Securing the Digital Realm

This book, *Securing the Digital Realm: Advances in Hardware and Software Security, Communication, and Forensics*, is a comprehensive guide that explores the intricate world of digital security and forensics. As our lives become increasingly digital, understanding how to protect our digital assets, communication systems, and investigate cybercrimes is more crucial than ever. This book begins by laying a strong foundation in the fundamental concepts of hardware and software security. It explains the design of modern computer systems and networks to defend against a myriad of threats, from malware to data breaches, in clear and accessible language.

One of the standout features of this book is its coverage of cutting-edge technologies like blockchain, artificial intelligence, and machine learning. It demonstrates how these innovations are used to enhance digital security and combat evolving threats.

Key features of the book include:

- Comprehensive coverage of digital security, communication, and forensics
- Exploration of cutting-edge technologies and trends
- Emphasis on digital forensics techniques and tools
- Coverage of ethical and legal aspects of digital security
- Practical guidance for applying cybersecurity principles

Additionally, the book highlights the importance of secure communication in the digital age, discussing encryption, secure messaging protocols, and privacy-enhancing technologies. It empowers readers to make informed decisions about protecting their online communications. Written by experts in the field, this book addresses the ethical and legal dimensions of digital security and forensics, providing readers with a comprehensive understanding of these complex topics. This book is essential reading for anyone interested in understanding and navigating the complexities of digital security and forensics.

Securing the Digital Realm

Advances in Hardware and Software Security, Communication, and Forensics

Edited by
Muhammad Arif, M. Arfan Jaffar, Oana Geman
and Waseem Abbasi

CRC Press
Taylor & Francis Group
Boca Raton London New York

CRC Press is an imprint of the
Taylor & Francis Group, an **informa** business

Designed cover image: © Shutterstock

First edition published 2025
by CRC Press
2385 NW Executive Center Drive, Suite 320, Boca Raton FL 33431

and by CRC Press
4 Park Square, Milton Park, Abingdon, Oxon, OX14 4RN

CRC Press is an imprint of Taylor & Francis Group, LLC

ISBN: 9781032802305 (hbk)
ISBN: 9781032806419 (pbk)
ISBN: 9781003497851 (ebk)

DOI: 10.1201/9781003497851

Typeset in Times
by Newgen Publishing UK

Contents

About the Editors

Dr. Muhammad Arif is an associate Professor at Superior University Lahore, Pakistan. His research interests include Artificial iIntelligence, big data, cloud computing, and cyberspace security, data mining, image processing, medical image processing, Privacy, Security, and E-learning. Currently, he is working privacy and security of vehicular networks. Previously he was a lecturer at the University of Gujrat, Gujrat, Pakistan. He completed masters and bachelor degrees in Pakistan. He received his BS degree in Computer Science from the University of Sargodha, Pakistan in 2011. He obtained his MS degree in Computer Science from COMSATS Islamabad 2013 Pakistan. He completed his PhD degree from Guangzhou University, China. He is the author of more than 220 SCIE journal publications, and more than 30 conference publications and a number of Best Paper Awards in the international conferences, including iSCI 2019. He has four SCI highly cited papers. He has more than 4,500 citations, according to Google Scholar. His H-Index is 39 and the i10-Index is 85. He is the Editorial board member of *International Journal of Advanced Intelligence Paradigms* (IJAIP) and to *International Journal of Computational Systems Engineering* (IJCSysE). He has participated in many international conferences as Program committee member, Session Chair, Technical Program Committee, or International Program Committees. He is the program chair of The 2020 International Workshop on Smart Technologies for Intelligent Transportation and Communications (SmartITC 2020 and 2021).

Prof. Dr. M. Arfan Jaffar received the M.Sc. degree in computer science from Quaid-i-Azam University, Islamabad, Pakistan, in March 2003, and the M.S. and Ph.D. degrees in computer science from the FAST National University of Computer and Emerging Sciences, in 2006 and 2009, respectively. He was an Assistant Professor with Al-Imam Mohammad Ibn Saud Islamic University, Riyadh, Saudi Arabia, from March 2013 to August 2018. He is currently the Dean with the Faculty of Computer Science and Information Technology, Superior University, Lahore, Pakistan, where he is also the Director of intelligent data visual computing research (IDVCR). He received a Postdoctoral Research Fellowship from South Korea and carried out research at the top raking Korean university, such as the Gwangju Institute of Science and Technology, Gwangju, South Korea, from 2010 to 2013. His research interests include image processing, data science, machine learning, computer vision, artificial intelligence, and medical images. He is a Reviewer of 30 reputed international journals, such as *IEEE Transactions on Pattern Analysis and Machine Intelligence, IEEE Transactions on Image Processing, IEEE Transactions on Industrial Electronics, Pattern Recognition, Knowledge, and Information Sciences.* Dr. Arfan is serving the educational sector of Pakistan for the past 20 years. With over 130 publications and one book to his credit, he is undoubtedly the most experienced scholar of CS and IT in Pakistan.

Oana Geman (Senior Member, IEEE) received the Ph.D. degree in electronics and telecommunication, in 2005. She is currently an Associate Professor with the Human and Health Development Department, Stefan cel Mare University of Suceava, Romania. Her current research interests include the non-invasive measurements of biomedical signals, wireless sensors, signal processing, artificial intelligence, nonlinear dynamics, stochastic networks, neuro-fuzzy methods, bioinformatics, biostatistics, biomedicine, the detection of neurological disorders, and rehabilitation. She has served as a member for the Technical Committees and the chair for several international conferences. She is a Reviewer of many journals, including IEEE T ransactions, IEEE Access, IEEE Internet of Things Journal, Sensors, and Symmetry.

Dr. Waseem Abbasi is an accomplished academic and researcher currently serving as an Associate Professor and Head of Department Computer Science at The University of Lahore, Sargodha Campus, Pakistan. With a profound dedication to the field of robotics and autonomous systems, Dr. Abbasi has established a prominent presence in academia and research. Dr. Abbasi is recognized as an esteemed academic professional enlisted in the Higher Education Commission of Pakistan (HEC) approved supervisors, contributing significantly to the development and growth of future talents in the realm of computer science and technology. In 2020, Dr. Abbasi completed a prestigious Postdoctoral Fellowship at the National Cheng Kung University, Taiwan. During this time, their research primarily focused on the intricate domain of Robotics and Autonomous Systems, garnering valuable insights and expertise in the field. In 2018, Dr. Abbasi achieved a milestone in their academic journey by obtaining a doctorate degree in Electrical Engineering, specializing in Robotics and Autonomous Systems, from the Capital University of Science and Technology, Pakistan. His academic excellence was further acknowledged through the receipt of the Indigenous PhD Scholarship from the Higher Education Commission of Pakistan, demonstrating their commitment to advancing knowledge and technology. Dr. Abbasi has authored and published more than 25 scientific papers in prestigious refereed and classified scientific journals and conferences, contributing significantly to the body of knowledge in their areas of expertise. Currently, Dr. Abbasi research is at the forefront of Robotics and Autonomous Systems, with a specific focus on Nonholonomic Systems, Nonlinear Control, Sliding Mode Control and Machine Learning algorithms for Autonomous Systems.

Contributors

Sajjad Abbas
Faculty of Computer Science & IT, Superior University Lahore, Pakistan

Waseem Abbasi
Department of Computer Science and IT, Superior University, Sargodha, Pakistan

Muhammad Adnan
Department of Computer Science, Superior University, Lahore, Pakistan

Amna Afza
Department of Information Technology, American Lycetuff, Lahore, Pakistan

Shehla Afzal
Department of Computer Science, Superior University, Lahore, Pakistan

Naseer Ahmad
Department of Information Technology, Superior University, Lahore, Pakistan

Waqar Ahmad
University of Engineering and Technology, Taxila, Pakistan

Abid Ali
Department of Computer Science, Lahore Leads University, Lahore, Pakistan

Muddassar Ali
Department of Information Technology, Tariq Glass Industries, Sheikhupura, Pakistan

Rashid Ali
Department of Engineering, Invozone Pvt. Ltd., Lahore, Pakistan

Ghadah Naif AlWakid
College of Computer and Information Sciences, Al Jouf University, Sakakah, Saudi Arabia

Muhammad Nabeel Amin
Superior University, Lahore, Pakistan

Muhammad Arif
Department of Computer Science, Superior University, Lahore, Pakistan

Muhammad Waleed Arif
Department of Information Technology, American Lycetuff, Lahore, Pakistan

Hinna Arqam
Superior University, Gold campus, Lahore, Pakistan

Arfan Arshad
School of Informatics and Robotics, Institute for Arts and Culture, Lahore, Pakistan

Masood Ashiq
Department of Computer Science, Lahore Leads University, Lahore, Pakistan

Abrar Ashraf
Department of Technology, The University of Lahore, Pakistan

Muhammad Ashraf
Department of Avionics Engineering, Air University, Islamabad, Pakistan

Muhammad Azam Khan
Department of Information Technology, American Lycetuff, Lahore, Pakistan

Sannia Bibi
University of Engineering and Technology, Taxila, Pakistan

M. Iram Baig
University of Engineering and Technology, Taxila, Pakistan

Irfan Ud Din
Department of Computer Science, Superior University, Lahore, Pakistan

Muhammad Ejazulghaffar
Department of Computer Science, Superior University, Lahore, Pakistan

Umar Farooq
Superior University Gold Campus, Lahore, Pakistan

Mehak Fatima
Superior University, Lahore, Pakistan

Noroze Fatima
Superior University, Lahore, Pakistan

Yazeed Yasin Ghadi
Department of computer science and software engineering, Al Ain University, Abu Dhabi, UAE

Hinna Hafeez
Department of Computer Science, Superior University, Lahore, Pakistan

Muhammad Ameer Hamza
Department of Computer Science, Superior University, Lahore, Pakistan

Ehtisham Ul Haque
Department of Computer Science, Muslim Youth University, Islamabad, Pakistan

Ommair Hameed
Superior University Gold Campus Raiwind Road, Lahore, Pakistan

Ahmed Hassan
Department of Computer Science, Superior University, Lahore, Pakistan

Moodser Hussain
Department of Information Technology, University of the Punjab, Gujranwala Campus, Gujranwala, Pakistan

Mamoona Humayun
College of Computer and Information Sciences, Al Jouf University, Sakakah, Saudi Arabia

Fatima Ijaz
COMSATS University Islamabad, Sahiwal Campus, Sahiwal, Pakistan

Aleema Imran
COMSATS University Islamabad, Sahiwal Campus, Sahiwal, Pakistan

Ali Imran
Superior University, Lahore, Pakistan

Ahsan Imtiaz
Department of Computer Science and IT, Superior University, Lahore, Pakistan

Ahsan Iqbal
University of Engineering and Technology, Taxila, Pakistan

Junaid Iqbal
Department of Computer Science, Lahore Leads University, Lahore, Pakistan

Maria Iqbal
Department of Computer Science, Superior University, Lahore, Pakistan

Muhammad Iqbal
Institute of Computing and Information Technology, Gomal University, Dera Ismail Khan, Pakistan

Danish Irfan
Faculty of Computer Science & IT, Superior University, Lahore, Pakistan

Faiqa Irum
Superior University, Gold campus, Lahore, Pakistan

Taimoor Hassan Jabbar
Department of Computer Science, Superior University, Lahore, Pakistan

M. Arfan Jaffar
Department of Computer Science, Superior University, Lahore, Pakistan

Muhammad Jameel
Department of Computer Science, Superior University, Lahore, Pakistan

Aetsam Javed
Department of Computer Science, Superior University, Lahore, Pakistan

Mannan Javed
Department of Avionics Engineering, Air University, Islamabad, Pakistan

Mubasher Khalid
Department of Computer Science, Superior University, Lahore, Pakistan

Hisham Khalil
Embedded System and Robotics Lab, Department of Technology, The University of Lahore, Pakistan

Farrukh Aslam Khan
Center of Excellence in Information Assurance, King Saud University, Riyadh, Saudi Arabia

Hamayun Khan
Superior University, Lahore, Pakistan

Naveed Ali Khan Kaim Khani
TU-Berlin, Germany. and Riphah Institute of System Engineering, Islamabad, Pakistan

Nouman Mabood
Department of Avionics Engineering, Air University, Islamabad, Pakistan

Tehseen Mazhar
Department of Computer Science, Virtual University of Pakistan, Lahore, Pakistan

Johar Mumtaz
Department of Computer Science Faculty of CSIT Superior University, Lahore, Pakistan

Hirra Mustafa
Department of Computer Science, Superior University, Lahore, Pakistan
and
Department of Computer Science, Sharif College of Engineering and Technology, Lahore, Pakistan

Syed Asad Ali Naqvi
Faculty of Computer Science & IT, Superior University Lahore, Pakistan

Muhammad Waqas Nasir
Department of Communication Technology, BZU, Multan, Pakistan

Nasir Nauman
Department of Computer Science, Superior University, Lahore, Pakistan
and
Embedded System and Robotics Lab, Department of Technology, The University of Lahore, Pakistan

Usman Nawaz
Department of Computer Science, Lahore Leads University, Lahore, Pakistan

Zaib un Nisa
Department of Computer Science, Superior University, Lahore, Pakistan

Asim Noor
Department of Avionics Engineering, Air University, Islamabad, Pakistan

Hajira Noor
Superior University, Lahore, Pakistan

Muhammad Ahmad Pasha
COMSATS University Islamabad, Sahiwal Campus, Sahiwal, Pakistan

Muhammad Yasir Qadri
Centers of Excellence in Science & Applied Technology Islamabad, Pakistan

Farhan Qamar
University of Engineering and Technology, Taxila, Pakistan

Tariq Qayyum
Department of Computer Science, Superior University, Lahore, Pakistan

Furqan Rafique
Department of Computer Science, Lahore Leads University, Lahore, Pakistan

Usman Ahmed Raza
Faculty of Information Technology, University of Central Punjab, Lahore, Pakistan

Ateeq Ur Rehman
Department of Electrical Engineering, Government College University, Lahore, Pakistan

Attiqur Rehman
Department of Computer Science Faculty of CSIT Superior University, Lahore, Pakistan

Faisal Rehman
Department of Computer Science, Lahore Leads University, Lahore, Pakistan
and
Department of Statistics and Data Science, University of Mianwali, Pakistan

Ahmad Shaf
COMSATS University Islamabad, Sahiwal Campus, Sahiwal, Pakistan

Hanan Sharif
Department of Computer Science, Lahore Leads University, Lahore, Pakistan

Danish Shehzad
Department of Computer Science, Superior University, Lahore, Pakistan

Farah Taj
Department of Computer Science,
Superior University, Lahore, Pakistan

Asadullah Tariq
Department of Computer Science, Superior University, Lahore, Pakistan

Imran Tariq
Department of Computer Science Faculty of CSIT Superior University, Lahore, Pakistan

Noshina Tariq
Department of Avionics Engineering, Air University, Islamabad, Pakistan

Muhammad Imran Tariq
Superior University, Lahore, Pakistan

Kainat Azmat Ullah
Department of Computer Science, Lahore Leads University, Lahore, Pakistan

Sahibzada M. Waqas
Department of Avionics Engineering, Air University, Islamabad, Pakistan

Haroon Waris
Centers of Excellence in Science & Applied Technology, Islamabad, Pakistan

Muhammad Waseem
Department of Software Engineering, Superior University, Lahore, Pakistan

Muhammad Younas
University of Engineering and Technology Taxila, Pakistan

Uns Bin Younas
Superior University Gold Campus Raiwind Road, Lahore, Pakistan

Muhammad Yousaf
TU-Berlin, Germany. and Riphah Institute of System Engineering, Islamabad, Pakistan

Fatima Zahid
Superior University Gold Campus Raiwind Road, Lahore, Pakistan

Muhammad Zakwan
Department of Avionics Engineering, Air University, Islamabad, Pakistan

Arif Zafar
University of Engineering and Technology Taxila, Pakistan

1 Performance Analysis of PSK-Based Advanced Modulation Formats for Free Space Optical Communication System

Sannia Bibi, M. Iram Baig, and Farhan Qamar

1.1 INTRODUCTION

A wireless system that uses light to transmit data is called a Free Space Optical (FSO) communication system [1]. This system employs laser technology to transfer data from one point to another. These systems are particularly useful in situations where traditional wired or wireless communication methods are not feasible or practical, such as in remote areas or where a secure and fast network connection is required. FSO technology can be installed in license-free environments. Furthermore, FSO systems are ideal for point-to-point (P2P) communication, particularly in outdoor or industrial environments. FSO communication systems possess several advantages over traditional wireless communication methods, including higher bandwidth, lower latency, and improved security. FSO technology has numerous practical applications, including disaster recovery, medical imaging, video surveillance, broadcasting, and even ship-to-ship communication, to name a few. However, FSO communication systems also have some limitations. For example, the weather conditions, absorption, scattering, and physical obstruction affect FSO communication [2]. FSO communication systems represent an innovative and effective way to transmit data wirelessly and are being used increasingly in various industries and applications. They offer substantial advantages over traditional RF systems, including higher bandwidth, lower latency, and improved security. This means the system may not be suitable in specific environments, such as urban areas with buildings or other obstacles that can block the light beam.

In conclusion, FSO communication systems represent an innovative and effective way to transmit data wirelessly and are being used increasingly in various industries and applications. They offer substantial advantages over traditional RF systems, including higher bandwidth, lower latency, and improved security. The performance of FSO can be defined using some characteristics such as (i) for long range, FSO system is able to operate at high power level; (ii) overall design of FSO system is compact; (iii) for high data rate and long range, modulation formats can be used [3–6]. The performance of FSO communication system in terms of link range, atmospheric condition, high data rates, and available hardware is improved using different modulation formats. Depending upon different weather conditions and desire results, we have used different modulation formats in this paper.

In the FSO system, the transmitter encodes data into a beam of light and transmits it through the air. The transmitter typically comprises several components, including a laser source, a modulator, and an optical assembly [7]. The laser source generates the light beam that carries the data, which is often produced by a laser diode. The modulator encodes the data into the light beam, using different modulation schemes that depend on the application and data requirements. Finally, the optical

DOI: 10.1201/9781003497851-1

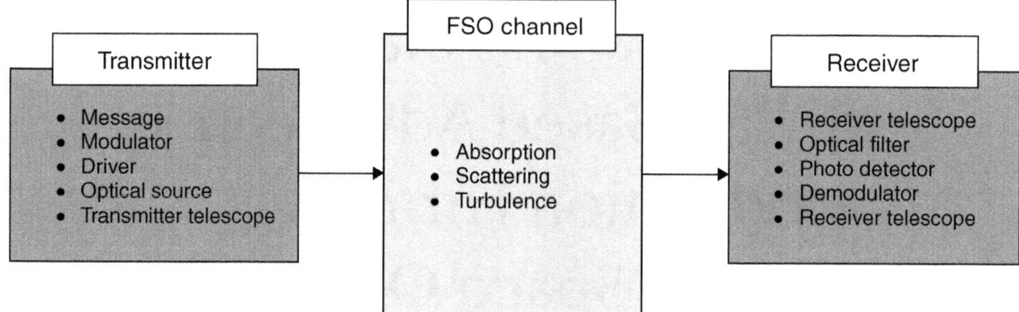

FIGURE 1.1 Component of FSO system.

assembly directs the light beam toward the receiver using lenses or mirrors and focuses it to maintain alignment over long distances. It is important to note that the transmitter also requires a control system to manage transmitted laser power, signal modulation, and proper positioning between the transmitter and receiver. This control system can be implemented using software or a combination of software and hardware.

The transmit telescope, an optical apparatus located at the transmission end, collects and steers the modulated light along its atmospheric journey. This telescope ensures that the beam of light is directed towards the receiving end. However, as beam travels through the atmospheric channel, it may experience a decrease in intensity due to various factors such as atmospheric disturbances, absorption of light, fluctuations in intensity, and deviation of the beam and background noise. Optical amplifier and detector are present at the receiver side of FSO, which amplify the incoming signal and convert the optical signal into electrical signal, respectively. A basic FSO system is shown in Figure 1.1.

In this chapter, we have enhanced the performance of the FSO system using advanced modulation formats such as DP-BPSK, DP-QPSK, DP-8PSK, and DP-16PSK, and compared the performance of these modulation formats on the bases of link range, weather turbulence, transmitting power, and at different aperture size. This chapter is organized as follows: Section 1.2 covers the advanced modulation formats. Section 1.3 shows the simulation and layout of the FSO system. In Section 1.4 results are reported and, finally, Section 1.5 concludes the chapter.

1.2 ADVANCED MODULATION FORMATS IN FSO COMMUNICATION

The method by which data is encoded onto the optical carrier signal and transmitted through the air is defined by modulation formats. The utilization of various modulation formats can enhance the performance of the FSO system by enabling longer link ranges, higher data rates, minimized interference, and superior outcomes in atmospheric turbulence. The selection of different modulation formats depends on various factors, such as atmospheric conditions, data rates, and available hardware. Comparison of FSO communication with different advanced modulation formats is described in Table 1.1.

1.2.1 Dual Polarization Binary Phase Sift Keying (DP-BPSK)

DP-BPSK is the combination of binary phase shift keying with dual polarization. By altering the phase of the transmitted signal, BPSK encrypts data. Two orthogonal polarizations of light are separately modified using dual polarization modulation to transport two distinct information streams. DP-BPSK has several benefits over conventional BPSK modulation. By enabling the transmission

TABLE 1.1
Comparison of FSO with Modulation Formats

Reference	Year	Modulation	System	BER	Q-Factor	Weather
[8]	2017	RZ, NRZ, CSRZ	FSO	✓	✓	✓
[9]	2018	MDRZ. DRZ, CSRZ	FSO	✓	✓	✓
[10]	2021	PPM	FSO	✓		
[11]	2019	PDM/OFDM	FSO	✓	✓	✓
[12]	2020	OOK, BPSK, DPSK, 8PSK, QPSK	FSO	✓		✓
[13]	2021	OOK, DPSK	FSO	✓	✓	✓
[14]	2022	4-QAM, 16-QAM,64 -QAM, 128-QAM, 256 -QAM	DWDM-FSO	✓		✓
[15]	2013	QPSK	FSO	✓	✓	✓
[16]	2020	DP-QPSK,16-QAM,	OWC	✓	✓	✓

of two separate data streams on the same carrier frequency, it first doubles the data rate [17]. Second, it offers superior resistance to polarization mode dispersion (PMD), which can distort the transmitted signal because of changes in the signal's polarization state as it travels down the optical fiber. The symbol rate in DP-BPSK can be calculated by the following equation:

$$Symbol\,rate = \frac{Data\,rate}{2} \tag{1.1}$$

In BPSK we deal with two phases so Bits per symbol is 1.

1.2.2 Dual Polarization Quadrature Phase Sift Keying (DP-QPSK)

DP-QPSK modulation format is the combination of two QPSK formats in dual polarization state. The difference between the current and previous values of the data symbols determines how the carrier signal's phase changes in DP-QPSK. Although it may be utilised in FSO communication systems, DP-QPSK is frequently employed in optical fibres. Its spectral efficiency and the durability of DP-QPSK to phase noise make it a preferred option for FSO communication systems in difficult situations. In [16], Tobias Siegel and Shun-Ping Chen compare the BER performance of DP-QPSK and 16-QAM with respect to different parameters. By using these two modulation schemes, the transmitted data rate is comparatively high but signals travel at only certain distance. By increasing the distance, signal quality is reduced. The symbol rate for DP-QPSK is calculated by the following formula:

$$Symbol\,rate = \frac{Data\,rate}{4} \tag{1.2}$$

Bits per symbol rate for DP-QPSK is 2 because in QPSK format, data is divided in four phases.

1.2.3 Dual Polarization – 8 Phase Sift Keying (DP-8PSK)

With 8PSK, the data is encoded by shifting the transmitted signal's phase to one of eight potential phase angles, each of which corresponds to three bits of information. Two orthogonal polarizations of light are modulated separately to transport two distinct streams of information in DP-8PSK, which also modulates the phase of the signal in addition to the polarization of the signal. DP-8PSK is used to improve the system transmission speed and DP-8PSK is sensitive to noise and shortens the link range [18–19]. The symbol rate for DP-QPSK is calculated by the following formula:

$$Symbol\,rate = \frac{Data\,rate}{8} \tag{1.3}$$

1.2.4 Dual Polarization – 16 Phase Sift Keying (DP-16PSK)

Dual polarisation – 16PSK has the fundamental benefit of enabling simultaneous transmission of two separate streams of data, thereby doubling the transmission's data throughput. This is accomplished by transmitting two distinct streams of data over the same frequency range utilizing two orthogonal polarizations without interfering with one another. In DP-16PSK, the data is encoded by shifting the transmitted signal's phase to one of 16 potential phase angles, each of which corresponds to four bits of information. The symbol rate for DP-QPSK is calculated by the following formula:

$$Symbol\,rate = \frac{Data\,rate}{16} \tag{1.4}$$

1.3 SIMULATION

In this section, we have implemented the advanced modulation formats in FSO communication systems. Table 1.2 shows the parameter of different components used in the FSO system. Table 1.3 describes the values of attenuation for different weather conditions [16].

Figure 1.2 shows the FSO communication system. In this diagram, we have designed separate transmitters and receivers for each modulation format and kept the remaining components the same in all types of modulation formats. In this system, transmitter transmits light over the FSO channel. The output signal is received by the receiver, which converts the light signal into an electrical signal. The DSP component ensures signal processing to obtain accurate signal. A PSK sequence

TABLE 1.2
Operating Parameters used in the FSO System

Parameter	Value
Data Rate	100 Gbps
Operating Frequency	1550 nm
Number of Samples	262144
Sequence Length	65536 bits
Beam divergence	1 mrad
Responsivity	1 A/w
Range	80 km

TABLE 1.3
Values of Attenuation for Different Atmospheric Turbulences

Atmospheric turbulence	Attenuation value (dB/ km)
Clear	0.17
Haze	4
Rain	6.23
Fog	21

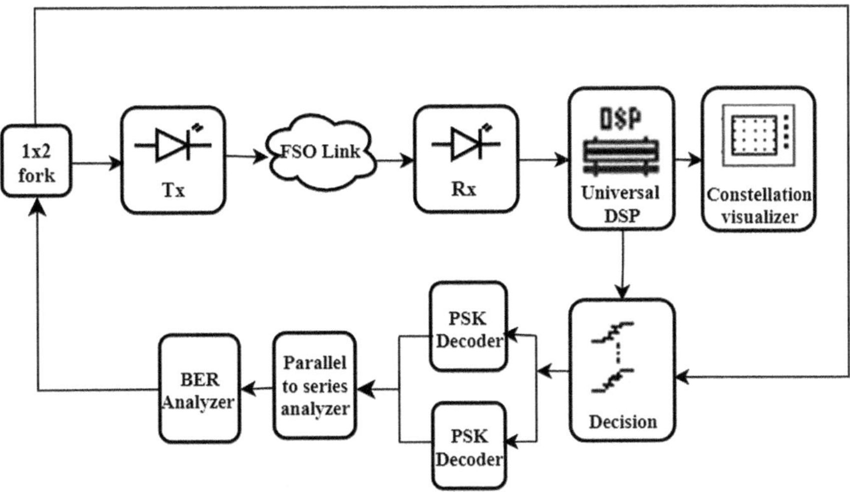

FIGURE 1.2 FSO communication system with advanced modulation formats.

decoder is used for decoding the parallel sequence into a binary signal. A BER analyzer is used to analyze the signal by comparing the transmitted signal to the received signal and providing the BER value.

1.4 RESULT AND ANALYSIS

In this section, we have analyzed the performance of FSO system with advanced modulation formats (DP-BPSK, DP-QPSK, DP-8PSK, and DP-16PSK) under different weather condition like Clear weather, Rain, Haze, and Fog, and presented the results by varying different parameters of the proposed system.

1.4.1 BER vs. Link Range for Clear Weather

The distance between FSO transmitter and receiver is called link range of FSO system. Link range is varied from 0 to 80 km for 100 Gbps data rate, while the other parameters mentioned in Table 1.1 are kept constant in this case. Figure 1.3 shows the value of BER against link range for different advanced modulation formats. The results show that DP-BPSK and DP-QPSK gives better performance up to 80 km as compared to DP-8PSK and DP-16PSK. Further, DP-8PSK performance is better up to 60 km while DP-16PSK gives acceptable BER value up to 20 km.

1.4.2 BER at Different Weather Conditions

FSO performance is affected by the change in weather conditions. Table 1.2 shows the attenuation values for different weather conditions. Figure 1.4 shows the BER vs. Link range for different modulation formats in the presence of rain, haze, and fog. This shows the performance of advanced modulation formats decreases with an increase in link range.

DP-BPSK performs well up to 10 km in the presence of rain after that performance is degraded. The link range achieved is 4 km and 2 km for haze and fog, respectively.

DP-QPSK performance is good up to 8 km in the presence of rain but after that performance is degraded. Similarly, the link range achieved is 6 km and 2 km for haze and fog, respectively.

DP-8PSK performs well up to 6 km in the presence of rain but after that performance is degraded. Link range achieved 4 km for haze while in the presence of Fog, signal is not properly retrieved.

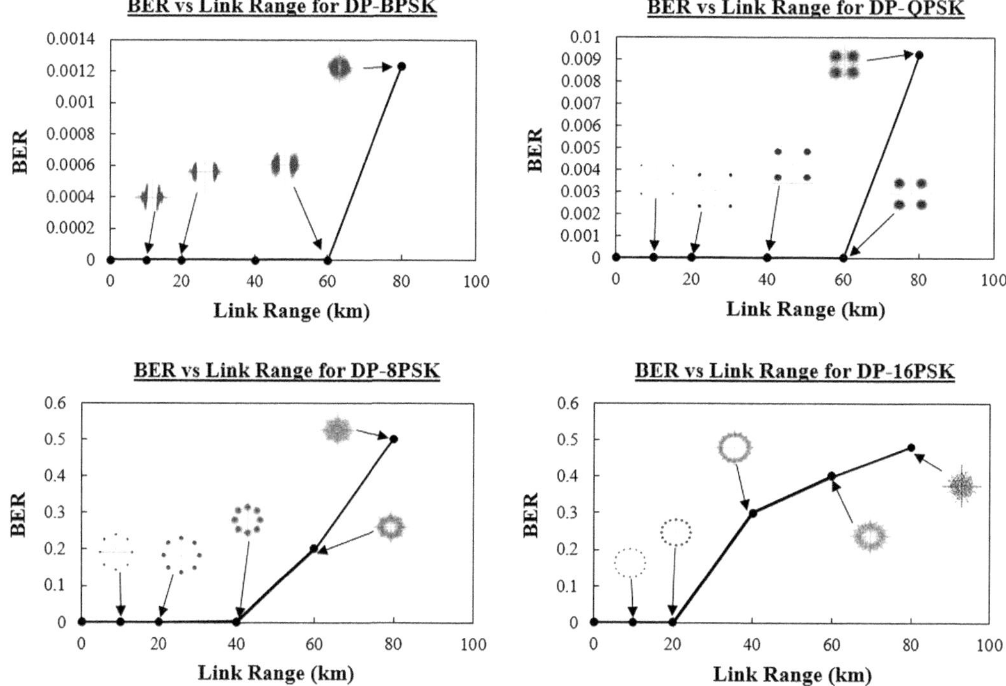

FIGURE 1.3 BER value vs. link ranges in clear weather using different modulation formats.

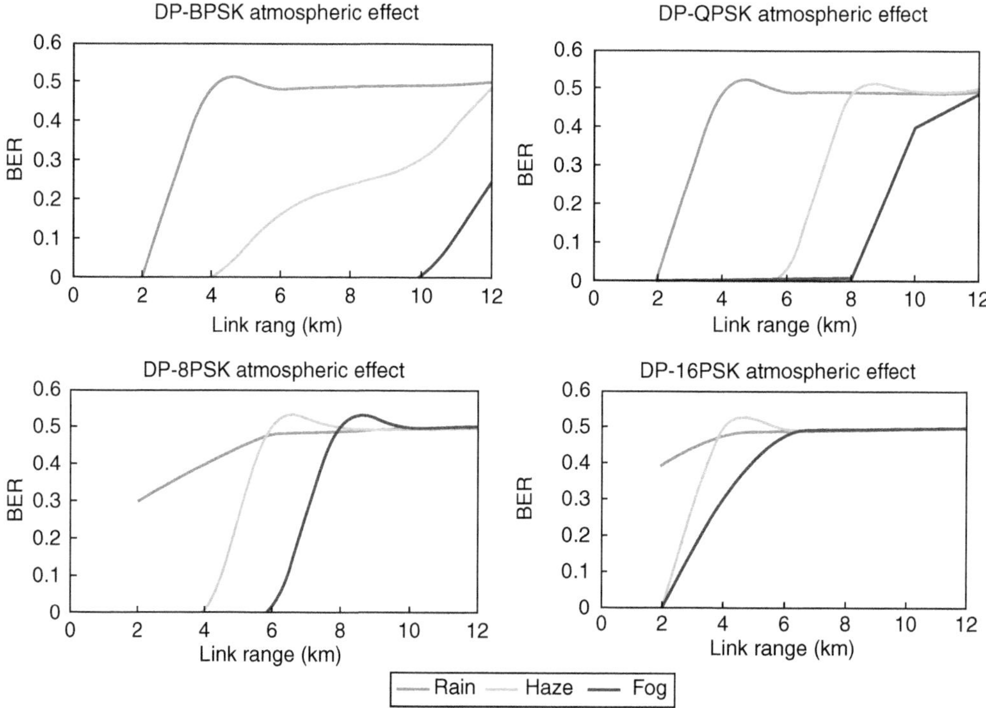

FIGURE 1.4 BER vs. link ranges in different weather conditions using different modulation formats.

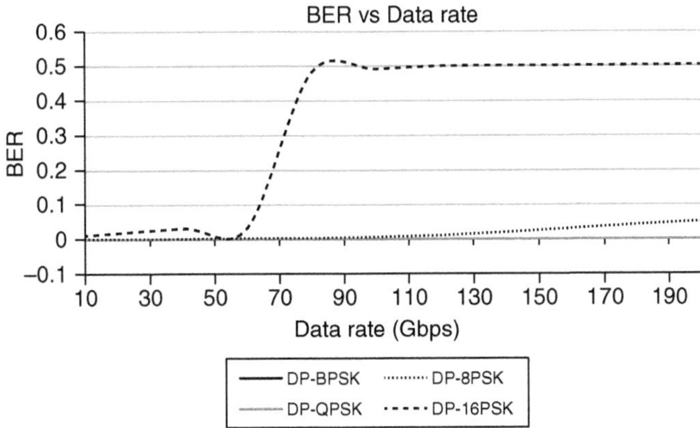

FIGURE 1.5 BER value vs. data rates using different modulation formats.

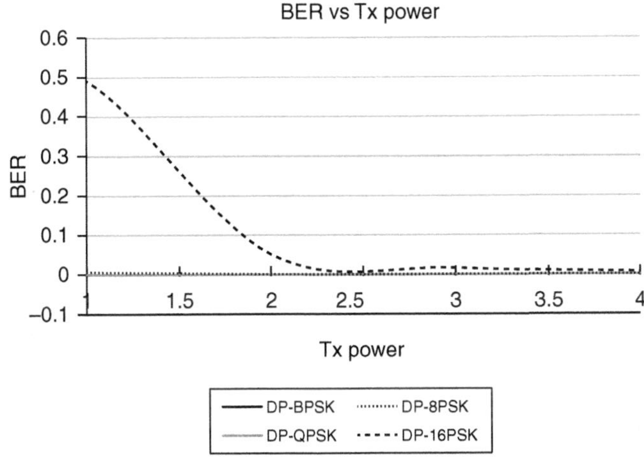

FIGURE 1.6 BER value vs. Transmitter power using different modulation formats.

In DP-16PSK, the signal travels up to 2 km only in the presence of rain and haze while no signal is retrieved in the presence of fog.

1.4.3 BER vs. Data Rate

Figure 1.5 shows the value of BER for different modulation formats at different data rates up to 190 Gbps. The link range is set to 40 km in this case. Results show the performance of DP-BPSK and DP-QPSK is better even at higher data rates. DP-8PSK gives high value of BER after 150Gbps, while DP-16PSK show sudden increase in BER value after 60 Gbps.

1.4.4 BER vs. Transmitter Power

Transmitter power is an important factor to check the performance of FSO communication. Figure 1.6 shows the effect of transmitter power on the performance of FSO system at data rate of 100 Gbps and link range 40 km. DP-BPSK, DP-QPSK, and DP-8PSK modulation formats gives acceptable BER value for all the power ranges. While DP-16PSK modulation format performance is improved at high value of transmitter power.

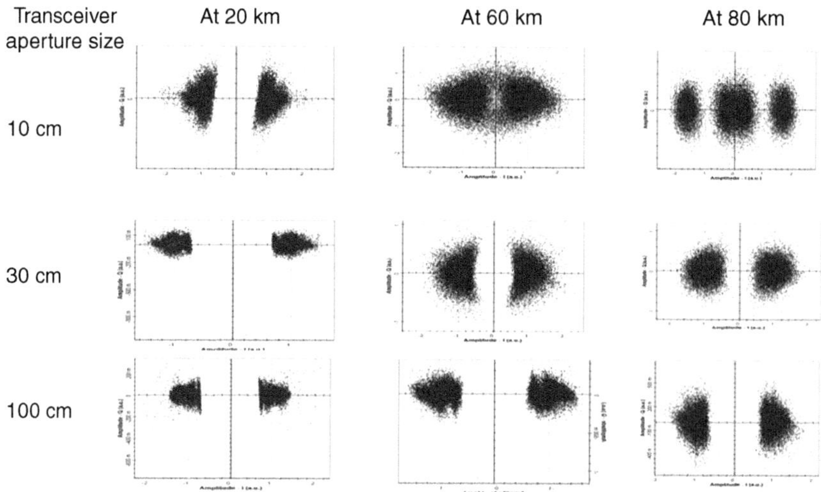

FIGURE 1.7 Constellation diagrams at different transceiver aperture sizes for DP-BPSK.

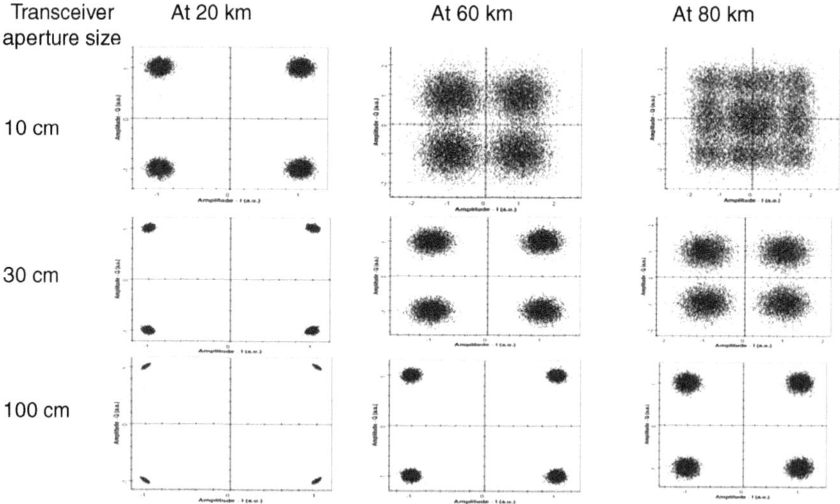

FIGURE 1.8 Constellation diagrams at different transceiver aperture sizes for DP-QPSK.

1.4.5 Signal Quality vs. Aperture Size

Figure 1.7 shows the constellation diagrams of DP-BPSK modulation format for different transceiver aperture sizes at different link ranges. At 10 cm DP-BPSK performance is degraded after 60 km. while at 30 cm and 1 m DP-BPSK gives prominent constellation diagrams, which shows signal quality is good at 30 cm and 1 m aperture sizes.

Constellation diagrams of DP-QPSK are shown in Figure 1.8 for different aperture sizes at different link ranges. At 10 cm of transceiver aperture size, DP-QPSK performance is good up to 20 km; after that signal quality is not acceptable. DP-QPSK performance is good at 30 cm and 1m of aperture sizes. Figures 1.9 and 1.10 shows the constellation diagram of DP-8PSK and DP-16PSK with the variation in aperture sizes and link ranges. Performance of DP-8PSK increases at 1m aperture size and up to 80 km of link range, while performance of DP-16PSK increases up to link range of 60 km and at aperture size 100 cm.

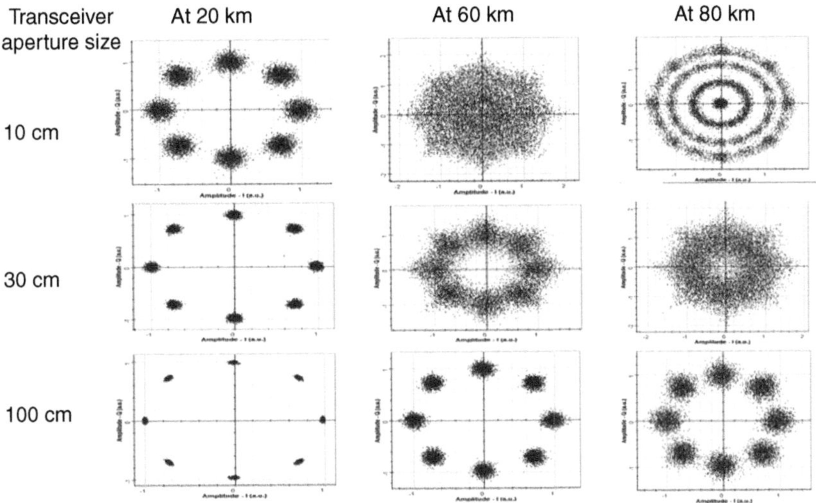

FIGURE 1.9 Constellation diagrams at different transceiver aperture sizes for DP-8PSK.

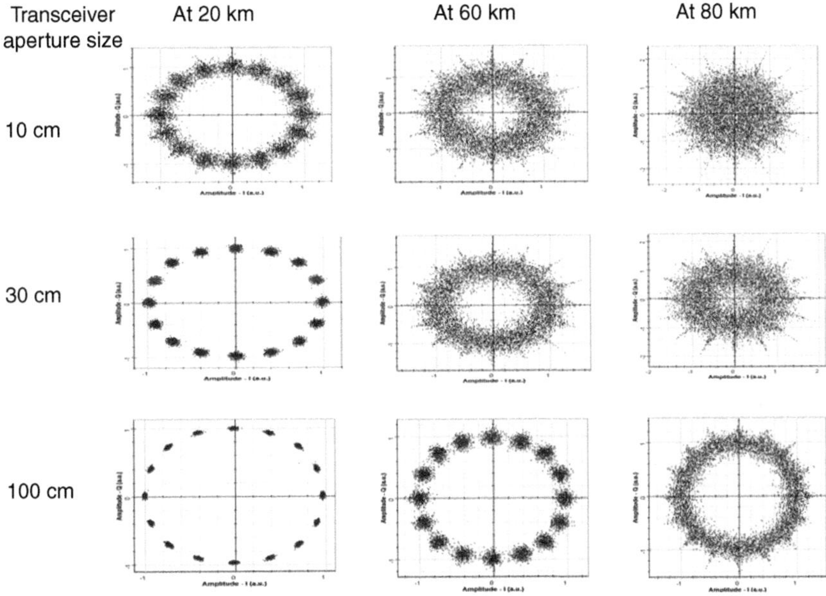

FIGURE 1.10 Constellation diagrams at different transceiver aperture sizes for DP-16PSK.

1.5 CONCLUSION

This chapter analyzes the performance of the FSO communication system using different modulation formats like DP-BPSK, DP-QPSK, DP-8PSK, and DP-16PSK by varying the link range, attenuation, data rate, transmission power and transceiver aperture size. In clear weather, DP-BPSK and DP-QPSK perform better while DP-8PSK and DP-16PSK links degrade after some distance. Further, due to increases in attenuation in case of rain, haze and fog, the performance of FSO communication system is limited up to certain distance. Performance of modulation formats increases with the increase in the transmission power and transceiver aperture size. The BER and constellation diagrams show that DP-BPSK and DP-QPSK outperformed DP-8PSK and DP-16PSK modulation formats in most of the cases.

REFERENCES

1. Kaur A, Sheetal A. Performance analysis of 16× 2.5 GB/s FSO system for the most critical weather conditions. *International Journal of Advance Research in Computer and Communication Engineering*. 2015 May;4(5):612–8.
2. Kaushal H, Kaddoum G. Optical communication in space: Challenges and mitigation techniques. *IEEE Communications Surveys & Tutorials*. 2016 Aug 26;19(1):57–96.
3. Khalil H, Qamar F, Ali M, Shahzadi R, Khan MFN, Qamar N. FSO communication: Benefits, challenges and its analysis in DWDM communication system. Sir Syed University Research *Journal of Engineering & Technology*. 2019;9(2).
4. Sadiq, N, Hussain A, Qamar F, Shahzadi R, Ali M, Qamar N, Nadeem MF, and Masud U. Performance analysis of NRZ and RZ variants for FSO communication system under different weather conditions. *Journal of Optical Communications*. 2020;1, no. ahead-of-print.
5. Niaz A, Qamar F, Ali M, Farhan R, Islam MK. Performance analysis of chaotic FSO communication system under different weather conditions. *Transactions on Emerging Telecommunications Technologies*. 2019 Feb;30(2):e3486.
6. Shahid T, Khalid F, Qamar F, Shahzad A, Shahzadi R, Ali M, Qamar N. Performance analysis of WDM based FSO communication with advance modulation formats. In *2020 IEEE 23rd International Multitopic Conference (INMIC)* 2020 (pp. 1–6). IEEE.
7. Mikołajczyk J, Bielecki Z, Bugajski M, Piotrowski J, Wojtas J, Gawron W, Szabra D, Prokopiuk A. Analysis of free-space optics development. *Metrology and Measurement Systems*. 2017;24(4):653–74.
8. Nadeem L, Saadullah Qazi M, Hassam A. Performance of FSO links using CSRZ, RZ, and NRZ and effects of atmospheric turbulence. *Journal of Optical Communications*. 2018 Apr 25;39(2):191–7.
9. Baiwa R, Verma P. Performance analysis of FSO system for advanced modulation formats under different weather conditions. In *2018 Second International Conference on Intelligent Computing and Control Systems (ICICCS)* 2018 Jun 14 (pp. 1490–1495). IEEE.
10. Yu N, Wang P, Zhuang Z. Design of digital pulse-position modulation system. In *Journal of Physics: Conference Series* 2021 Nov 1 (Vol. 2093, No. 1, p. 012030). IOP Publishing.
11. Kaur G, Srivastava D, Singh P, Parasher Y. Development of a novel hybrid PDM/OFDM technique for FSO system and its performance analysis. *Optics & Laser Technology*. 2019 Jan 1;109:256–62.
12. Sharoar J, Choyon AK, Chowdhury R. Performance comparison of free-space optical (FSO) communication link under OOK, BPSK, DPSK, QPSK and 8-PSK modulation formats in the presence of strong atmospheric turbulence. *Journal of Optical Communications*. 2024 Mar 31;44:s763-s769.
13. Xu Z, Xu G, Zheng Z. BER and channel capacity performance of an FSO communication system over atmospheric turbulence with different types of noise. *Sensors*. 2021 May 15;21(10):3454.
14. Sinha S, Kumar C. Performance Analysis of Hermite Gaussian 8× 40 Gb/s DWDM-FSO System using Advance Modulation Schemes.
15. Patnaik B, Sahu PK. Design and study of high bit–rate free–space optical communication system employing qpsk modulation. *International Journal of Signal and Imaging Systems Engineering*. 2013 Jan 1;6(1):3–8.
16. Siegel T, Chen SP. Investigations of free space optical communications under real-world atmospheric conditions. *Wireless Personal Communications*. 2021 Jan;116(1):475–90.
17. Li Y, Wang M. Simultaneous target positioning and velocity measurement using a DP-BPSK modulator with VLFM waveform modulation. In *Asia Communications and Photonics Conference* 2021 Oct 24 (pp. W1D–4). Optical Society of America.

18. Kojima K, Koike-Akino T, Millar DS, Pajovic M, Parsons K, Yoshida T. Investigation of low code rate DP-8PSK as an alternative to DP-QPSK. In *Optical Fiber Communication Conference* 2016 Mar 20 (pp. Th1D–2). Optica Publishing Group.
19. Sillekens E, van Uden RG, van Weerdenburg JA, Kuschnerov M, de Waardt H, Koonen AM, Okonkwo CM. Experimental demonstration of 8 state turbo trellis coded modulation employing 8 phase shift keying. In *2015 European Conference on Optical Communication (ECOC)* 2015 Sep 27 (pp. 1–3). IEEE.

2 Volatile Kernel Rootkit Hidden Process Detection View (VKRHPDV) Approach for Detection of Malware in Cloud Computing Environments

Danish Shehzad, Hinna Hafeez, Farah Taj, and Zaib un Nisa

2.1 INTRODUCTION

Via distributed computing clouds, customers can access distinctive organizations like SaaS, PaaS, and IaaS. Attackers using rootkits are tarnishing Windows instances (AWS) all over the internet [1]. Rootkit Convenience foresaw a crucial measure to conceal multiple malware testing, including network access, keylogging, recordings, and process [2, 3], the aggressors' behaviors. This subject has been discussed for a very long time. Malware attacks like rootkit [4] assaults can grant hazardous access to a machine as well. It can hide its proximity and the specific poisonous operation in the meantime by indicating some upgrade and change to the framework. As a result, it is more challenging to distinguish a rootkit from other infections.

Cloud is an approach in which multiple servers are used to provide numerous services to the client machines over the internet; rootkits are malware software / processes which enter in the cloud environment to get access of kernel and manipulate the system environment. Different unique organizations like SaaS, PaaS, and IaaS are offered to customers by distributed cloud computing environment. The attackers are dealing with rootkit to get access to kernel in cloud environment.

2.2 SIGNIFICANCE OF THE STUDY

Kernel rootkits are hidden processes that can be loaded into kernel spaces dynamically. Due to the lack of security mechanisms and no protection of kernel memory this process can work maliciously. These processes can access data section, code section and can also access the registers. It is important to find the techniques to detect this hidden process and control their execution. Many approaches have been used for detection, but some approaches are platform dependent, some are efficient when the system is online but unable to execute while the system is in offline mode. Some approaches cause significant performance overhead. In this chapter we have discussed several popular approaches, compared the approaches and discussed the limitation of the proposed approached.

DOI: 10.1201/9781003497851-2

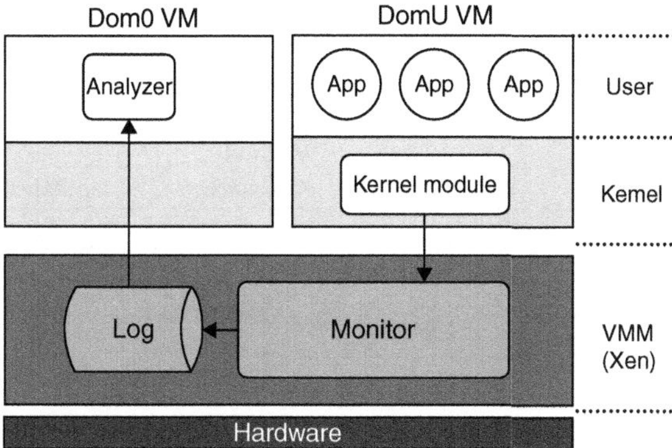

FIGURE 2.1 The figure highlights the key techniques used by the specified five papers.

2.3 BACKGROUND

Various techniques have been used to detect the kernel-level rootkit malware (Figure 2.1, Table 2.1). Some of them are:

- Information extraction at hypervisor-level through deep learning.
- An online cloud anomaly detection approach.
- Hypervisor-based malware detection.
- Meta statistics approach, derived from packet header and volumetric information which is packet and byte etc. count.
- The approach is based on the concept of giving the same access to hypervisor as the VM has.
- PoKeR(Profiler for Kernel Rootkits).
- Static analysis-based technique for detecting.
- X-Anti approach. X-Anti is based on the overall rootkit detection program.
- Host-based and virtualization-based approach.

Because of some limitations and performance overheads, another approach, which is VKRD is considered much more efficient.

The key concept of VKRD is to isolate targeted kernel module and observe its runtime behavior. To achieve this, the approach baseline is:

1. Memory isolation
2. Register isolation
3. Features extraction

The VKRD also has its own limitations and overhead.

2.4 PROBLEM STATEMENT

The existing approaches to detect the kernel-level malware are platform and architecture dependent; there is a need to tackle the rootkit kernel-level attacks when the system is in offline mode.

A popular way to compromise OS kernel is through a kernel rootkit (i.e., malicious kernel module).

TABLE 2.1
Different Approaches Used to Detect Malware in Kernel Level

Paper	Journal	Year	Area	Method/ Approach	Limitation	Efficiency	Platform
A Review on Learning-based Detection Approaches of the Kernel-level Rootkit By Mohammad Nadim; David Akopian; Wonjun Lee	IEEE	2021	Rootkit detection	A learning-based detection. Virtualization-based Kernel Rootkit Detection system.	Cant handle VM based operations	high accuracy	Window
Detection pf Malware and Kernel-level Rootkits in cloud Computing Environments By: Win, Thu Yein and Tianfield, Huaglory and Mair, Quentin		2016	Rootkit detection	Approach used system call hashing and SVM together in VM.I.	use single point of control for the attack detection, it need signature database.	90%	Window
Malware Detection in Cloud Computing Infrastructures [b] By: Michael R. Watson, Noor-ul-hassan Shirazi, Angelos K. Marnerides, Andreas Mauthe and David Hutchison	IEEE Transactions on Depend able and Secure Computing	2015	Rootkit detection	approach is based on per flow meta statistics, derived from packet header and volumetric information which is packet and byte etc count.	The paper has discussed an online cloud anomaly detection approach but it didn't discuss off line mode detection	moderate accuracy	Window
Rootkit Detection on Virtual Machines through Deep Information Extraction at Hypervisor-level By: Xiongwei Xie & Weichao Wang	IEEE-	2013	Rootkit detection	the view difference between hypervisor and VM, paper proposed to give the same access to hypervisor at the VM has.	The proposed solution is not executed on windows	very low	Linux

2.5 OBJECTIVES

- Find the efficient algorithm to detect rootkit kernel-level attacks in cloud computing.
- Find efficient approach to prevent cloud environments from the rootkit kernel level attacks in the most efficient way.

2.6 RESEARCH QUESTIONS AND HYPOTHESIS

The following are research questions:

1. How to prevent cloud computing environment from kernel-level rootkit attacks in an efficient way.

2. Which are exiting methods to detect rootkit kernel level attacks and what are their limitations, strengths and weaknesses.
3. How to ensure the security of cloud computing environment in both online as well as offline mode.

The following are hypothesis:

H0: The defection of malware is possible in an efficient way when the system is online.
H1: The performance cost related to detection of malware is platform dependent.
H2: The detection approach shows different results on different hardwares.

2.7 LITERATURE REVIEW

The key concept of the security of cloud is that it should not only detect net-based threats but also make cloud capable of handling new challenges that targets cloud infrastructures. The paper has discussed an online cloud anomaly detection approach, compared detection components etc. The paper claims detection of malware and DoS attacks, authors not only evaluated the system-level data but also covered the network-level data depending on attack type and showed that their detection approach based on components monitoring per VM is applicable to cloud and its flexible detection system which can detect malware without any knowledge of their functionalities or underlying instructions.

Cloud datacenters are used at private, public, and commercial levels so should be able to handle all type of cyber-attacks; the properties of cloud, e.g. transparency and elasticity of its services, make it vulnerable. Cloud's dependency on IP networks makes it vulnerable . The current approach is based on resource-intensive deep packet inspection (DPI), which relies on payload information whereas the proposed approach is based on per flow meta statistics, derived from packet header and volumetric information which is packet and byte etc. count. Their approach targets the cloud and also integrates with infrastructure for detection and also for remediation. At infrastructure level they targeted the cloud nodes and network infrastructure that provides the connectivity within the cloud and with external services, cloud services are provided with one or more VMs which are interconnected. Cloud services are divided into three categories [5].

Software as a service: it has most control and gives limited access to the user.

Platform as a service (PaaS): it gives the choice to the user of execution environment, deployment tools but not the ability to be the administrator of their own operating system.

Infrastructure as a service (IaaS): it provides the most control, the user has the ability to install and administer their own choice of OS and run anything on the provided virtualized hardware, it is more sensitive and bit difficult to secure IaaS. The paper mainly focused this cloud service and the techniques are also applicable on the other services.

The paper discussed the approach that uses one class Support Vector Machine (SVM) algorithm and provides its effectiveness, tested this approach on malware and DoS in controlled environment.

Used malware samples were Kelihos and Zeus.

The experiments are performed on cloud. These experiments have used the implementation of the concepts which are based on Virtual machine. The results of the experiments have shown that online detection of the anomaly takes less time for the bulk of the data per VM with the help of SVM approach. The accuracy rate of these approaches is more than 90%.

The detection of malware in actual cloud is related to VM live-migration; SVM specific parameter estimation is used for better detection; they evaluated overall system, network-based or joint datasets.

Some other findings are:

- Computational cost is not very high for this approach.
- They used the sub modules of architecture's cloud resilience managers which is used in the detection at the end system.

The limitations are:

The methodology overall worked well and also improved the efficiency of the cloud, but it worked more efficiently on joint datasets than the other case.

Following are the recommendations:

- This approach is good when the system is online; offline mode techniques should also be secure and efficient.
- For joint dataset it needed to be improved.
- Datasets are basically under processed so data mining techniques can made to be more efficient [a].

Another popular approach is hypervisor-based malware detection. Virtual machines is used to store or process data of client machines; these virtual machines are targeted by cyber-attacks, e.g. VENOM (use to access hypervisor).

Approaches to malware detection is classified into distributed and hypervisor-based malware detection, distributed is VM agent running into the guest VM, remote monitoring server is monitoring its behavior, to use single point of control for the attack detection, it needs signature database.

Hypervisor-based malware detects malware within the guest by hypervisor; it protects the results, it makes it infeasible for deployment in production house.

The paper presented novel virtualization security system, combined system call monitoring and system call hashing in guest kernel with SVM-based external monitoring on host. With this approach malware and rootkit detection protect the guest against attacks without any extra burden [6].

The paper presented rootkit and malware detection system to protect the virtualization infrastructure against cyber-attacks in cloud environment.

- Approach used system call hashing and SVM together in VMI.
- It makes sure that the internal guest VM state can be accurately achieved.
- It also handles the offline SVM case.
- Offline SVM classifies allows quick attack classification.

Some recommendations to improve this approach are:

Some additional call can be added for achieving accuracy of attach detection in guest VM.

Data mining algorithms can also be applied for detection and tracking.

Attacks should be logged for avoiding in future.

Artificial techniques can improve the efficiency of the given approach. [b]

2.8 VKRD APPROACH

The paper discussed the kernel rootkit detection techniques previously used and then find the related limitation. The two most common methods previously used were classified into two categories

2.9 STATIC METHOD

2.9.1 Dynamic Method

Static method uses statistical analysis to differentiate malicious and legitimate kernel modules, but the limitation of this method is if there is any encryption technique implemented in kernel module this method is not capable to analysis the encrypted code.

To deal with this limitation the other approach is dynamic method. The key concept is the execution of kernel level modules in real environment and for later observation monitor its runtime behavior. The emulation technique used in this method is QEMU.

The limitation of this technique is performance cost, secondly some malicious function change their behavior in emulation environment, thirdly as this approach uses emulator therefor all kernel modules doesn't perform accurately in the emulator environment as they perform in actual hardware. On the basis of these limitations the paper proposed VKRD (visualization-based kernel rootkit detection system) [7]. The basic idea of this approach is to run targeted kernel level module & analyze its runtime behavior. The environment which VKRD provides kernel level target module is transparent & efficient. The approach used kernel memory & registers isolation approach with this approach transparency is ensured.

The visualization technique is hardware assisted for kernel data, registers and kernel code. Supervised machine learning technique is also implemented. A prototype system based on Xen hypervisor is designed & implemented. VKRD is online system which works on dynamical approach.

In this approach the targeted kernel module is isolated from OS kernel then dynamically intercept the interaction btw targeted module and OS kernel, when the behavior is recorded then the next step is to find is it regular process or its malicious process. With the help of hardware assisted visualization many features are extracted dynamically then these features are used to train detection model. These features are further classified into data access, code access & register access operations.

VKRD which is prototype based on Xen, separate targeted kernel module then in hardware assisted visualization monitors in VMM space. The approach intercept kernel modules that try to interact with memory or hardware registers. Analyzer is a detection component that runs in user space. Analyzer is use to detect the running process is malicious process of legitimate process. DomO VM is user space which is for VM. The log files which are resided in VMM logs all the operations, the analyzer has access of these logs. A timer in VMM is used to notify the analyzer to analyze the logs. If the analyzer detects any operation which needs to access kernel and is different than the regular access demanding operations if will generate an alert.

In regular operating system, kernel module & OS kernel both share the same memory region which makes it difficult to detect kernel module behavior to address this the following approach used Intel Extended page Table (EPT) technology. VMM also ensures that kernel modules memory area is not in the access of other data and code session. MmLoadsystemImage is kernel function which is used to use return addresses of kernel modules functions. If the kernel module tries to access kernel data this access will result kernel Exit call. Emulation technique is also used to emulate common data access operation.

The kernel module also access the hardware registers, intel VT technology is used to monitor the access. VM execution control field in VMCS is configured, with this configuration (28th & 31th bit of processor based VM) the access to the registers can be monitored which can detect the malicious operations. With this configuration any access to the specific registers will result into VM Exit.

Whenever there is any access VMM needs to check whether the faulting address is within the code memory region if the address isn't in the region it indicates that the access operation is from other kernel module.

Another important part is feature extraction, with this the runtime behavior of the kernel is observed. For feature extraction important operations are:

1. Code access operations
2. Data access operations
3. Register access operations

The related features of these operations shows the runtime behavior of kernel module. Each of them is either a binary flag or a counter number.

Flag checks the presence of an attribute while the counter shows the occurrence of a particular event.

For the features extraction machine learning based detection approach is used. Following machine learning algorithms have been used.

1. Decision tree
2. KNN
3. SVM
4. Random Forest

As compared to SVM & KNN the performance of random forest & Decision tress is 96.74 % & 95.11%.

Findings:

- Accuracy of random forest & decision tress is much better than KNN & SVM.
- DECAF shows significant performance overhead as compared to VKRD.

Limitation:

- Some kernel rootkits exploit the VM aware technique and can hide their malicious behavior which can lead to bypass the VKRD detection approach.
- Memory isolation technique can result considerable performance overhead.

Previous Approaches:

1. A learning-based detection [8]
2. Information extraction and reconstruction techniques at the hypervisor level [9]
3. Online cloud anomaly detection approach [10]
4. VKRD Method [11]

Limitation:
To defeat kernel rootkits, many approaches have been proposed in the past few years (Figure 2.2, Table 2.2). However, existing methods suffer from some limitations:

1. Most methods focus on user-mode rootkit detection
2. Some methods are limited to detect complicated kernel modules
3. Some methods introduce significant performance overhead

Selected Method:
Based on comparison of different approached used previously the Volatile Kernel Rootkit Hidden Process Detection View (VKRHPDV) is effective solution of kernel level rootkit detection. This procedure accuracy is great and the rootkit detection time is faster as compared to other approaches.

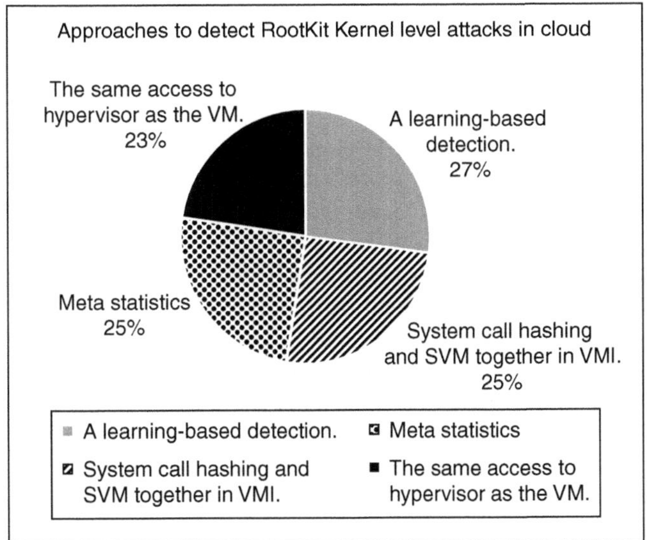

FIGURE 2.2 Comparison btw different approaches.

TABLE 2.2
Other Approaches to Detect Rootkit

Paper	Approach
Alhassan, J. K., S. O. Subairu, and S. Misra. "Evaluating Capabilities Of Rootkits Tools." [7]	PoKeR(Profiler for Kernel Rootkits)
Kruegel, Christopher, William Robertson, and Giovanni Vigna. "Detecting kernel-level rootkits through binary analysis." 20th Annual Computer Security Applications Conference. IEEE, 2004. [6]	static analysis-based technique for detecting
Liu, Leian, et al. "Research and design of rootkit detection method." Physics Procedia 33 (2012): 852–857. [12]	called X-Anti. X-Anti is based on the overall rootkit detection program
Bowman, Michael, Heath D. Brown, and Paul Pitt. "An undergraduate rootkit research project: How available? How hard? How dangerous?." Proceedings of the 4th annual conference on Information security curriculum development. 2007. [4]	Latter based approach which was able to get complete, stealthy, access, and control of targeted machine
Joy, Jestin, Anita John, and James Joy. "Rootkit detection mechanism: a survey." International Conference on Parallel Distributed Computing Technologies and Applications. Springer, Berlin, Heidelberg, 2011. [13]	host-based & virtualization-based approach

This method uses the approaches as cross view and clean boot based and it defines a process monitoring frame work that continuously monitors of all running processes and can recognize any unidentified rootkit with the less amount of performance overhead as compare to other techniques.

The three components of VKRHPDV are:

i. Process Monitor
ii. Comparing Process Analyzer
iii.Contaminated Process List

Different system calls are used to interpret the process activities; process activities & sequences are monitored. For this monitoring system calls are interpreted by different OS libraries. Process monitoring also manage process cleaning as in this step all the activities are monitored including starting and finishing activity of all the processes. A clean process list is created and maintained by process monitoring phase in following steps:

- Initialize the clean process list in the boot step of the server
- Every new process will create new procedure for the collection of information
- The process list will be updated & clean whenever there is any new instant/ event is generated

The creation of new process list is dynamically maintained. System calls differentiates new and existing functions. The system calls are important in process monitoring, with system class each process is monitored. System call itself can be attacked which is treat to this approach. By attacking system calls any malicious process can breach this monitoring process. To avoid this risk hash tables can also be used. It will add the performance overhead but with this two way authentication the risk can be avoided or minimized. The system calls and hash tables both verify the process and then updated the process cleaning list. In hash tables each process is logged. Process id is assigned to process, the process is logged into hash table. Whenever a process is initiated the system calls examines the process and the table is also used to check whether it's a new process or normal existing process.

Another concern is hidden rootkit processes. One approach can be list these processes in cleaning process list to minimize the performance overhead but this is not an efficient way as these hidden processes need to be monitor separately [12, 13]. The behavior of hidden process is different than the other malicious process. For this another list is created to log theses hidden processes. So basically, two lists are created:

- lean process list
- idden process list

Process analyzer is basically comparing generated lists, the analyzer is confirming if the process is regular process, if the analyzer stops generating confirmation notification it means the process on which it stopped is attacked process. The basic working algorithm is following:

Step 1: create clean process list
Step 2: create hidden process list
Step 3: compare process lists
Step 4: if the comparison generates alert (malicious process detection) terminate the loop
Step 5: log the process in log file as tainted process

To make the process efficient the log files can be used for the detection of malware while the system is in offline mode. The treats can be harmful for the targeted kernel as well as the client machines associated with targeted machine. The timer can also be effective to generate alerts when the analyzer detects any attack. The extract of the method is the analyzer analyzes every process, if the process is harmless then the clean process list is updated otherwise an alert is generated, the system call terminates the running process.

These lists can be beneficial to train the system as well. By extracting the features of a regular process & tinted process the system can be trained to detect the malware by itself. Different machine learning algorithms can be used for the training, KNN, SVM, decision tress etc. algorithms can classify the processes on the basis of its extracted features.

The flow of a regular process & a tainted process is given in Figures 2.3 and 2.4.

Identifying open ports on a target system is the next step to defining the attack surface of a target. The general protocols used for port scanning are TCP (transmission control protocol) and UDP (user datagram protocol) (Figure 2.5). They are both data transmission methods for the internet but have different mechanisms.

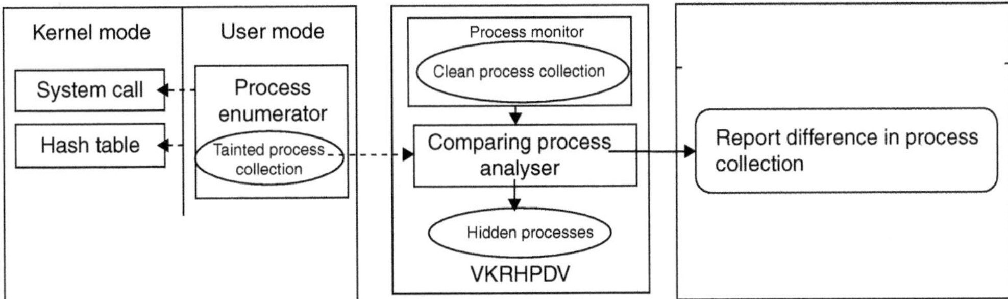

FIGURE 2.3 Tainted process detection.

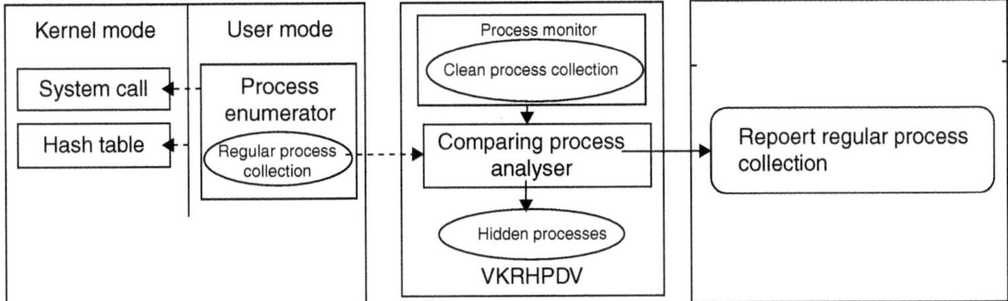

FIGURE 2.4 Clean process.

Both TCP and UDP are transport protocols. Transmission Control Protocol (TCP) is the more commonly used of the two and provides connection-oriented communication. User Datagram Protocol (UDP) is a non-connection-oriented protocol that is sometimes used with services for which speed of transmission is more important than data integrity. The penetration testing technique used to enumerate these services is called port scanning.

While TCP is a reliable, two-way connection-based transmission of data that relies on the destination's status in order to complete a successful send, UDP is connectionless and unreliable. Data sent via the UDP protocol is delivered without concern for the destination; therefore, it is not guaranteed that the data will even make it.

Using these two protocols, there are several different techniques for performing port scans.

- Ping scans
- Half-open or SYN scans
- XMAS scans

Port scan results reveal the status of the network or server and can be described in one of three categories: open, closed, or filtered.

Open ports: Open ports indicate that the target server or network is actively accepting connections or datagrams and has responded with a packet that indicates it is listening. It also indicates that the service used for the scan (typically TCP or UDP) is in use as well.

Finding open ports is typically the overall goal of port scanning. There are many port detection methods some are:

- Ping Scan
- ACK Scan
- Nmap's (network mapping) scan method

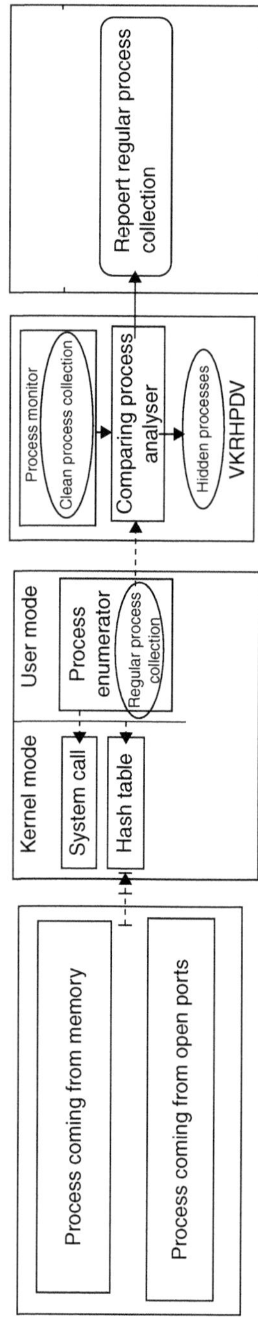

FIGURE 2.5 Process entry point source detection.

These are Nmap's workhorse scans, and they're the default scan methods because they can identify open ports in almost any situation. The next step is to identify from which path the process is coming. The process entering into the system can come from following:

1. Memory
2. Open ports

The process which is coming from memory needs to be compared with the cleaning list, memory originated processes are less likely to be the threat for the system. If the process is coming from open ports then it needed to be scan more carefully as compared to the process coming from memory.

2.10 FUTURE WORK /RECOMMENDATIONS

Future work will be done on memory levels and port numbers identification for the detection of process source.

REFERENCES

[1] CloudMon. Monitoring virtual machines in clouds. *IEEE Transaction on Computers*, 65, Dec 2016.
[2] Eresheim, Sebastian. "The Evolution of process Hiding Techniques in Malware-Current Thread and Possible Countermeasures." *Journal of Information Processing*, 25 (866–874), Sep 2017.
[3] Win, Thu Yein, Huaglory Tianfield, and Quentin Mair. "Detection of Malware and Kernel-level Rootkits in Cloud Computing Environments". IEEE, NY –USA, 2015.
[4] Bowman, Michael, Heath D. Brown, and Paul Pitt. "An undergraduate rootkit research project: How available? How hard? How dangerous?" *Proceedings of the 4th Annual Conference on Information Security Curriculum Development*, 2007.
[5] Infrastructures." *IEEE Transactions on Dependable and Secure Computing*, 13 (192–205), 2015.
[6] Kruegel, Christopher, William Robertson, and Giovanni Vigna. "Detecting kernel-level rootkits through binary analysis." *20th Annual Computer Security Applications Conference*. IEEE, 2004.
[7] Alhassan, J. K., S. O. Subairu, and S. Misra. "Evaluating capabilities of rootkits tools." 2016.
[8] Singn, Baljit, and Dmitry Evtyushkin. "On the detection of kernel level rootkits using hardware performance counters." Asia CCS, April 2017.
[9] Xie, Xiongwei, and Weichao Wang. "Rootkit detection on virtual machines through deep information extraction at hypervisor-level." 2013.
[10] Watson, Michael R, Noor-ul-hassan Shirazi, Angelos K. Marnerides, Andreas Mauthe, and David Hutchison. "Malware Detection in Cloud Computing
[11] Tian, Donghai, Rui Ma, Xiaoqi Jia, and Changzhen Hu. "A kernel rootkit detection approach based on virtualization and machine learning." *IEEE Access*, 7 (91657–91666), 2019.
[12] Liu, Leian, et al. "Research and design of rootkit detection method." *Physics Procedia*, 33 (852–857), 2012.
[13] Joy, Jestin, Anita John, and James Joy. "Rootkit detection mechanism: a survey." *International Conference on Parallel Distributed Computing Technologies and Applications*. Springer, Berlin, Heidelberg, 2011.

3 Privacy Preserved Federated Learning with Optimized Differential Privacy (FL-ODP)

Maria Iqbal, Asadullah Tariq, Muhammad Adnan, Irfan Ud Din, and Tariq Qayyum

3.1 INTRODUCTION

Machine learning (ML) has gained significant attention in both academic and industrial settings, with applications spanning diverse domains [1,2]. A decentralized learning system enables a group of individuals to collaboratively participate in training an ML model without compromising the privacy of their respective training datasets [3–7]. Privacy regulations, such as the General Data Protection Regulation (GDPR) and the California Consumer Privacy Act (CCPA), have mandated strict privacy standards for handling personal data [3,8]. These regulations preserve the person's privacy rights by verifying that their user data is collected, accessed, and stored securely. Privacy preservation is crucial for protecting sensitive information and preventing data leakage. In response to concerns regarding personal data privacy, a decentralized approach has been introduced for collaborative learning [9,10]. Federated Learning (FL) [11,12] enables scalable and collaborative machine learning while preserving data privacy. FL enables individuals to narrowly convey a model and contribute only model parameters with others, rather than sensitive training data. Despite its privacy-preserving benefits, FL is not immune to privacy risks, as highlighted by various studies. These include inference attacks during the learning phase, deriving private information from trained models, and model inversion attacks, as outlined in [12], [13], and [14] respectively. To mitigate privacy risks associated with FL, Differential Privacy (DP) [5] has been presented as a learning framework [3,4]. A particular approach involves a trustworthy aggregator managing the privacy exposure to assure the model's performance privacy. This paper presents a practical framework, named Privacy Preserved Federated Learning with Optimized Differential Privacy (FL-ODP), which addresses the issue of private information leakage under FL scenarios. FL-ODP aims to fully prevent privacy breaches during the FL process.

Our main contributions are:

1. The application of DP with FL to enhance privacy.
2. DP with different parameters is tested to optimize results.
3. We illustrate the quantitative results of this study about the trained models' accuracy and their corresponding privacy guarantees achieved.
4. The optimized results show that with constant values of epsilon, different noise levels and delta result in improved privacy protections.

3.2 LITERATURE REVIEW

Privacy preservation is a crucial concern in machine learning [3,5], particularly in the generation of big data where a huge amount of individual data is possessed and analyzed [16,17]. There have been

 DOI: 10.1201/9781003497851-3

extensive research efforts in developing privacy-preserving techniques [18,19] for machine learning, including differential privacy [18,20–22], homomorphic encryption [23,24], and secure multi-party computation [25,26]. FL and DP have received considerable attention among these techniques due to their strong privacy promises and versatility in various machine-learning settings [3,20].

FL is a distributed machine learning paradigm that enables multiple devices to train a global model collaboratively without the need for sharing their raw data. This approach allows for the collaborative training of a model while preserving the privacy of individual devices' data. Federated learning ensures user privacy through encryption, differential privacy, and security integration, preventing breaches of regulations and leakage of user data. It offers several advantages over traditional centralized machine learning approaches [27,45], such as improved scalability, reduced communication overhead, and the ability to handle data that is not centralized [5,43,44]. FL has been successfully applied in various domains [28]. However, FL also poses several challenges, including privacy and security concerns, communication efficiency, and heterogeneous data distribution [43]. FL also requires specific architectural patterns for effective deployment [11]. The architectural design of an FL system should consider several factors, including the number of participating devices, their computing and communication capabilities, and the heterogeneity of the data. Several architectural patterns, like FL as a service (FLaaS), edge learning, and federated transfer learning, have been proposed to address these challenges [12,27].

DP is a preferred approach for improving privacy in FL [5]. DP provides a mathematical framework that guarantees the assurance of personal data, like in the existence of a malicious aggregator [28]. DP achieves this by adding a controlled amount of noise to the data before it is transmitted to the aggregator. The noise level is controlled by a privacy parameter, which gets to decide the trade-off between privacy and utility [29]. Certain DP-based FL approaches have been introduced in recent years, similarly DP-SGD [30], DP-FedAvg [31], and DP-FedSGD [32]. The aggregation process is a critical component of FL, which involves connecting the model to modernize from a couple of devices to create a global model [33]. Aggregation methods can significantly affect the performance of FL in terms of convergence speed, accuracy, and communication efficiency. The scientific literature has proposed several aggregation methods for FL, including averaging-based techniques, weighted averaging, and secure aggregation [34–36]. To ensure strong privacy guarantees, secure aggregation methods such as secure multi-party computation (MPC) and homomorphic encryption (HE) enable devices to jointly compute the global model without disclosing their updates [36,37]. According to federated optimization, these properties are critical:

$$\forall s, u_{r+1}^{s} \leftarrow u_r - \mu \nabla g\left(u_r\right) \tag{3.1}$$

The parameters of client s, denoted as u_{r+1}^{s}, and the learning rate μ at which the parameters are computed are utilized in computing an updated global model. To improve the global model, the central server aggregates the parameters u_{r+1}^{s} by determining a weighted average of the updates from each client. This can be mathematically expressed as:

$$u_{r+1} \leftarrow \sum_{s=1}^{L} \frac{m_s}{m} u_{r+1}^{s} \tag{3.2}$$

Though u_{r+1} is the latest version of the global model, m_s Is the count of the data points at client s and m is the number of clients.

Recent research has proposed several approaches related to FL with DP. In [38], the authors introduce FL algorithms that incorporate DP and provide a performance analysis. In [39], the authors suggest a decentralized deep learning approach that utilizes differential privacy to enhance privacy and security. The authors of [40] propose a multi-objective Federated Learning framework that incorporates differential privacy to balance the trade-off between privacy and accuracy. Furthermore, [41] presents FedDP, a comprehensive framework for privacy-preserving Federated Learning with differential privacy that guarantees privacy and performance across various scenarios.

The main inspiration of this study is to analyze the usage of federated learning with DP to achieve high privacy in machine learning. Specifically, varying the noise and delta values while keeping the epsilon value constant is expected to identify the optimal configuration that maximizes privacy without compromising model accuracy.

3.3 METHODOLOGY

3.3.1 Data Interpretation

The methodology of this study involves analyzing the MNIST database [6], a commonly used dataset for recognizing handwritten digits and assigning them their corresponding numerical values. The training dataset comprises 50,000 examples, with each example being a gray-level image of size 28 by 28 pixels. These images represent handwritten digits. We are utilizing a basic feed-forward neural network that includes Rectified Linear Unit (ReLU) activation functions and a SoftMax output layer consisting of 10 classes, representing each of the 10 digits. A training procedure was run for 50 epochs with $C = 1.0$, $\sigma = 1.2, 2.0, 4.0, 8.0$, $\delta = 1e - 5$ and $1e - 3$, and a learning rate of 0.01, corresponding to an overall target privacy loss of = 0.5.

3.3.2 Research Problem

This research aims to address privacy concerns related to user-generated data in a federated learning environment. The proposed solution is a differential privacy model that locally stores data on the device and adds noise before transmitting it to the server. By doing so, personal data can be secured, and the problems associated with dispersed data can be reduced. The decentralized methodology using differential privacy techniques aims to provide a secure environment for federated learning, addressing privacy concerns and contributing to privacy-preserving machine learning.

3.3.3 Method of Analysis and Classification

3.3.3.1 Preprocessing

The MNIST dataset is a collection of images of handwritten digits that are utilized for training and testing ML models. The images are pre-processed through various techniques such as grayscale conversion, pixel value normalization, and data augmentation to ensure consistency and accuracy, and improve the robustness and generalization performance of the models. The pre-processed images are subsequently employed in training and evaluating machine learning models, which renders the MNIST dataset a highly preferred option for assessing the efficacy of various machine learning algorithms.

3.3.3.2 Hyperparameter

As a component of this investigation, a sequence of experiments was carried out employing the MNIST dataset, with non-independent and identically distributed (non-iid) data, to analyze the influence of modifications to the hyperparameters of noise and delta on the effectiveness of federated learning. The experiments involved ten clients, each with a batch size of 64 and an output size of 10. For the MLP model, I set the value of epsilon to 0.5 and used five local iterations (E) and a sampling rate (Q) of 0.1. I conducted the experiments for 50 rounds to evaluate the convergence of the algorithm. In each experiment. The value of the Laplace noise 1.2, 2.0, 4.0, and 8.0, which were included in the model's output, was modified during the experiments. I also tested the impact of different delta values, setting it to 1e-5 in some experiments and 1e-3 in others. By varying these hyperparameters, I sought to explore their influence on the adjustment between privacy and accuracy in FL.

3.3.3.3 (ε)-Differential Privacy and (ε, δ)-Differential Privacy

DP is a technique for preserving privacy that is defined by two terms: (ε)-DP and (ε,δ)-DP. The former is a stricter definition that does not allow any leakage, whereas the latter is a more practical definition that permits a small amount of accidental leakage. For this study, we utilized the MNIST dataset, which encompasses 50,000 images for training and 10,000 images for testing, and was segregated into 10 separate clients. To uphold privacy, we introduced Laplace noise into the average value of the dataset, denoted as x_i. To incorporate Laplace noise, the following equation was utilized: $y = \mu + Lap(b)$, where μ denotes the actual average height of the dataset, and $b = 1/\varepsilon$. Here, Lap(b) denotes a random variable, drawn from the Laplace distribution, where the scale is b. The value of δ should be chosen such that it is less than the inverse of the size of the dataset, to prevent privacy violations. The (ε)-DP term cannot guarantee privacy when the Gaussian mechanism is used, which is why (ε,δ)-DP was introduced. The possibility of any specific data point x_i being disclosed is given by: Pr[|y - x_i| ≤ t] = 2 * exp(-ε * t) <= δ, where t is the sensitivity of the data and δ is the desired privacy parameter. The Laplace noise added to the average height can be varied in different experiments, such as 1.2, 2.0, 4.0, and 8.0 while keeping ε fixed at 0.5. Moreover, the value of δ can be altered between 1e-5 and 1e-3. If ε=0.5 and δ=10^-6, the value of b would be b = 2 * 10^4. The Laplace noise added to the average height would then be within the range of ±b, ensuring that the probability of any individual data point being disclosed is less than 10^-6.

3.3.3.4 Classification

In this research, FL and FL with DP have been implemented using MLP models on the MNIST dataset, where Laplace noise [42] has been added to the model updates for ensuring privacy and security. The SGD optimizer and FedAvg aggregation method have been used with hyperparameters specified in Table 3.1. The study shows the effectiveness of Federated Differential Privacy in achieving better privacy and security while maintaining model accuracy. The use of the FedAvg aggregation method has also helped in achieving better concurrence of the models. Overall, the results demonstrate the effectiveness of Federated Learning with Federated Differential Privacy using Laplace noise for training models in a distributed and privacy-preserving manner.

3.3.3.5 Privacy Analysis

This experiment demonstrates privacy preservation in federated learning. I employed differential privacy by introducing varying noise and delta values while keeping the epsilon value fixed. DP is a privacy-preserving method that computes random noise to the data to protect sensitive information, ensuring that the statistical analysis is still accurate while opposing the privacy of the user's data points. By varying the noise and delta values, efforts are made to strike a balance between accuracy and privacy. To ensure proper privacy preservation, I used statistical measures to analyze the amount of privacy loss and ensured that the delta values were set appropriately to maintain the desired level of privacy. Overall, my research focused on developing effective privacy-preserving techniques for federated learning, enabling the analysis of sensitive data while opposing the privacy of particular users.

3.4 EXPERIMENTS AND RESULTS

The following section will provide a comprehensive evaluation of both the quantitative and qualitative outcomes of each experiment. The quantitative analysis will include statistical measures such as accuracy, and loss, whereas the qualitative analysis will focus on assessing the model's efficacy in answering the research questions. The objective of the experiments was to assess the effect of several factors, including the number of clients, dataset size, and training round count, on the model's performance. Therefore, the results will provide valuable insights into the optimal conditions for deploying federated and differential privacy learning.

TABLE 3.1
Hyperparameter Federated Learning and FL with Differential Privacy

	Federated Learning	FL with Differential Privacy
Hyperparameter	**MNIST (IID and Non-IID)**	**MNIST(Non-IID)**
Number of clients	100	10
Batch Size	32	64
Epoch	5	5
No of rounds	50	50
Learning rate	0.1	0.01
Optimizer	SGD	SGD
Model	MLP	MLP
Output size	x	10
Delta	x	1e-5,1e-3
Epsilon	x	0.5
Q (sampling rate)	x	0.1
Noise	x	1.2,2.0,4.0,8.0
Aggregation method	FedAvg	FedAvg

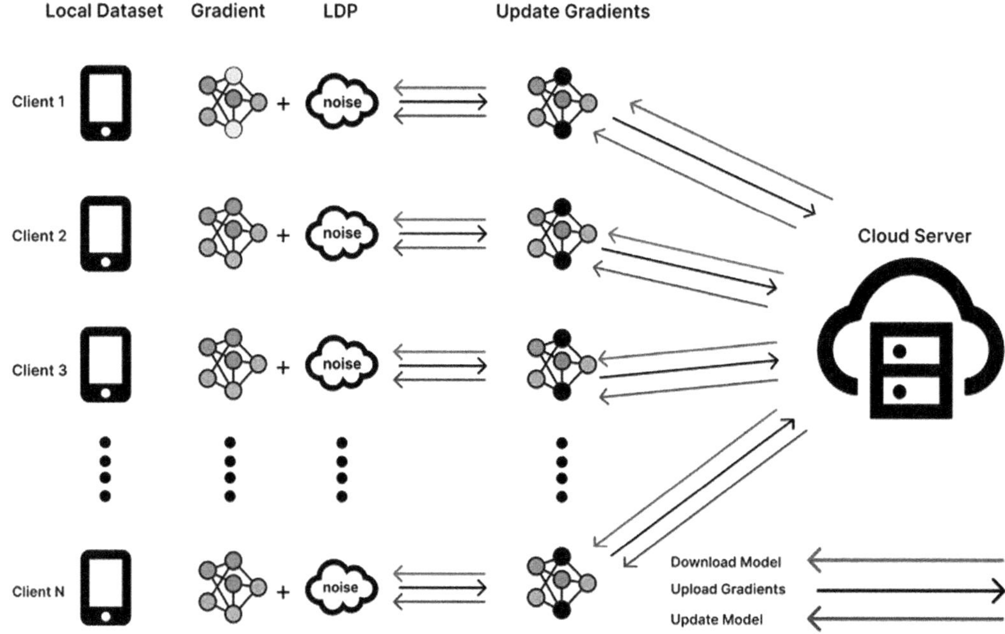

FIGURE 3.1 FL-ODP architecture.

3.4.1 Simple Federated Learning

In the initial phase of my federated learning experiment, I used the MNIST dataset and a multi-layer perceptron (MLP) model. Two types of experiments were conducted: IID and non-IID. Figures 3.1 and 3.2 present the findings of the experiments. These figures demonstrate the performance of the model in both IID and non-IID scenarios. The results suggest that the model performs better in the IID scenario as compared to the non-IID scenario.

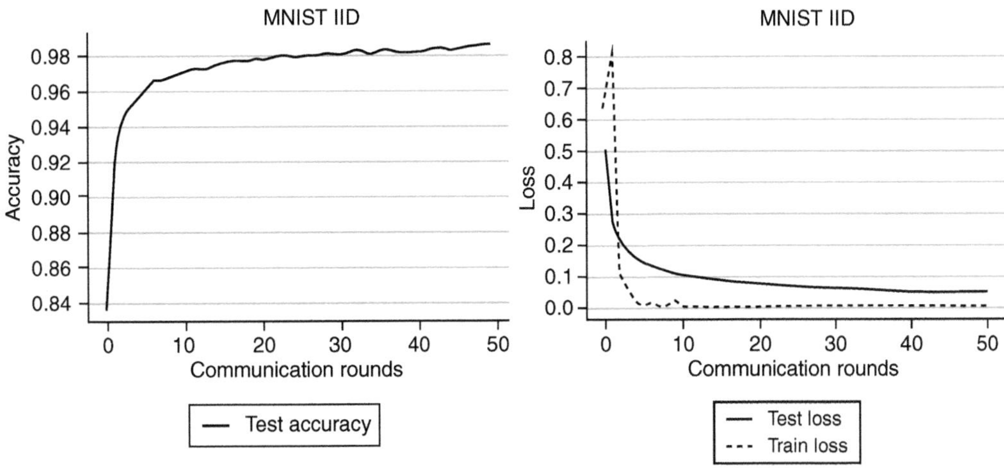

FIGURE 3.2 Test set accuracy and loss (MNIST IID) setting within federated learning. (a) Accuracy with IID setting (b) Loss with IID setting.

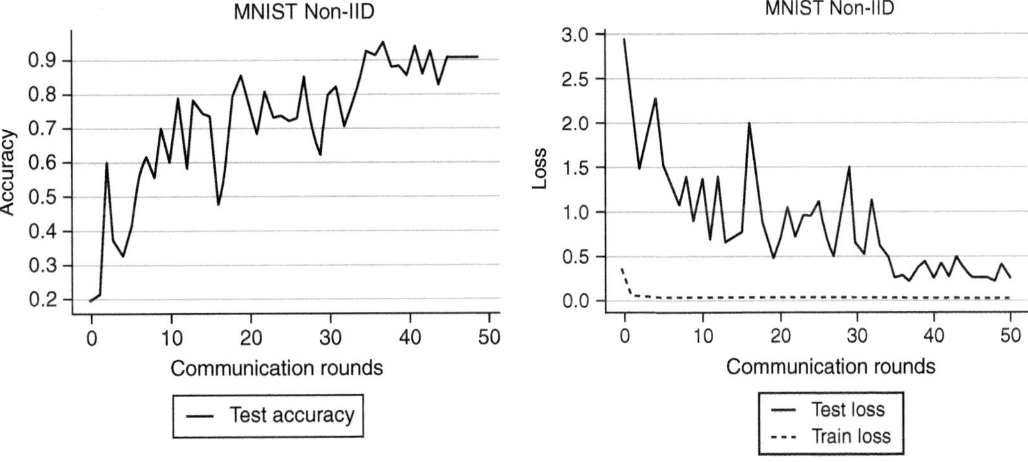

FIGURE 3.3 Test set accuracy and loss (MNIST Non-IID) setting within federated learning (a) Accuracy with non-IID setting (b) Loss with non-IID setting.

Figure 3.2(a) displays the accuracy achieved over 50 epochs using MNIST IID data. The hyperparameters mentioned in Table 3.1 were used. Accuracy gradually increased after each round and reached 98% in my experiment using an MLP model with a batch size of 32 and 100 clients. With each round of federated learning, the model becomes more accurate and better able to generalize to new data, as it is trained on a larger and more diverse set of data. This can lead to significant improvements in the overall performance of the model, even when the individual data samples are not sufficient for training a high-quality model on their own. In the 4.1(b) figure the training loss started at 0.6 and then increased to 0.8 initially. However, after reaching a certain point, it began to gradually decrease and eventually reached a plateau where it neither increased nor decreased significantly. In contrast to the training loss, the test loss showed a different pattern. The test loss started at 0.5 and then decreased in every epoch until it reached a point where it leveled off into a straight line with no significant change.

Figure 3.3(a) illustrates the accuracy achieved on MNIST non-IID data. Initially, the accuracy was low and started at only 0.20 percent. It fluctuated in every epoch and decreased or increased at multiple points. However, after reaching 40-plus epochs, the accuracy became more consistent and eventually reached 0.87 percent. The iterative process of federated learning endeavors to enhance the accuracy of a machine learning model by training it on data that is dispersed across multiple servers or devices. During each round of federated learning, a subset of servers or devices, commonly referred to as clients, engage in the training process by executing local computations on their respective data sets and sharing the updated model parameters with a central server. The central server subsequently consolidates the revised parameters from all clients and computes a novel collection of model parameters, which are then communicated back to the clients for the next round of training. Figure 3.3(b) shows that the training loss is consistently lower than the test loss, despite using the same hyperparameters of 100 clients and a batch size of 32 for both. The training loss steadily decreases over time, while the test loss exhibits fluctuations with some epochs showing an increase and some showing a decrease.

3.4.2 Federated with Differential Privacy (FDP)

The second phase of the federated learning experiment incorporated differential privacy to address privacy concerns that remained unresolved by FL alone. A non-IID dataset was used with specified parameters listed in Table 3.2. The experiment aimed to evaluate the effectiveness of differential privacy by varying the noise and delta values while keeping a constant epsilon value of 0.5. The objective was to identify the optimal noise-delta combination that would maintain a suitable level of privacy while maximizing the model's accuracy.

In the current experiment, the objective is to evaluate the effectiveness of per-example privacy outside the federated learning framework with non-independent and identically distributed (non-iid) samples. The MLP is trained for numerous epochs until it attains a predefined upper limit of (ε,δ)-differential privacy (DP) guarantee deemed acceptable. For the training procedure, a batch size of 64, a clipping threshold of 5, and a learning rate of 0.01 are utilized. The runs are terminated once the lowest accepted δ value of 1e-5 has been reached. This value serves as the threshold for terminating the training process. The MLP is trained using a batch size of 64, which represents the number of training examples processed in each iteration of the training algorithm. A clipping threshold of 5 is also used, which limits the magnitude of the gradients to prevent the model from overfitting the training data. Additionally, a learning rate of 0.01 is employed, which determines the step size at which the model updates its parameters during training. As the epsilon value increases, the model becomes less private but more accurate. Conversely, as the noise value increases, the model becomes more private but less accurate. Table 3.3 shows that as the noise value increases from 1.2 to 8.0, the model's accuracy decreases from 76.0% to 56.6%. This demonstrates the trade-off between privacy and accuracy, where higher privacy levels come at the cost of reduced accuracy. Additionally, as the epsilon value decreases from 2.47 to 0.25, the model's accuracy also decreases, showing that lower privacy levels also come at the cost of reduced accuracy. The summary can be seen in Table 3.4.

3.4.3 Federated with Optimized Differential Privacy (FL-ODP)

In the third phase of the study, Federated Learning with optimized Differential Privacy was implemented, keeping the noise and all other parameters at the same values as in Phase 2, but changing the delta value. The results of this experiment are presented in Table 3.2 and Figure 3.4.

TABLE 3.2
Shows the Results of IID and Non-IID

Model	IID (%)	Non-IID (%)	Rounds
MLP	98	87	50

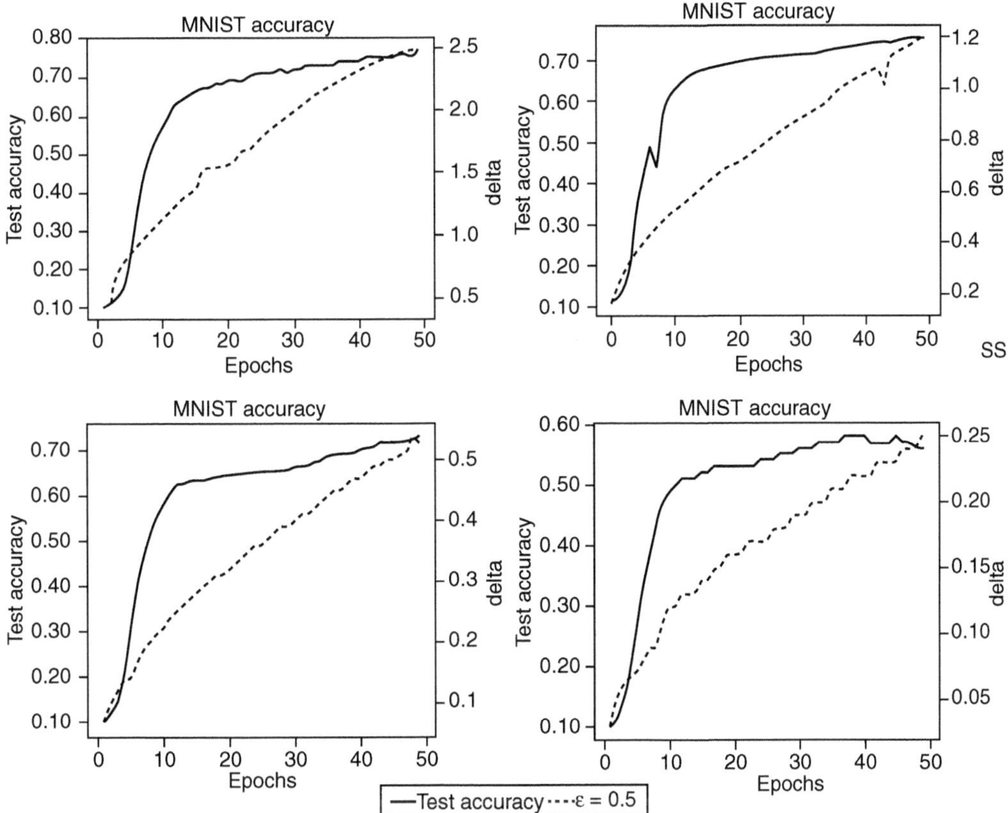

FIGURE 3.4 An illustration of the accuracy and epsilon results with delta(1e-5) for various noises on the MNIST non-IID dataset (a) noise, σ = 1.2 (b) noise, σ = 2.0 (c) noise, σ = 4.0 (d) noise, σ = 8.0.

TABLE 3.3
FL with DP Delta Value (Delta = 1e-5)

DP Parameters	FL with DP (Delta = 1e-5)			
σ (noise)	1.2	2.0	4.0	8.0
ε (epsilon)	0.5	0.5	0.5	0.5
Rounds	50	50	50	50
Accuracy	76.5	75.5	72.0	56.6
Epsilon	2.46	1.20	0.53	0.25

The experiment results show that the accuracy of the model remained consistent with the results from Phase 2 while improving the privacy of the model (Figure 3.5 and Table 3.4). These findings suggest that changing the delta value (1e-3), better privacy can be achieved without sacrificing the accuracy of the model.

3.5 CONCLUSION

In conclusion, this research highlights the effectiveness of Federated Learning with Optimized Differential Privacy (FL-ODP) techniques in addressing privacy issues in machine learning. The

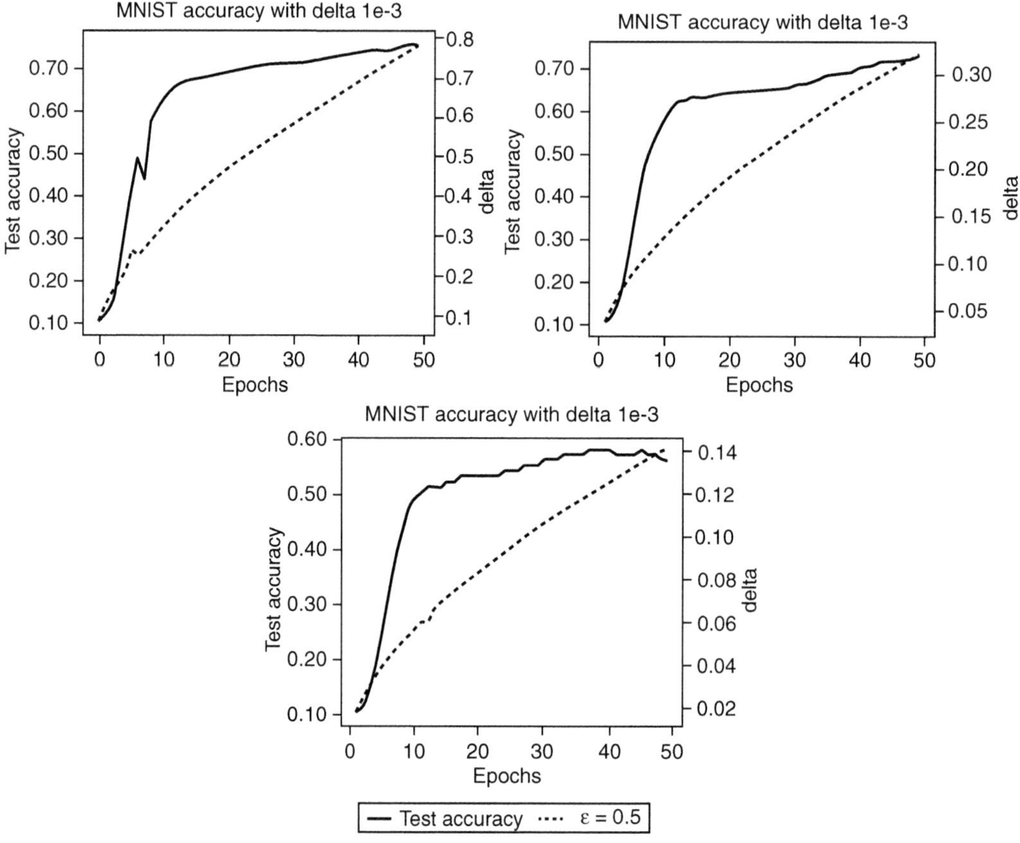

FIGURE 3.5 An illustration of the accuracy and epsilon results with delta(1e-3) for various noises on the MNIST non-IID dataset noise, σ = 2.0 noise, σ = 4.0 noise, σ = 8.0.

TABLE 3.4
FL with DP Delta Value (Delta = 1e-3)

DP Parameters	FL with ODP (Delta = 1e-3)			
σ (noise)	1.2	2.0	4.0	8.0
ε (epsilon)	0.5	0.5	0.5	0.5
Rounds	50	50	50	50
Accuracy	76.5	75.5	72.0	56.6
Epsilon	1.07	0.77	0.31	0.14

experiments conducted using the Opacus library demonstrate that adding noise and changing the delta value can significantly improve user privacy without negatively impacting the accuracy of the model. The findings also indicate that the level of noise added affects the accuracy of the model, with lower levels of noise resulting in a greater increase in accuracy compared to higher levels of noise. Moreover, the experiments show that changing the delta value can further enhance privacy without compromising the accuracy of the model.

3.6 FUTURE WORK

Future research in the field of differential privacy should aim to improve the mechanisms used to achieve privacy while maintaining utility. One of the challenges in determining an acceptable value of ε lies in the difficulty of quantifying which results are satisfactory and which are not. Additionally, in some cases, arbitrary values are chosen in the hopes that they will be suitable. It is therefore important to explore different values of ε to gain a better understanding of the consequences of using differential privacy mechanisms. Furthermore, research has shown that the choice of ε that achieves an acceptable level of privacy depends heavily on the specific implementation details. To address this, future research could focus on developing more robust and customizable differential privacy mechanisms that can better balance privacy and utility trade-offs. By doing so, we can advance the field of differential privacy and ensure that it is a reliable tool for protecting individuals' privacy in various applications.

REFERENCES

[1] Alpaydin, E. (2010). *Introduction to Machine Learning* (2nd ed.). Cambridge, MA: MIT Press.

[2] Jordan, M. I., & Mitchell, T. M. (2015). Machine learning: Trends, perspectives, and prospects. *Science*, 349(6245), 255–260. doi: 10.1126/science.aaa8415

[3] Abadi, M., Chu, A., Goodfellow, I., McMahan, H. B., Mironov, I., Talwar, K., & Zhang, L. (2016). Deep learning with differential privacy. In *Proceedings of the 2016 ACM SIGSAC Conference on Computer and Communications Security* (308–318). ACM, ACM, Vienna, Austria.

[4] Papernot, N., Song, S., Mironov, I., Raghunathan, A., Talwar, K., & Erlingsson, Ú. (2018). Scalable private learning with PATE. In *Proceedings of the 2018 Sixth International Conference on Learning Representations*. arXiv preprint arXiv:1802.08908.

[5] Dwork, C., & Lei, J. (2009). Differential privacy and robust statistics. In *STOC*, Vol. 9. ACM, 371–380.

[6] LeCun, Y., Cortes, C., & Burges, C. (1998). The MNIST database of handwritten digits.

[7] Sarathy, R., & Muralidhar, K. (2011). Evaluating Laplace noise addition to satisfying differential privacy for numeric data. *Transaction On Data Privacy*, *4*(1), 1–17.

[8] Hitaj, B., Ateniese, G., & Perez-Cruz, F. (2017). Deep models under the GAN: Information leakage from collaborative deep learning. arXiv preprint arXiv:1702.07464.

[9] Chen, M., et al. (2019). Blockchain-based federated learning: A privacy-preserving solution for industrial IoT. *IEEE Network*, 33(5), 211–216.

[10] Kairouz, P., et al. (2019). Advances and open problems in federated learning. arXiv preprint arXiv:1912.04977.

[11] Jakub, K., McMahan, H. B., Yu, F. X., Richtárik, Suresh, A. T., & Bacon, D. (2016). Federated learning: Strategies for improving communication efficiency. arXiv preprint arXiv:1610.05492 (2016).

[12] McMahan, H. B., Moore, E., Ramage, D., Seth H., et al. (2016). Communication-efficient learning of deep networks from decentralized data. arXiv preprint arXiv:1602.05629.

[13] Shokri, R., Stronati, M., Song, C., & Shmatikov, V. (2017). Membership inference attacks against machine learning models. In *2017 IEEE Symposium on Security and Privacy (SP)*.

[14] Zhang, W., Li, Y., Zhou, Z., Yang, X., & Yang, Y. (2021). CIDA: Certifiable Inversion Detection in Autonomous Systems. arXiv preprint arXiv:2108.10962. IEEE, 3–18.

[15] Shokri, R., & Shmatikov, V. (2015). Privacy-preserving deep learning. In *Proceedings of the 22nd ACM SIGSAC Conference on Computer and Communications Security* (pp. 1310–1321). ACM.

[16] Hasan, M. M., Ahsan, M. N., & Haque, A. T. (2022). A survey on big data security and privacy challenges and solutions. *Journal of Big Data*, 9(1), 1–31.

[17] Chen, W., Ma, J., Li, X., Li, J., & Li, X. (2021). A survey on big data-driven anomaly detection. *IEEE Access*, 9, 37563–37579.

[18] McMahan, B., Ramage, D., Talwar, K., & Zhang, L. (2017). Learning differentially private recurrent language models. In *Proceedings of the 2017 Conference on Empirical Methods in Natural Language Processing* (pp. 2854–2864). Association for Computational Linguistics.

[19] Bonawitz, K., Ivanov, V., Kreuter, B., Marcedone, A., McMahan, H. B., Patel, S., ... & Shin, H. (2019). Practical secure aggregation for privacy-preserving machine learning. In *Proceedings of the 2017 ACM SIGSAC Conference on Computer and Communications Security* (pp. 1175–1191). ACM.
[20] Dwork, C. (2008). Differential privacy: A survey of results. In *International Conference on Theory and Applications of Models of Computation* (pp. 1–19). Springer.
[21] Song, S., Shu, R., Bai, Y., Zhou, Z., Jiang, X., & Zou, J. (2019). Federated learning with differential privacy: Algorithms and experiments. *IEEE Journal on Selected Areas in Communications*, 37(6), 1205–1218.
[22] Yang, Q., Liu, Y., Chen, T., & Tong, Y. (2019). Federated machine learning: Concept and applications. *ACM Transactions on Intelligent Systems and Technology*, 10(2), Article 12, 1–19.
[23] Chen, C., Li, X., & Yang, J. (2021). Fully homomorphic encryption: A comprehensive survey. *Journal of Network and Computer Applications*, 181, 103074.
[24] Kim, J., Ju, H., Kim, S., & Paik, J. (2020). A survey on homomorphic encryption schemes: Theory, implementation, and applications. *Journal of Information Processing Systems*, 16(4), 836–870.
[25] Lindell, Y., & Pinkas, B. (2017). A brief survey of multi-party computation. In *Advances in Cryptology–CRYPTO 2017* (pp. 388–408). Springer.
[26] Zhang, Y., Huang, H., Liu, J., & Li, J. (2021). Secure multi-party computation: A comprehensive survey. *IEEE Communications Surveys & Tutorials*, 23(2), 1042–1081.
[27] Lo, S. K., Lu, Q., Zhu, L., Paik, H. Y., Xu, X., & Wang, C. (2022). Architectural patterns for the design of federated learning systems. *Journal of Systems and Software*, 191 (p111357), 2022.
[28] Dwork, C., Rothblum, G. N., & Vadhan, S. (2010). Boosting and differential privacy. In *2010 IEEE 51st Annual Symposium on Foundations of Computer Science* (51–60). IEEE, IEEE.
[29] Bonawitz, K., Ivanov, V., Kreuter, B., Marcedone, A., McMahan, H.B., Patel, S., Ramage, D., & Seth, K. (2017). Practical secure aggregation for privacy-preserving machine learning. In *Conference on Computer and Communications Security (CCS)* (pp. 1175–1191).
[30] Zhang, M., Wang, Y., Chen, X., Sun, Y., & Yang, Q. (2021). A survey on federated learning: Challenges, methods, and future directions. *IEEE Transactions on Knowledge and Data Engineering*, 33(7), 2895–2917.
[31] Kairouz, P., McMahan, H. B., Avent, B., Bellet, A., Bennis, M., Bhagoji, A. N., … & Song, D. (2021). Advances and open problems in federated learning. *Foundations and Trends® in Machine Learning*, 14(1), 1–161.
[32] Chen, X., Liu, C., Wang, Y., Zhang, M., Sun, Y., & Yang, Q. (2021). Learning with differential privacy: A survey. *IEEE Transactions on Knowledge and Data Engineering*, 33(9), 3574–3594.
[33] McMahan, B., Moore, E., Ramage, D., Hampson, S., & Arcas, B. A. (2017). Communication-efficient learning of deep networks from decentralized data. In *Artificial Intelligence and Statistics* (pp. 1273–1282).
[34] Konečný, J., McMahan, H. B., Ramage, D., & Richtárik, P. (2016). Federated optimization: Distributed machine learning for on-device intelligence. arXiv preprint arXiv:1610.02527.
[35] Bonawitz, K., Salehi, F., Konečný, J., McMahan, B., & Gruteser, M. (2019, November). Federated learning with autotuned communication-efficient secure aggregation. In *2019 53rd Asilomar Conference on Signals, Systems, and Computers* (pp. 1222–1226). IEEE.
[36] Elkordy, A. R., & Avestimehr, A. S. (2022). Heterosag: Secure aggregation with heterogeneous quantizetion in federated learning. *IEEE Transactions on Communications*, 70(4), 2372–238
[37] Acar, A., Aksu, H., Uluagac, A. S., & Conti, M. (2018). A survey on homomorphic encryption schemes: Theory and implementation." *ACM Computing Surveys (Csur)*, 51(4), 1–35.
[38] Wang, Y., Chen, J., Liu, Y., & Tong, H. (2020). Federated learning with differential privacy: Algorithms and performance analysis. *IEEE Journal on Selected Areas in Communications*, 38(8), 1717–1728.
[39] Bagdasaryan, E., Veeling, B. S., & Van Der Schaar, M. (2020). Differential privacy for decentralized deep learning. *Journal of Machine Learning Research*, 21(91), 1–36.
[40] Liu, C., Wang, T., & Song, L. (2021). Multi-objective federated learning with differential privacy. *IEEE Transactions on Information Forensics and Security*, 16, 3277–3291.
[41] Cai, R., Chen, Z., Huang, S., & Yang, W. (2021). FedDP: A general framework for privacy-preserving federated learning with differential privacy. In *Proceedings of the 2021 IEEE International Conference on Communications (ICC)* (pp. 1–6). IEEE.

[42] Sarathy, R., & Muralidhar, K. (2011). Evaluating Laplace noise addition to satisfy differential privacy for numeric data. *Transaction on Data Privacy*, *4*(1), pp.1–17.

[43] Mahlool, D. H., & Abed, M. H. (2022). A Comprehensive Survey on Federated Learning: Concept and Applications. arXiv preprint arXiv:2201.09384.

[44] Mammen, P. M. (2021). Federated learning: Opportunities and challenges. arXiv preprint arXiv:2101.05428.

[45] The White House. (2013). Consumer data privacy in a networked world: A framework for protecting privacy and promoting innovation in the global digital economy. *Journal of Privacy and Confidentiality*, 2013.

[46] Nasr, M., Shokri, R., & Houmansadr, A. (2019). Comprehensive privacy analysis of deep learning: Stand-alone and federated learning under passive and active white-box inference attacks. In *2019 IEEE Symposium on Security and Privacy (SP)*. IEEE, ACM.

4 Cyber Physical Systems Having Nodes/Layers with Trustworthy Data

A Review

Muhammad Ejazulghaffar, Muhammad Arif, M. Arfan Jaffar, and Ahmed Hassan

4.1 INTRODUCTION

In order to construct a special system that can interact with the physical environment in real-time, a cyber-physical system, or CPS, combines physical and cyber segments, including software, hardware and networks. CPS is used in various fields, such as transportation, healthcare, manufacturing, and energy. CPS relies on sensors, actuators, and communication networks to collect and transmit data between the physical and cyber components. The system analyzes this data to control physical processes and enable automated decision-making. Examples of CPS include autonomous vehicles, smart cities, and medical monitoring systems. CPS is essential for creating efficient, reliable, and secure systems that can operate in complex environments. As technology continues to advance, CPS will become more prevalent in everyday life and play an increasingly important role in shaping the future of our world.

The cornerstone of Industry 4.0, as promoted by the Government of Germany for the construction of smart factories to usher in the industrial revolution 4.0, is the cyber physical system (CPS). When a workshop under National Science Foundation (NSF) convened in Austin, Texas, in 2006, the word "CPS" was first used. "A system made up of cooperative entities, capable of computation, and agents with a close relationship to the surrounding physical environment and phenomena, employing and simultaneously delivering network services for the treatment and transfer of data" was how it was described. Storage, processing, and network management functions have increasingly moved to central data centers, IP mainstay networks, and primary networks over time. The amount of data being gathered from Internet of Things (IoT) gadgets is growing at an exponential rate, necessitating careful processing of the data to reduce reaction time and efficiently disperse network burden. Several research have revealed the big-data handling limitations of the cloud. The number of "smart" and "dumb" objects linked to the Internet is projected to reach 50 billion by the year 2020 [1]. A common example of a CPS is a production system, which integrates a number of machines, actuators, sensors, and control systems to manufacture goods as quickly as feasible [2]. Cyber-Physical Production Systems (CPPSs) are sophisticated production platforms that strive to synchronize and integrate the physical world of machines with the digital computational realm. Nevertheless, for the application of CPPSs in the real world, extensive interconnection and a computing platform are essential [3]. The future of the Internet is the Internet-of-Things (IoT), in which everything will be interconnected. Research have shown that by performing intermediate functions at the network's edge, fog/edge computing-based operations will play a significant part in expanding the cloud [4]. Most of today's processors are implemented as components that are

 DOI: 10.1201/9781003497851-4

integrated into different CPS, using actuators and sensors to sense and amend the environment, and they are becoming more and more linked to one another through the Internet. There is much more to this dispersed and linked strategy than merely using information and communication technologies (ICT). To regulate the CPS, a variety of multidisciplinary knowledge is incorporated for pertinent physical sensing and cognition [5].

Computer experts and network specialists must work closely with specialists in a variety of fields, including digitization and control, mechanical engineering, civil engineering, and biology. This is because CPS is a new area of research that includes the overlapping and incorporation of many science and engineering fields. As a result, current descriptions of CPS are primarily provided by various academics from their respective views. Cyber-physical systems, as per the description of E. A. Lee, are those that combine working out with physical procedures while using embedded machines and networks to observe and manage those procedures. Through feedback loops, physical processes influence computations, and the reverse is true [6]. Professor J. F. views CPS as scalable, credible, and controllable interconnected physical tools systems that deeply integrate computing, control, and communication capabilities based on perceptions of the environment. To boost or enlarge the functionality of linkages and physical structures, and to observe or control physical objects in a safe, dependable, proficient, and simultaneous manner, in-depth integrating and real-time communication are made possible through the feedback chain of mutually beneficial special effects between computer procedures and physical procedures.

4.1.1 Motivation

The interoperability, connectedness, networking capacity, communication, decentralized management, autonomy, and modularity of the CPS are what make it useful. A CPS can communicate data that is mutually understood between its components when its components are connected, as opposed to when the components are connected via different communication protocols. One distinguishing feature of the CPS is its modularity that permits a CPS to adapt and rearrange its rejoinder to the demands and ease of users. To do this, a CPS a network of interconnected modules or subsystems must possess networking capabilities that allow all subsystems to communicate and interact with one another concurrently while maintaining their autonomy and decentralized control capabilities. The distinctiveness of CPS in contrast to conventional critical infrastructure systems is characterized by the direct communication of several components among each other enabling real-time decision making. The main structural overview of cyber physical system is described in Figure 4.1.

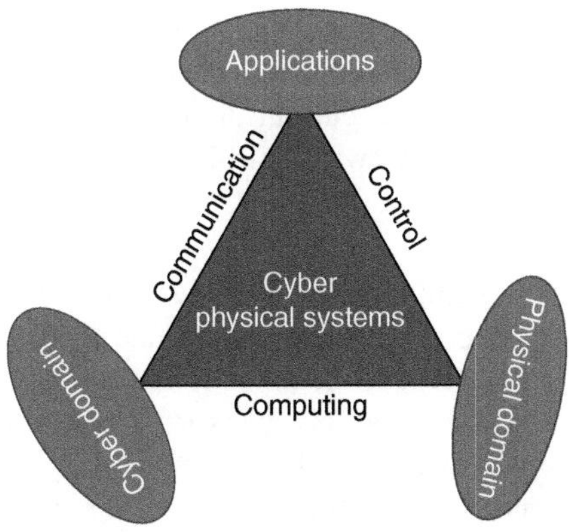

FIGURE 4.1 Cyber-physical system.

Mortality ratio, other problems and losses due to fire eruption are increasing with time due to many faults in cyber-physical systems components which are employed in fire detection systems. Early fault detection can reduce the losses rate with proper detection. The proposed work intends to apply privacy protection methods at upstream nodes level to transmit trustworthy data and also to indicate early fault detection in the authentication process. The study will attempt to provide a better error detection technique and increase the chances of avoidance. This work will support and provide new horizons for further research and the betterment in data protection of cyber physical architecture.

4.1.2 Scope

This study will provide an overview of the cyber physical system, its applications and issues. It will also give an overview of data protection at the level of sink/gateway (before the aggregation of data) and at fog and cloud level (after data accumulation and before data storage or further processing) using IGTDC and gate level modeling and encoding using gray coding and k-mapping. The techniques beyond this scope will not be addressed by this architecture.

4.1.3 Background

By using high-fidelity networks, cyber physical systems link actuators, sensors, and systems together [1]. This network provides system functionalities, like data transmission in real time and status observing, enabling communication with other systems. CPS develops a collaborative infrastructure similar to a digital twin to facilitate the electronic demonstration of information all over the product and procedure life cycle [2, 3]. The combination of feedback control with processing, communication, and control tools makes these systems effective [4]. More recent advancements have enhanced the interaction between people and CPS in the loop. For activities requiring a human operator to collaborate with a robot, for instance, human CPS, which entails CPS with close monitoring to the human components to avoid mishaps, is utilized. In general, there is a wealth of literature on the design of efficient techniques to assure and enhance the reliability of data in diverse CPS applications [5]. According to the literature, comprehensive solutions must be created by integrating various strategies and tactics to guarantee data trustworthiness throughout the lifecycle of data of sensors; for example, during the process of data gathering, broadcast, accumulation, and cloud storing. However, maintaining data dependability throughout the sensors' data lifecycle is difficult in the diverse CPS domain, and there isn't a perfect solution to the problem [6].

4.1.3.1 Major Research Contribution

Computer experts and network specialists must work closely with specialists in a variety of fields, including digitization and control, mechanical engineering, civil engineering, and biology. This is because CPS is a new area of research that involves the overlying and incorporation of many engineering and science fields. As a result, the current definitions of CPS are primarily provided by various academics from their respective views (Table 4.1).

4.1.3.2 Data Trustworthiness

The sensor modules in CPS are able to transfer only trustworthy data and reject the inaccurate or corrupted data before delivering it to the upstream nodes due to in-network data manipulation and coordination amongst sensors. As a result, the amount of data transmitted to the upstream component is greatly decreased. Meanderingly, this helps to solve the network deferral communication issue as the sensor's modules filter out a lot of unnecessary data. We expect it to alleviate/improve latency, battery, and bandwidth problems. But still there is a danger at upstream units such as gateway/sink in CPS that data may greatly be affected. So our main focus is to protect data before sending it to upstream components.

TABLE 4.1
The Most Important Attributes of CPS

Attributes	Statement	Ref	Year
Adaptive control	A CSP has the capacity to alter its mode of operation in response to shifting environmental circumstances in order to produce optimal results.	[7]	2021
Human Computer Interaction	The CSP's agent-based design permits its many agents to interact openly with humans to receive directions for carrying out specific activities.	[8]	2021
Data-Driven	The interconnections between both physical and cyberspace systems, that are made possible by data flow and data authenticity, are the basis for a CSP's operation.	[9]	2021
Real Time Monitoring and Control	A CSP can continuously monitor the performance of its many subsystems, promptly recognize changes in the ecological framework, and make aware and wise judgements in response.	[10]	2020
Predictive Maintenance	In order to detect irregularities in the procedure of various sub-systems and pieces of tools and rectify them prior to botch, a CSP executes predictive maintenance using data analysis techniques and tools.	[11]	2021
Internet of Things	Internet of Things (IoT) gadgets and protocols enable the interconnection and data flow between various CSP subsystems.	[12]	2021
Inter-connected system	A CSP's interconnectivity between its many subsystems enables its various agents to cooperate and work independently by exchanging information and making wise decisions.	[13]	2020
Privacy and Security	CSP's top priorities these days are keeping the system secure and protecting consumers' privacy.	[14]	2020

4.2 OVERVIEW OF CYBER-PHYSICAL SYSTEMS

Investigations on CPS involve a number of overlapping scientific and engineering disciplines as an emerging field of study. As a result, the official description of the CPS has also changed as a result of various viewpoints from various academics. According to [15], CPS is a procedural framework where physical procedures have a positive feedback loop on computations and vice versa. According to the authors of [16], a CPS is an interconnected physical engineering system that uses computations to monitor and manage the performance of physical systems.

4.2.1 Historical Perspective

The development of digital control systems began roughly 30 to 40 years ago. This was the preliminary phase regarding current Industry 4.0 because they had the idea of integrating thousands of sensors. Then, in the early 2000s, there was a change brought on by the creation of good open-source software. Humanity has been able to utilize the data gathered from the sensors in control systems with greater effectiveness because of the development of application packages. Another substantial wave of advantages for the sector resulted from this.

The term "Industry 4.0" was first used in 2011 by Klaus Schwab. Industry 4.0, in his opinion, is not a collection of novel technology but rather "a completely new approach to production." The phrase "means of increasing the competitiveness of the German manufacturing industry through the enhanced integration of cyber-physical systems into factory processes" [17] is how Industry 4.0 was initially defined in Germany.

The virtual and physical worlds come together, resulting in the creation of new cyber-physical complexes. In addition, the goal was to create an automated, in-depth interface that would allow the product's consumer to virtually watch while his purchase was produced [18].

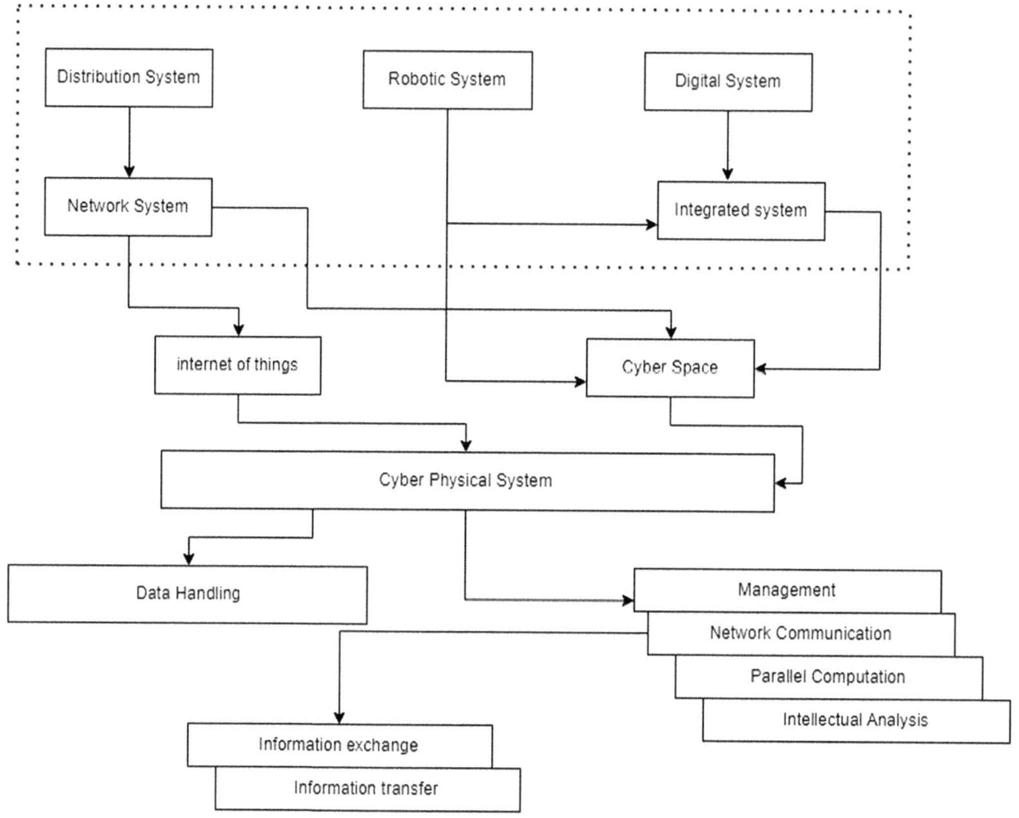

FIGURE 4.2 The development of cyber physical systems [19].

Cyber-physical systems are being used in many different aspects of human activity, including transportation, energy, life support, and medical. Embedded real-time systems, distributed computing systems, robotic systems of control for technical and industrial processes, networks, and integrated networks are examples of "cyber-physical systems" that came before them [19] (Figure 4.2).

4.2.2 CPS Layers and Components

The framework of CPS systems is made up of various layers and components that communicate with one another using various communication protocols and technologies [20].

4.2.2.1 Layers of CPS

The three primary layers of the CPS architecture, the transmission layer, perception layer, and application layer, are shown and described in Table 4.2.

4.2.2.2 CPS Components

Components of the CPS are employed to control signals (Figure 4.3) or to sense information [21]. Accordingly, the components of the CPS are divided into two groups: sensing components (SC), which gather and detect information, and controlling components (CC), which watch over and regulate signals.

TABLE 4.2
CSP Layers

Layer Name	Objectives	Targets
Perception Layer	Data and information Collection	Confidentiality, Privacy, Authentication
Transmission Layer	Data and information Transmission	Confidentiality, Integrity, Authentication, Availability
Application Layer	Data and information Analysis and Decision making	Privacy, Security, Safety, Authentication

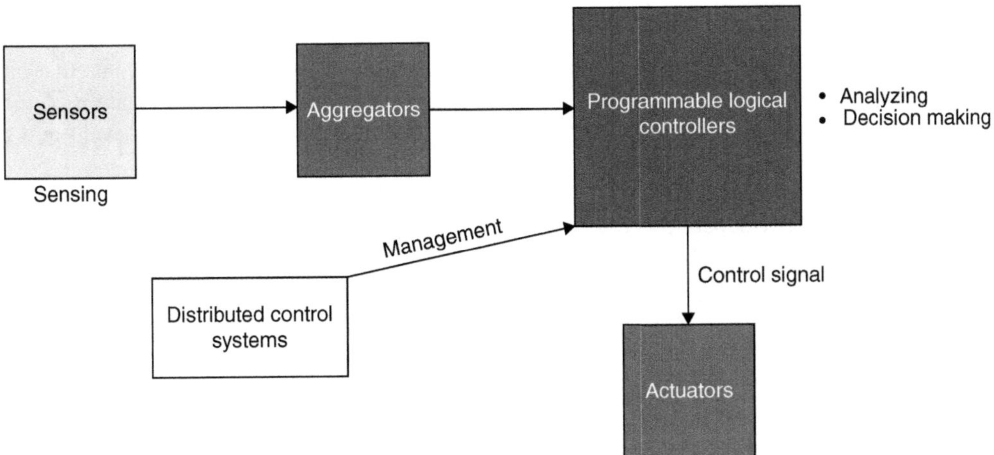

FIGURE 4.3 CSP components.

Sensors that gather data and send it to aggregators make up sensing constituents, which are predominantly found at the perception layer. The actuators will then get this data and information for additional processing to ensure proper decision-making. The main CPS sensor components are listed below.

- Sensors
- Aggregators
- Actuators

In order to obtain greater stages of precision and safeguarding against nefarious attacks or accidents, primarily noise, interference, and signal jamming, controlling components are used to control signals. They play an important role in signal control, monitoring, and management. As a result, it has become crucial to use Programmable Logic Controllers and Distributed Control Systems (DCSs) and all of associated parts, including PAC [22], OT/IT [23], Server/Control Loop and (HMI)/(GUI) [24]. Following that, we outline the various control system types utilized in CPS systems:

- Programmable Logic Controllers (PLC)
- Distributed Control Systems (DCS)
- Remote Terminal Units (RTU)

4.2.2.3 Types of Models of CPS

Models of CPS fall under the following categories:

CPS Timed Actor: Geilen et al. [25] proposed a theory with a functional and classical modification that limits a specific behavioral set, increasing efficiency while lowering complexity. In fact, these time predictable models may more easily derive analytical bounds and are less prone to state expansion issues [26].

Event-Based CPS: In these kinds of models, actuation decisions cannot be made until an event has been sensed and detected by the appropriate CPS components. As a result of the various CPS actions, the non-deterministic system delay affects the individual component time restrictions which can be managed, according to Hu et al. in [27], by using an event-based strategy that relies on CPS events.

Model of Lattice-Based: Events pursuant to the type of event, as well as the events that have internal and external properties, the CPS events are depicted in. The combination of these events enables the definition of a spatio-temporal attribute of every specified event and the identification of all the observers.

CPS Model Based on Hybrid: The physical dynamical systems and the discrete computing systems are the two separate interactive system types that make up hybrid CPSs, which are the heterogeneous systems [28].

4.2.3 Characteristics of Cyber Physical Systems

A system must possess a number of qualities in order to be classified as a CPS. The production processes can become more efficient and flexible when there is high autonomy and quick decision-making procedures, which are fundamental criteria [29]. The key characteristics of the cyber physical system are shown in Table 4.3.

4.2.4 Design of CPS

The components of the CPSs are positioned at an advanced degree of collaboration than other information concepts [20].

4.2.4.1 Layers of Cyber Physical Systems

The primary layers can be described as follows:

TABLE 4.3
Characteristics of CPS

Characteristics	Description
Reliable	Consistent operation without failures
Stability	Consistent operations
Accurate	High precision data analysis
Efficiency	No waste operation
Communicative	Data connection and sharing with all systems
Scalability	On-demand system expansion
Autonomy	Decision-making capacities
Safe Trustworthy system	Avoiding accidents
Robustness	Tolerant to perturbations in the system

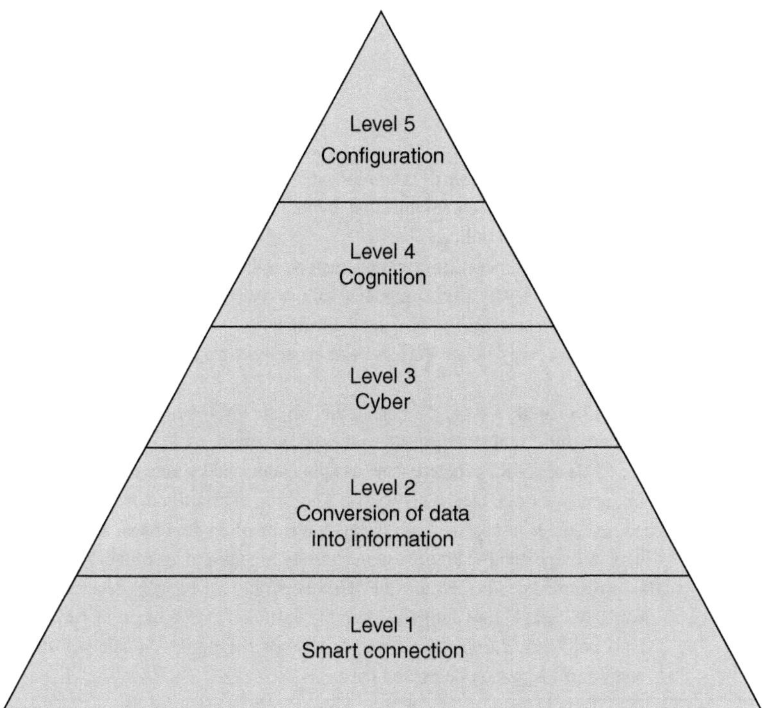

FIGURE 4.4 Communication levels of cyber-physical systems.

- First is the physical layer, which consists of actual objects in the real world.
- Second is the digital layer, which consists of algorithms for handling actual objects.
- The interface that allows for the communication of the physical layers and the digital layers with people, which consists of control mechanisms and detectors.
- The interface between the physical and digital layers with actual objects, which consists of augmented reality technologies.

The following levels of communication between cyber-physical system items exist [20], about which Figure 4.4 displays a graphical perspective.

4.3 DISCUSSION

A CPS has various difficulties that aren't necessarily present in an embedded or traditional corporate information system [30]. The key challenges faced in most of the cyber physical systems are described in Table 4.4.

Different characteristics allow CPS to function in many domain-related applications. Depending on the uses, there are many CPS kinds that work either self-sufficiently or cooperatively to provide services that are secure, trustworthy, dependable, efficient, agile, and affordable. The main applications are discussed in Table 4.5.

4.4 CONCLUSION

CPS will address various facets of social and economic life, have a significant impact, and drive the all-encompassing advancement of computer science as well as other disciplines. The development of CPS, however, also faces significant difficulties because of the limitations imposed by the

TABLE 4.4
Challenges of CPS

Challenge	*Description*
Data heterogeneity	Platforms must be capable to accommodate a wide range of unique uses and gadgets. Different protocols related to communication, including Bluetooth, RF, Zigbee, and infrared, are utilized by digital policies.
Reliability	Due to the environmental impact of actuators, reliability and safety are fundamental considerations. Any CPS component failure can result in system degradation, which could seriously affect people's lives and their property.
Data management	Big data from many linked devices must be stored, analyzed, processed, and displayed in real-time.
Privacy	Since CPSs handle a lot of data, including private information about people's health, gender, and religion, significant privacy concerns are raised.
Security	Since CPSs are built on heterogeneous applications and wireless communications, there are frequently serious security concerns. The issue of security is now a global one.
Real-time	Because failures to take actions on time might result in permanent harm, CPSs must make sure they have the bandwidth or system capacity required to accomplish time-critical functions.
Data Trustworthiness	The sensor modules in CPS are able to transfer only trustworthy data and reject the inaccurate or corrupted data before delivering it to the upstream nodes due to in-network data manipulation and coordination amongst sensors. As a result, the amount of data transmitted to the upstream component is greatly decreased [31].

TABLE 4.5
Applications of CPS

Application	Description
Smart Buildings	Smart or intelligent buildings, commonly referred to as green buildings, place a strong emphasis on sustainability and are outfitted with cutting-edge technology like 5G, cloud computing, the Internet of Things (IoT), edge computing, and artificial intelligence which renders them human- and energy-friendly [32].
A Smart Grid and Smart Cities	A smart grid, often referred to as a smart energy grid, is divided into seven categories, including users, service providers, markets, procedures, generation in bulk, transmission, and then distribution, as defined by the National Institute of Standards and Technology (NIST) [33]. As the name suggests, a smart city is an urban region that uses technology to become connected and adaptable in order to handle the environmental, social, and economic concerns of the twenty-first century and enrich the eminence of life for its occupants [34]
Intelligent Transportation Systems	Today, vehicles can avoid traffic jams, consume less land and energy, improve operating control, security and safety, act as distributed resources of energy to maintain a balance between supply of electricity and demand, lower costs (or increase benefits) for those who own them, and enhance environmental sustainability [35].
The Tele-Medical Systems	The Tele-medical systems are interconnected medical equipment systems that offer context aware, services relating to life critical situations in hospitals to provide patients with high-quality continuous care. [36]. As a result, patients can now receive care both at home and in a hospital.
Autonomous Robotic Systems	Unmanned ground-based vehicles (UGV) and unmanned aerial vehicles (UAV) are examples of autonomous robotic systems that are increasingly in demand as the CPS of the future. A UAV, also referred to as a drone, is a vehicle that can fly like a plane without a human pilot or other passengers. UGVs, are autonomous system of robots that fly over the land without a human pilot [37].

TABLE 4.5 (Continued)
Applications of CPS

Application	Description
Smart Agricultural Systems	The population of the planet will surpass 9.0 billion people by 2050. Under the stringent constraints of scarce resources, unskilled labor, a lack of agricultural land, and climate change, it would be difficult to achieve the necessary level of agricultural output. One potential solution to this significant issue is precision agriculture, a clever CPS-based farming technique [38].
Industry 4.0	Industry 4.0 has already begun to take shape thanks to CPS's integration of cutting-edge functions through the IoT, cloud manufacturing, big data analytics, IoE, and the fog-based computing [39].

theory and technology of computation, communications, and control that are already in use. Lead the world in CPS development with a groundbreaking advancement in a critical CPS technology.. CPSs combine physical and computational processes. CPSs have much higher economic potential than has been realized, and significant investments are being undertaken all around the world to further the technology. The capabilities, adaptability, scalability, resilience, safety, security, and usability of CPS will greatly outpace those of today's straightforward embedded systems. This work will support and provide new horizons for further research and the betterment in data protection cyber physical architecture. This will provide a better understanding and conducive decision support system for effective and efficient transmission of trustworthy data from upstream nodes.

New studies should definitely be done to emphasize and utilize the potential of CPS. A trustworthy and secure environment for technological growth could be created through CPS technology, but this will take time and much research. Future research should concentrate on creating standards and best practices for CPS implementation while addressing trustworthiness of data, reliability, scalability, standardization, integration, and complexity.

REFERENCES

1. Serpanos, D.: The cyber-physical systems revolution. *Computer*. 51, 70–73 (2018).
2. Colombo, A.W., Karnouskos, S., Kaynak, O., Shi, Y., Yin, S.: Industrial cyberphysical systems: A backbone of the fourth industrial revolution. *IEEE Industrial Electronics Magazine*. 11, 6–16 (2017).
3. Cogliati, D., Falchetto, M., Pau, D., Roveri, M., Viscardi, G.: Intelligent cyber-physical systems for industry 4.0. In: *2018 First International Conference on Artificial Intelligence for Industries (AI4I)*. pp. 19–22. IEEE (2018).
4. Jamaludin, J., Rohani, J.M.: Cyber-physical system (cps): State of the art. In: *2018 International Conference on Computing, Electronic and Electrical Engineering (ICE Cube)*. pp. 1–5. IEEE (2018).
5. Kumar, R., Kumar, P., Tripathi, R., Gupta, G.P., Gadekallu, T.R., Srivastava, G.: SP2F: A secured privacy-preserving framework for smart agricultural Unmanned Aerial Vehicles. *Computer Networks*. 187, 107819 (2021).
6. Hai, T., Bhuiyan, M.Z.A., Wang, J., Wang, T., Hsu, D.F., Li, Y., Salih, S.Q., Wu, J., Liu, P.: DependData: Data collection dependability through three-layer decision-making in BSNs for healthcare monitoring. *Information Fusion*. 62, 32–46 (2020).
7. Hao, L., Sun, B., Li, G., Guo, L.: The eco-driving considering coordinated control strategy for the intelligent electric vehicles. *IEEE Access*. 9, 10686–10698 (2021).
8. Wilhelm, J., Petzoldt, C., Beinke, T., Freitag, M.: Review of digital twin-based interaction in smart manufacturing: Enabling cyber-physical systems for human-machine interaction. *International Journal of Computer Integrated Manufacturing*. 34, 1031–1048 (2021).
9. Zhang, X., Wang, J., Zhang, H., Li, L., Pan, M., Han, Z.: Data-driven transportation network company vehicle scheduling with users' location differential privacy preservation. *IEEE Transactions on Mobile Computing*. 1–8 (2021).

10. Guddanti, K.P., Matavalam, A.R.R., Weng, Y.: PMU-based distributed non-iterative algorithm for real-time voltage stability monitoring. *IEEE Transactions on Smart Grid*. 11, 5203–5215 (2020).
11. Tang, Y., Dananjayan, S., Hou, C., Guo, Q., Luo, S., He, Y.: A survey on the 5G network and its impact on agriculture: Challenges and opportunities. *Computers and Electronics in Agriculture*. 180, 105895 (2021).
12. Pivoto, D.G., de Almeida, L.F., da Rosa Righi, R., Rodrigues, J.J., Lugli, A.B., Alberti, A.M.: Cyber-physical systems architectures for industrial internet of things applications in Industry 4.0: A literature review. *Journal of Manufacturing Systems*. 58, 176–192 (2021).
13. Ning, Z., Dong, P., Wang, X., Hu, X., Guo, L., Hu, B., Guo, Y., Qiu, T., Kwok, R.Y.: Mobile edge computing enabled 5G health monitoring for Internet of medical things: A decentralized game theoretic approach. *IEEE Journal on Selected Areas in Communications*. 39, 463–478 (2020).
14. Wang, J., Wu, L., Wang, H., Choo, K.-K.R., He, D.: An efficient and privacy-preserving outsourced support vector machine training for internet of medical things. *IEEE Internet of Things Journal*. 8, 458–473 (2020).
15. Lee, E.A.: Computing foundations and practice for cyber-physical systems: A preliminary report. University of California, Berkeley, Tech. Rep. UCB/EECS-2007-72. 21, (2007).
16. Gonzalez, G.R., Organero, M.M., Kloos, C.D.: Early infrastructure of an internet of things in spaces for learning. In: *2008 Eighth IEEE International Conference on Advanced Learning Technologies*. pp. 381–383. IEEE (2008).
17. Hecklau, F., Orth, R., Kidschun, F., Kohl, H.: Human resources management: Meta-study-analysis of future competences in Industry 4.0. In: *Proceedings of the International Conference on Intellectual Capital, Knowledge Management & Organizational Learning*. pp. 163–174 (2017).
18. Plakhotnikov, D.P., Kotova, E.E.: Design and analysis of cyber-physical systems. In: *2021 IEEE Conference of Russian Young Researchers in Electrical and Electronic Engineering (ElConRus)*. pp. 589–593. IEEE (2021).
19. Kudzh, S., Cvetkov, V.: YA. Setecentricheskoe upravlenie i kiberfizicheskie sistemy. *Obrazovatel'nye resursy i tekhnologii*. 19 (2017).
20. Plakhotnikov, D., Kotova, E.: The Use of Artificial Intelligence in Cyber-Physical Systems. In: *2020 XXIII International Conference on Soft Computing and Measurements (SCM)*. pp. 238–241. IEEE (2020).
21. Yaacoub, J.-P.A., Salman, O., Noura, H.N., Kaaniche, N., Chehab, A., Malli, M.: Cyber-physical systems security: Limitations, issues and future trends. *Microprocessors and Microsystems*. 77, 103201 (2020).
22. Mazur, D.C., Quint, R.D., Centeno, V.A.: Time synchronization of automation controllers for power applications. In: *2012 IEEE Industry Applications Society Annual Meeting*. pp. 1–8. IEEE (2012).
23. Morelli, U., Nicolodi, L., Ranise, S.: An open and flexible cybersecurity training laboratory in IT/OT Infrastructures. In: *Computer Security: ESORICS 2019 International Workshops, IOSec, MSTEC, and FINSEC*, Luxembourg City, Luxembourg, September 26–27, 2019, Revised Selected Papers 2. pp. 140–155. Springer (2020).
24. Ardanza, A., Moreno, A., Segura, Á., de la Cruz, M., Aguinaga, D.: Sustainable and flexible industrial human machine interfaces to support adaptable applications in the Industry 4.0 paradigm. *International Journal of Production Research*. 57, 4045–4059 (2019).
25. Geilen, M., Tripakis, S., Wiggers, M.: The earlier the better: A theory of timed actor interfaces. In: *Proceedings of the 14th International Conference on Hybrid Systems: Computation and Control*. pp. 23–32 (2011).
26. Canedo, A., Schwarzenbach, E., Al Faruque, M.A.: Context-sensitive synthesis of executable functional models of cyber-physical systems. In: *Proceedings of the ACM/IEEE 4th International Conference on Cyber-Physical Systems*. pp. 99–108 (2013).
27. Hu, F., Lu, Y., Vasilakos, A.V., Hao, Q., Ma, R., Patil, Y., Zhang, T., Lu, J., Li, X., Xiong, N.N.: Robust cyber–physical systems: Concept, models, and implementation. *Future Generation Computer Systems*. 56, 449–475 (2016).
28. Yang, Y., Zhou, X.: Cyber-physical systems modeling based on extended hybrid automata. In: *2013 International Conference on Computational and Information Sciences*. pp. 1871–1874. IEEE (2013).

29. Rojas, R.A., Rauch, E., Vidoni, R., Matt, D.T.: Enabling connectivity of cyber-physical production systems: a conceptual framework. *Procedia Manufacturing*. 11, 822–829 (2017).
30. Alwan, A.A., Ciupala, M.A., Brimicombe, A.J., Ghorashi, S.A., Baravalle, A., Falcarin, P.: Data quality challenges in large-scale cyber-physical systems: A systematic review. *Information Systems*. 105, 101951 (2022).
31. Rahman, H.U., Wang, G., Bhuiyan, M.Z.A., Chen, J.: In-network generalized trustworthy data collection for event detection in cyber-physical systems. *PeerJ Computer Science*. 7, e504 (2021).
32. Li, P., Lu, Y., Yan, D., Xiao, J., Wu, H.: Scientometric mapping of smart building research: Towards a framework of human-cyber-physical system (HCPS). *Automation in Construction*. 129, 103776 (2021).
33. Abrahamsen, F.E., Ai, Y., Cheffena, M.: Communication technologies for smart grid: A comprehensive survey. *Sensors*. 21, 8087 (2021).
34. Habibzadeh, H., Nussbaum, B.H., Anjomshoa, F., Kantarci, B., Soyata, T.: A survey on cybersecurity, data privacy, and policy issues in cyber-physical system deployments in smart cities. *Sustainable Cities and Society*. 50, 101660 (2019).
35. Massar, M., Reza, I., Rahman, S.M., Abdullah, S.M.H., Jamal, A., Al-Ismail, F.S.: Impacts of autonomous vehicles on greenhouse gas emissions—positive or negative? *International Journal of Environmental Research and Public Health*. 18, 5567 (2021).
36. Gupta, A., Singh, A.: A Comprehensive Survey on Cyber-Physical Systems Towards Healthcare 4.0. *SN Computer Science*. 4, 199 (2023).
37. Asadi, K., Suresh, A.K., Ender, A., Gotad, S., Maniyar, S., Anand, S., Noghabaei, M., Han, K., Lobaton, E., Wu, T.: An integrated UGV-UAV system for construction site data collection. *Automation in Construction*. 112, 103068 (2020).
38. Torky, M., Hassanein, A.E.: Integrating blockchain and the internet of things in precision agriculture: Analysis, opportunities, and challenges. *Computers and Electronics in Agriculture*. 178, 105476 (2020).
39. Lee, J., Azamfar, M., Singh, J.: A blockchain enabled Cyber-Physical System architecture for Industry 4.0 manufacturing systems. *Manufacturing Letters*. 20, 34–39 (2019).

5 Analysis of Device-to-Device Communication in IoT System

Usman Nawaz, Faisal Rehman, Kainat Azmat Ullah, Usman Ahmed Raza, Hanan Sharif, Furqan Rafique, Abid Ali, Masood Ashiq, and Junaid Iqbal

5.1 INTRODUCTION

Sending and receiving messages through language, expression, or some other medium is referred to as communication. The major feature of the Internet of Things (IoT) is device-to-device communication which is peer-to-peer communication. Working outcomes, enhanced efficiency, and resource utilization can be improved with the help of the D2D [1] communication approach. There are multiple wireless or wired methods for IoT-based networks but we lack in the field of communication between IoT devices while they are the primary client in IoT systems. Gadgets will autonomously communicate with each other without centralized power and pre-collect, distribute, and save in a multi-tabbed manner. Facts may be transformed into intelligence as a way to facilitate the creation of a wise environment. The main benefit of IoT systems is that they can gather data in real time. The collection of data in real time and the speed of devices to manage the time-critical environment shows that the devices are more clever than our imaginations [2]. Furthermore, speaking devices working under extraordinary requirements for connectivity, will have a connection that comes and goes and would be limited in terms of resources in many cases. Such traits are there to create a slew of networking problems that typical routing techniques can't tackle. Mostly as a reason to gain a competitive advantage devices would need astute network parameters.

The communication network is required to meet the Internet of Things needs. Many companies and standards organizations have demonstrated a significant interest in implementing the D2D technique in wireless networks. People are becoming increasingly connected via the internet, resulting in exponential development in the number of related gadgets. Several networks such as Bluetooth connectivity, Wi-Fi, RFID, ZigBee, etc. are used for short-range communication. They're made up of record-processing equipment. The method of creating, sharing, and altering content necessitates a connection between technologies, whether or not they require processing time. Some gadgets can be any type of devices or objects (for example, smartphones, sensors, electrical devices, autos, equipment for the house, or RFID tags) that are equipped with intelligent skills in outcome, calculation, and public speeches. As a result, not only are the most effective persons interconnected, but so are the most effective gadgets. From this paradigm shift, the concept of the Internet of Things (IoT) was born. The Internet of Things (IoT) is an extraordinary extension of the current Internet that has evolved from an interconnected community of people to an interconnected community of gadgets. Therefore, humans can collect data from their environment, and these devices interact with the body's world through network protocols and standards.

The IoT, along with machine and deep learning, has emerged to be a recent research area while talking about networks and networks security, and various types of research have also been proposed in this domain, for instance, [3–10].

DOI: 10.1201/9781003497851-5

The IoT will make it possible to transform perception or information gathering into intelligent knowledge and successfully integrate intelligence into the surrounding environment. In addition, the IoT will consist of billions of gadgets that can save their space, identity, and information over Wi-Fi connections. In 2003, the world's population was estimated to exceed 6.3 billion and all 500 million of the simplest devices were connected to the internet. By 2010, the types of internet-related devices had increased to 12.5 billion and the population of the arena to 6.8 billion, making the number of related devices equal to one floor for the first time.

As a consequence, there was no IoT in 2003 because there were several types of connected devices. After creating the first computer system for smartphones and pill makers in 2003, the number of connected devices gradually increased. According to a recent prognosis, the number of connected devices would quadruple by 2013 when compared to the total number of people on the planet. It is estimated that it will reach 25 billion in 2015 and the world population will reach 7.2 billion. It is also estimated that there will be around 50 billion connectable devices by 2020. The number of gadgets will eventually exceed the industry's population by four times. Improvements in the tools used to plan and control human physical activity daily could contribute to this expansion. IoT devices, including private devices on the IoT, are non-standard computing devices that can connect to the community wirelessly and have the ability to tell the truth.

5.1.1 D2D Communication

Device-to-device (D2D) communication is a cutting-edge data transmission method designed to boost network efficiency. D2D communication-capable devices can communicate with each other utilizing a secure transmission protocol similar to how they communicate with the base station in LTE-Direct. D2D communication networks are divided into two categories.

There is an existing helping infrastructure, with network-assisted D2D communication. With the advancement in technology, the need for a D2D communication system is increasing. The communication between devices should be smooth, reliable, and time efficient, for the success of IoT in bigger societies.

5.1.2 D2D Communication Approach

In the Wi-Fi community, D2D stands for the location of the hobby because of several benefits. This strategy is ideal for reducing the stress on the network infrastructure caused by site traffic. Devices will communicate with one another via the D2D technique, which eliminates the need for BS manipulation. It also converts hop communications (through BS) into a single hop spoken exchange. D2D generation is becoming increasingly significant to deliver improved facilities. For instance the event of an attack or tragedy or when internet work is unavailable or broken. D2D connectivity might reduce reliance on existing core infrastructure. D2D communication is primarily based on the mobile community and includes all three types of D2D communication. Essentially, D2D conversations can be classed as running in a certified or unauthorized band. The most extensively utilized wireless technologies are Wireless charging and Wi-Fi eras in the unlicensed spectrum for establishing tiny communities for information sharing.

Because it is publicly available it provides a vital platform for inventors. A comfortable network for quick-range communication can be created by commercial use of unlicensed spectrum, clinical or scientific services at any time and from any location. The community is constrained by its constraints due to the limited range of signals available. For a large community, many intermediate nodes are required, which results in higher bandwidth utilization or delay difficulties.

These gadgets are also susceptible to intervention. Once again, in the certified frequency band, the wireless community can transmit a density of 1,000 samples and a distance hundreds of times greater than the communication system [11]. It will improve the communications variety, throughput, and performance. Security concerns may likely be alleviated, and the system will be less susceptible to outside intervention. Furthermore, the use of green resources and interference control are key subjects to consider when evaluating network performance. In an IoT environment established with D2d users, a connection between unregistered and allowed bands is essential.

5.1.2.1 Cellular Mode

Devices speak in this mode as though they have been using a traditional cell telephone. The data is passed via the BS which serves as an intermediary point. From source to destination, a bridge connection is required (Uplink and downlink).

5.1.2.2 Dedicated Mode

A separate connection between the sender and receiver is formed under this state, allowing for an immediate spoken connection. There are no gaps between cellular and D2D voice communication.

5.1.2.3 Reuse Mode

In this method, D2D verbal interaction collaborates with cellular communication to share available resources. It allows you to repurpose a tunnel that is currently being used by mobile communication. This mode makes environmentally friendly usage with no trouble easily accessible communal sources. The main difficulties to overcome in the updated usable plan and intervention management are two reusable kinds of strategies.

5.1.3 Advantages of D2D Communication

- Increased efficiency in the community.
- Less power usage and a longer battery life.
- Controlling traffic loads efficiently.
- Even in the absence of a community, a connection can be established.
- Extending the network without the need for expensive infrastructure-like base stations is possible.
- Multiple applications of D2D communication.
- Public safety and security services.
- Peer-to-peer communication [12] between machines.
- Independent riding packages and vehicle-to-vehicle communication.
- Offloading of cells (to WLAN community for quicker records transfer).
- Gaming and social networking in the neighborhood.
- Content distribution, local advertising, and location-based services.
- Emergency broadcast and warning message for a specific place.

5.1.4 Types of D2D Connectivity Approach

The participation of cellular infrastructure in the setup of the direct connection, as well as the spectrum over which the direct speech exchange occurs, may be used to classify D2D (direct-to-device) connectivity cooperation between wireless connections as managed by the users in self-organized D2D communication. This method works in the same way as unregistered radio is used in classic wireless communication. It identifies software when the mobile network isn't always available, and it's frequently motivated by its low interaction and straightforward maintenance (e.g. in case of an herbal catastrophe). In community-assisted D2D communication, on the other hand, the base station (BS) aids immediate facts dissemination via command communication and aid maintenance [13]. In this situation, to limit the likelihood of interruption, computer systems may integrate any transmissions. One consequence of this cooperation is that it may necessitate a lot of signaling and utilization that is central. Multiple kinds of systems integration are also available that are expected such as the intention of striking a good balance of expense in terms of difficulty and transmission, as well as promised efficiency. As illustrated at some point within the investigation portion, D2D clients may be provided by the most effective network. They are in charge of their own operations and media sources after which they can plan accordingly D2D users accessing the range.

There are some main components that one needs.

- Sensor
- Gateway
- Connectivity
- Cloud
- Analytics
- User interface

5.1.4.1 Sensor

Sensors are the link between IoT devices and the outside world or people. It identifies changes and sends the data to the cloud for processing as the name implies. These sensors collect data from the environment constantly and send it to the next level. Sensors such as pressure sensors, temperature sensors, and light density detectors are examples.

5.1.4.2 Gateway

The gateway helps mobile data management and protocol layer transfer statistical information from one device to any other device. It interprets the device's community protocol and provides encryption for information flowing into the community. It is like a layer between the cloud and the device that can clean up statistics of network attacks and illegal logins.

5.1.4.3 Connectivity

Statistical data collected by the sensor should be sent to the cloud for analysis. Information needs the internet as a means of transferring facts from one device to any other device. Therefore, all IoT devices want to be connected to the internet at all times. Wi-Fi, Bluetooth, WAN, satellite networks, and other networks make it easy to stay in touch. Statistics can be sent to the cloud via these networks.

5.1.4.4 Cloud

IoT systems generate large volumes of data from devices that must be properly managed to produce useful outputs. The IoT cloud is used to store so much data. It provides tools for data collection, processing, and storage. Data can be easily accessed and accessed from anywhere via the internet. It can also be used as an analysis tool. The IoT cloud is a complex network of high-performance servers that can rapidly process large amounts of data.

5.1.4.5 Analytics

The process of transforming raw data into something useful is called analysis. IoT analytics supports real-time analytics to capture changes and anomalies in real time. The data is then converted into an easy-to-understand format for the end user. For the successful implementation of their ideas, users and businesses can use the reports to adapt trends, market predictions, and plans beforehand.

5.1.4.6 User Interface

The Internet of Things system has a visible and tangible user interface. It is the part where the machine communicates with the plug. Statistics can be provided in document format or in the form of actions such as alerting, or sending notifications. Also, the user can choose to perform some actions. It is very important to design a user-friendly interface that can be used without much effort or technical knowledge.

5.1.5 Characteristics of IoT

5.1.5.1 Connectivity

The primary purpose of the IoT is connectivity, which makes the devices send data and stay connected with other devices. It allows devices to connect to the network and collaborate on functions.

5.1.5.2 Intelligence

The IoT system is preferred by the market because of its intelligence. The combination of algorithms and computers enables the device to detect processes in the environment and take appropriate actions. For example, the system is smart enough to detect a sudden temperature rise and trigger an oven alarm.

5.1.5.3 Expressing

The IoT is about intelligent interaction with the outside world and people. This interaction is achieved through expression. Expression allows us to communicate the output to the real world, as well as the inputs of people and the environment.

5.1.5.4 Sensing

Sensitivity refers to being aware of the changes occurring in our environment. Sensor technology enables us to create an experience that reflects awareness of changes in the physical world and its inhabitants. This makes it easier for you to express yourself. This document records the introduction to the IoT system and provides a better understanding of the complex world around us.

5.1.5.5 Energy

Everything in this world is based on the use of force. The design of the IoT structure is smart enough to synthesize and store energy from the outside world. It has also been developed to be strong enough to paint for a long time.

5.1.5.6 Security

The most important feature of any device is its safety and security. Nobody will utilize a device if it is vulnerable to cyber-attacks and unauthorized access. IoT systems deal with personal information [5], so it's a must that every safety level be met on this machine. All IoT structures are secure enough to handle personal information. Aside from that, the equipment required is substantial. IoT networks may also be vulnerable because device security is critical and crucial.

5.1.6 Problem Statement

The voice D2D era will play a key role in the Internet of Things (IoT). The incredible rapid improvement of operators' internal levels necessitates big adjustments to the IoT tools' design, and community devices supply a set of processes to improve the element network. The system's physical restraint and the community's auxiliary restraint make for a difficult decision. The performance of language exchange will be greatly impacted by D2D pair interference, and the minimum distance requirement for D2D communication must assure SNR.

5.1.6.1 Research Objective

This research aims to study modern linguistic communication techniques and how we can follow these strategies to communicate in IoT. Our important goal is to provide readers with the latest cutting-edge material on the work (protocols and proposed solutions) done so far in D2D communication and uncover the issues that still need to be resolved.

5.1.7 Research Question

What is D2D communication and how does D2D communication work in IOT?

5.2 LITERATURE REVIEW

In this study [14] authors categorized the forms of interaction as well as the reasons for each form of oral exchange of information in IoT. In intra-domain networks, gadgets can also collect records through calls or record their kingdoms with each other. Across networks, devices can also communicate without or with manual intervention. Interacting with humans or issuing alarms for human decisions may require communication between devices through human intervention. Devices can also speak to perform actions or discover or find different devices without human intervention. Devices can communicate with humans without delay to bypass information about humans or immediately obtain information from humans to make decisions. Gadgets can also communicate with information garage agents to deliver captured statistics, update stored records, or retrieve saved records for computerized decision-making.

In this work [15] a complete end-to-end solution for each application is needed to be provided for the healthcare strategy to cover the technical details. From the standpoint of verbal communication technology, the survey looked at crucial application-specific criteria. Different approaches and protocols are provided in the studies which help to make these applications work, with the possible potential benefits and barriers of short- and long-distance communication methods. Finally, the poll identified important open research challenges and difficulties, particularly those concerning the future healthcare structure, which would be driven by the IoT. Optical sensors (blood sugar and strain, oxygen saturation, and other key symptoms), environment (temperature, pressure, humidity measurement, etc.), accelerometer, magnetometer, and other sensors are all included in today's smartphones (which can be used as a framework coordinator or an important unit for personalized health monitoring) (for measuring speed, route, and gravity results). Additionally, the Smartphone's built-in bag (such as S-Health) can be utilized to save daily fitness music. The widespread usage of wearable auxiliary generation and the open supply of general-purpose systems, on the other hand, has issues with dependability, statistical privacy and security, and value effectiveness. In general, cutting-edge language exchange technology can meet the requirements of dependability, connectivity, factual cost, and delay for most practical situations, including infectious diseases, musculoskeletal, and neuromuscular disorders. Converged technologies may, however, be required to deliver smoking cessation solutions for expanding healthcare application.

This research [16] states that to promote a structure with controlled structural resistance. A device-to-instrument (d-d) negotiating mechanism is proposed in this study that allows green to react at the point of consumption. The proposed method creates a neighborhood voice exchange community and a negotiation algorithm between the smart grids connected to the IOT gadget that communicates along with the intelligent machine's base structure. In addition, a mesh network to enable a dynamic and responsive network system amongst IoT devices and configuration is used. As a result, the language exchange system ensured that the suggested set of d-d energy settlement rules is highly reliable. Each IOT tool will receive statistics on a specified amount of electricity consumption reduction from the smart meter and send a backup record of its preparedness to perform the required electricity diminution. The d-d interaction and IOT gadget selection rule set are dependent on the operational priority of the device, power requirement as well as accessibility. Therefore, the proposed d-d trading algorithm guarantees a knowledge selection mode and reliable factual dialogue to promote green energy management.

In this work [17], one of the most advanced technologies for improving network performance is communication. Customers near the periphery or close to the insurance region can communicate through D2D linkages, allowing D2D communication to grow. Gadgets in the radio location are authorized to use D2D oral communication in the D2D communication device settings. When using a mobile network, however, several communication devices are not always in the same radio range. In this case, two communication parties exchange statistical data over a direct hyperlink without the need for any involvement from the cellular infrastructure, including the BS or both in the cellular mode.

Here, the two talking devices communicate utilizing mobile services and peer-to-peer transmission, as well as direct linkages between pals. It is possible to simulate and analyze D2D customer behavior, whether it be individual or collective, using a set of mathematical tools provided by the theory of many transmission systems. As a result, a new strategy is required in this location to address some issues connected to disconnection. These issues may need to be dealt with well in addition to bandwidth limits, high latency, and so on. Before structuring D2D type, the peer must go through the peer discovery process before the device can begin D2D communication. For two capacity devices to create a connection for D2D discussions on the direct link, peer discovery is required. D2D candidates are a pair of devices that can discover each other during P2P discovery. From the discovery phase until the two devices begin to communicate, this is the overall discovery process of D2D communication.

In this study [18], the reaction time of edge gadgets in an IoT system [5] is calculated by layout time. These partial devices continuously offer a flow of data to guarantee that real-time IoT units run smoothly. Facet devices, on the other hand, are prone to errors and must often address issues while attempting to maintain a decent level of vocal communication in the face of external interference. The absence of any conversation at the threshold tool level can cause the entire system to fail or deliver inaccurate information. Because IoT equipment contains a wide array of heterogeneous devices it's not easy to see all of these gadgets in one place or to look through the gadget logs to see if there's a problem with connectivity. As a result, a lightweight intelligent layer on the edge device, which may be modified according to internal changes, may be required to retain the maximum quality of first-class communication while being as close as feasible to the precise theoretical reaction time. Their proposed ASB can be implemented on IoT devices as a lightweight intelligence layer to track good communication and offer the next best exchange alternative to maintain a sufficiently high level of communication so that the device can communicate in real time.

In this work [19] in the IoT D2D newsletter they looked at the importance of security. Furthermore, we offer Secure Key Exchange with QR Code (SeKeQ), a single authentication mechanism that ensures automatic key assessment and uses the Diffie-Hellman key agreement with SHA-256 hash key to validate personal identity. We used a test mattress and a gadget with a Wi-Fi-Direct connection interface to evaluate SeKeQ's performance. The results suggest that our concept can provide the necessary security aspects, such as crucial backup, data secrecy, and integrity. Furthermore, our approach achieves the same level of security as MANA (Manual Authentication) and UMAC (Universal Hash Message Authentication Code), but with a 10x reduction in key computation and a smaller memory footprint. Currently, regardless of the tool's vicinity, all data traffic in a Wi-Fi cellular network creation must transit via a set of infrastructure. To minimize overburdening the infrastructure, it's ideal to allow direct device-to-device communication whenever possible. The problem of the radio access network load and intermediate network load will be discovered when the gadgets are close to each other.

This paper [20] is designed to increase the cumulative machine improvement system by determining the transmission rate of each d2d and mobile hyperlink.D2d customers considered three modes of communication: mobile mode, delivery function as well as reusable mode. Energy management and combined power management and connection allocation are two sub-problems of the optimization problem (Table 5.1).

5.3 MATERIALS AND METHODS

The generation of PCs has been upgraded in the middle of twentieth century and laid its foundation despite all the technical obstacles. The foundation for the idea of conversation from system to device (M2M) which progressively evolved collectively with connectivity improvements the answers initiated the concept of the Internet of Things as we understand it today. Factor Network (IoT) [34] is a system of virtual tools, machines, objects, or people connected by specific identifiers and can transmit statistics and percentages within the community without the need for people or persons with PC interactions. The Internet of Things has several goals to create intelligent humans and also groups of people who live in situations that are conducive to learning. Entire cultures could be capable of maintaining a more intelligent and calm state. As grandiose as it could appear, the Internet of Things (IoT) [35] has indeed proven its worth and emerged as a regular aspect of our lives. With that in consideration, let's take a look at the equipment that powers the global IoT and allows it to move spherically.

5.3.1 Research Methodology

IoT gadgets are composed of multiple additives, which interact with each other. It gives a conceptual model of the components that make up a typical IoT device, such as the relationship between the main IoT ideas. The user can be a human character or a software agent with a goal. Things are discrete and recognizable parts of the physical environment, and users may be interested in accomplishing their goals. Things can be any physical entity, including people, cars, animals, or computers.

The interaction between users and issues in the IoT is mediated through services, which constitute the interface between consumers and IoT devices. Services are associated with virtual entities, that is, virtual diagrams of physical entities [36]. Components can be represented in a virtual world by digital entities. Different types of numerical representations of factors can be used, such as gadgets, 3D models, avatars, or perhaps social community bills. Some virtual entities can also interact with different digital entities to meet their purposes.

A key element of the Internet of Things is that the adjustment of the problem location and its corresponding virtual entities need to be synchronized. This is usually achieved by devices that are embedded, connected to, or placed in the vicinity of the factor. Filters, identifiers, and actuators are the three categories of gadgets that can be identified in theory. Sensors are used to rank the countries and locations where things are filtered. The sensor turns a mechanical, optical, magnetic, or thermal signal into a voltage and tip. These details can then be analyzed and used to create a description of the sport in question. The tag is a device that aids in the analysis procedure, which normally involves the use of a specialized sensor called a reader. Special documentation, including optics, such as barcodes and QR codes, can be handled by the identifying system, which is solely based on the frequency range. Actuators are employed to transform or influence things once more.

5.3.2 Architecture of an IOT System

Companies and services that are one-of-a-kind outline, develop, and comprehend the structure of the IoT in a unique way. Regardless of the application or business strategy, the underlying framework of IoT devices remains the same.

The four-layer version of the IoT devices' basic structure can be understood as:

1. IoT devices and Gateway
2. Network of communication
3. Server or cloud
4. Programmed is the Internet of Things Utility Program

TABLE 5.1
Analysis of Literature Review Regarding D2D Communication Systems in IoT

Sr. No	Author	Year	Title	Objective	Tools or techniques	Short summary
1	F. Zenalden, et al[17]	2020	Peer Selection in Device-to-Device Communication Based on Multi-Attribute Decision Making	When a user moves between multiple network spots, it tries to find and select the optimal peer to initialize a D2D connection.	Using the Analytic Hierarchy Process and Hierarchical Adaptive [20] Weighting, the peer selection process makes decisions.	When the network verified the D2D connection, better results were observed.
2	S. Umrao et al [21]		Device-to-Device Communication from Control and Frequency Perspective	Issues like interference, mode decision-making, supplemental distribution, and security can all be resolved via D2D communication's native support.	Over the analysis technique in detail and ruled out any open research requirements.	discussed the core D2D architecture, protocols, and associated procedures
3	Y. Haung et al [22]	2016	Mode selection, resource allocation [23], and power control for D2D-enabled two-tier cellular network.	Two-tier mobile network mode selection with d2d enabled, payload allocation, and intensity control	It is suggested to use geometric vertex search technology to boost overall performance and address the power distribution issue.	The total cost of the system can be increased with the frequency-sharing technology.
4	A. Sultana et al. [24]	**2017**	Efficient resource allocation in device-to-device communication using cognitive radio technology."	Optimize the transmission cost of d2d users while limiting electricity costs, interference costs, and statistical charges.	He proposed a subcarrier allocation scheme using cognitive radio networks and also developed a power allocation scheme.	Determine the most suitable solution with less computational complexity to solve the nonlinear failure rule
5	M. Belleschi et al. [25]	2011	Performance analysis of a distributed resource allocation scheme for D2D communications.	Distributed rule sets require minimal reporting overhead and reduced computational complexity	Combined mode transmission and resource allocation for d2d communication	A two-step method of low complexity is proposed and a d2d communication in mixed mode with a useful resource allocation method is proposed.
6	Y. Zhang et al. [26]	2014	Social network aware device-to-device communication in wireless networks.	Optimize the stream download process in d2d communication	Indian Buffet gained knowledge on modeling methods of online social networking. Chernoff's limit is derived to assess the overall performance of the proposed algorithm.	A weird social awareness method is proposed to optimize traffic offload for d2d communication

7	J. Liu et al. [27]	**2015**	A stochastic geometry analysis of D2D overlaying multi-channel downlink cellular networks.	Improve insurance probability and cross-cost of multi-channel D2D mask	The random geometry method works on a complete cellular network based on d2d, which will improve the performance of community coverage	Enabling d2d links relies primarily on store connections, which can dramatically improve network assurance performance
8	Q. Ye et al. [28]	2014	Distributed resource allocation in device-to-device enhanced cellular networks.	d2d distribution assistance distribution in the largest mobile networks	Provides useful resource locks and power allocation and abandons the idea of going out of the transmission relay to improve the performance of the d2d	Achieve big productivity gains with low overheads
9	M. Tehrani et al. [29]	2014	Device-to-device communication in 5G cellular networks: Challenges, solutions, and future directions.	Discover cellular applications in D2D cellular communication	For d2d cellular communication, a code-based protocol is proposed, which uses a method entirely based on the hash and bloom clear characteristics.	The proposed protocol dealt with an excessive density of devices and quickly absorbed a small number of radio resources.
10	L. Wang et al. [30]	2015	Socially enabled wireless networks: Resource allocation via bipartite graph matching.	Maximize the overall performance of the entire system	Used to match the Hungarian rule set. Optimize functionality by maximizing total registration costs	In the event of maximization of conventional performance indicators, taking into account power and quality of service constraints and taking into account aid sharing problems
11	Q. Duong et al. [31]	2014	Location-dependent resource allocation for mobile device-to-device communications	Research on the allocation of useful green resources for D2D communication in the cellular network	The total allocation of resources based on distance based on outage probability assessment	A resource allocation plan that is beneficial for interference detection is suggested to increase the efficiency of the spectrum.
12	G. Yu et al. [11]	2014	Distributed resource allocation in device-to-device enhanced cellular networks.	Selection of common mode and auxiliary distribution of d2d communication	For light and medium load scenarios, a set of low-complexity heuristic rules is proposed	To get the most out of the system as a whole, low-complexity algorithms are designed.
13	Wang, F. et al. [32]	2012	Energy-efficient radio resource and power allocation for device-to-device communication underlying cellular networks.	The community management side chain communication scheme has a context-sensitive rule in place to ensure the effectiveness	Apply tool-to-device newsletters to decorate IOT products	A signaling system is employed to enable large-scale D2D communication and to configure linkages and cell links.
14	D. Feng et al. [11]	2013	Device to Device communication underlying cellular networks	Maximize the overall productivity of the community	Maximize overall throughput when the device and cellular users pay attention to uplink control	With the help of the suggested plan, d2d began to load and enhanced the total network throughput.
15	C. Xu et al. [33]	**2013**	Efficiency resource allocation for device-to-device underlay communication systems	Optimize the total cost of the machine by sharing useful resources in d2d and mobile mode	Public sale using the opposite iterative combination	Low complexity auxiliary distribution scheme is proposed to increase the total cost of the machine

Through IoT devices, data is generated, transported, processed, and transformed to provide meaningful insights.

5.4 DISCUSSION AND RESULTS

This chapter focuses on the analysis of empirical statistics gathered during the fieldwork phase of the study. After completing the interviews and surveys, review the company's statistics data to verify that any questionnaires that are invalid or incorrect are removed by looking for samples of the respondents' responses and adding the finishing touches. The data analysis and interpretation will be finished soon, and the research goals will be revealed. The empirical portion of this thesis focuses on device-to-device verbal communication, sequencing, blessings, and challenging scenarios, applications, issues, and safety, all of which are based only on IoT devices. Community social networks, on the other hand, have gotten a lot of attention from scholars in recent years. Cluster and alliance creation are the two main solutions recommended for resolving the privacy issue in this situation. In general, D2D communication [37] uses one-hop routing; however, in some instances (public safety, insurance growth, content distribution, and so on), multi-hop routing may be used. There is virtually no research on routing components in D2D communications in the literature. A lot of work needs to be done in terms of security factors, particularly to combat security issues associated with a lack of trust authorization. On the one hand, it's the topology's dynamic nature; on the other, it's to safeguard customers' security and privacy so that they can be observed. Without authorization, a large amount of sensitive data goes through a single node.

The IoT is a concept that has yet to be fully realized, despite extensive research. Many off-the-shelf products rely on effective ways to meet demand. The installation of the IoT is surely progressing with the technology supplied here. RFID allows gadgets to be automatically identified without the need for costly manual intervention. IPv6 enables IP processing to be offered for all projects. The IoT is a forward-thinking and innovative concept that is now gaining traction. The concept of connecting anything and everything at all times is incredibly appealing. Its actual implementation is tough to assume, therefore overcoming these problems may require a significant amount of responsibility. There may be demanding scale requirements in terms of IP addressability, privacy, protection, and information control and analysis.

5.5 CONCLUSION

The IoT needs to connect something, anytime, anywhere, build a bridge, and connect some real worlds with virtual worlds. Intelligent vehicles have autonomous functions under proper cooperation. Evolving equipment and new applications require wireless networks to deliver higher data costs, more powerful capacities, lower latency, and a better service experience. The d2d dialogue method meets these requirements but faces many difficult situations. The goal is to highlight the capabilities of the d2d method in improving the IoT environment. Also, it has an advertising value capable to causes an impact on the economy in the future and the domain. D2D conversation and IoT have performed a brilliant position in developing and improving the networks. This paper intends to study both technologies in the past, current cutting-edge techniques, and predicts destiny based on the previous studies and demonstrates the venture. IoT expanding to plenty of industries and offerings together with enterprise and healthcare safety will keep being extra vital to shield touchy records and devices in opposition to harm. IoT has many advantages, however is developing at a fast pace and some devices are stuck with old protection leaving them susceptible to assault and being restrained via the gadget's processing and energy consumption barriers. For a better, reliable, and user-friendly environment, D2D oral communication is becoming an important part of the IoT system that is growing rapidly. The most favorable paradigm for the future communication of devices is D2D conversation. We provided a comparative evaluation of the IoT society with supportive arguments from the conventional systems. This study highlights the usage, benefits, and challenges of the entire convergence of IoT.

REFERENCES

[1] P. Pawar, and A. Trivedi, "Device-to-device communication based IoT system: Benefits and challenges," *IETE Tech. Rev. (Institution Electron. Telecommun. Eng. India)*, vol. 36, no. 4, pp. 362–374, 2019, doi: 10.1080/02564602.2018.1476191.

[2] Y. S. Chen, and J. S. Hong, "Intelligent device-to-device communication in the internet of things," *IEEE Syst. J.*, vol. 7, no. 1, pp. 77–91, 2014.

[3] S. Ashraf, F. Rehman, H. Sharif, H. and Kirn, H. Arshad, "Fake reviews classification using deep learning," *2023 Int. Multi-disciplinary Conf. Emerg. Res. Trends*, (pp. 1–20). 2023, doi: 10.1109/IMCERT57083.2023.10075156.

[4] M. F. Safdar, H. Sharif, F. Rehman, Z. Bilal, and S. Ahmad, "A review of recent advances in deep learning for object detection," *in 3rd International Conference on Innovations in Computer Science & Software Engineering (ICONICS)*, (pp. 1–6), 2023, doi: 10.1109/ICONICS56716.2022.10100486.

[5] M. H. Khalid, H. Sharif, F. Rehman, M. N. Ullah, S. Shaukat, H. Maqsood, C. N. Ali, and A. Hussain, "A brief overview of deep learning approaches for IoT security," *in 4th International Conference on Computing, Mathematics and Engineering Technologies (iCoMET)*, (pp. 1–5), 2023, doi: 10.1109/iCoMET57998.2023.10099306.

[6] A. Hassan, F. Rehman, M. S. Ashraf, A. Ashfaq, H. Sharif, and R. Zeeshan, "Performance enhancement in agriculture sector based on image processing," *in 3rd International Conference on Innovations in Computer Science & Software Engineering (ICONICS)*, (pp. 1–6), 2022, doi: 10.1109/ICONICS56716.2022.10100392.

[7] I. Manan, F. Rehman, H. Sharif, N. Riaz, and M Atif, "Quantum computing and machine learning algorithms - A review,", *in 3rd International Conference on Innovations in Computer Science & Software Engineering (ICONICS)*, (pp. 1–6), 2023, doi: 10.1109/ICONICS56716.2022.10100452.

[8] N. Sarfraz, F. Rehman, H. Sharif, S. Akram, and B. H. Mughal, "Efficient energy storage systems management in power plants with artificial intelligence and price control,", *in 3rd International Conference on Innovations in Computer Science \& Software Engineering (ICONICS)*, (pp. 1–5), 2022, doi: 10.1109/ICONICS56716.2022.10100617.

[9] N. Riaz, S. I. A. Shah, F. Rehman, and S. O. Gilani, "An approach to measure functional parameters for Ball-Screw drives,", *in Intelligent Technologies and Applications: Second International Conference, INTAP*, (pp. 398–408), 2020, doi: 10.1007/978-981-15-5232-8_34.

[10] R. Alkurd, R. M. Shubair, and I. Abualhaol, "Survey on device-to-device communications: Challenges and design issues," in *IEEE 12th International New Circuits and Systems Conference (NEWCAS)*, (pp. 361–364), 2014, doi: 10.1109/NEWCAS.2014.6934057.

[11] D. Feng, L. Lu, Y. W. Yi, G. Y. Li, G. Feng, and S. Li, "Device-to-device communications underlaying cellular networks," *IEEE Trans. Commun.*, vol. 55, no. 8, pp. 9–7, 2013, doi: 10.1109/TCOMM.2013.071013.120787.

[12] J. Yan, D. Wu, C. Zhang, H. Wang, and R. Wang, "Socially aware D2D cooperative communications for enhancing Internet of Things application," *EURASIP J. Wirel. Commun. Netw.*, vol. 2018, 2018, doi: 10.1186/s13638-018-1127-0.

[13] F. Jameel, Z. Hamid, F. Jabeen, S. Zeadally, and M. A. Javed, "A survey of device-to-device communications: Research issues and challenges," *IEEE Commun. Surv. Tutorials*, vol. 20, no. 3, pp. 2133–2168, 2018, doi: 10.1109/COMST.2018.2828120.

[14] A. Hussain, S. V. Manikanthan, T. Padmapriya, and M. Nagalingam, "Genetic algorithm based adaptive offloading for improving IoT device communication efficiency," *Wirel. Networks*, vol. 26, no. 4, pp. 2329–2338, 2020, doi: 10.1007/s11276-019-02121-4.

[15] M. M. Alam, H. Malik, M. I. Khan, T. Pardy, A. Kuusik, and Y. Le Moullec, "A survey on the roles of communication technologies in IoT-Based personalized healthcare applications," *IEEE Access*, vol. 6, pp. 36611–36631, 2018, doi: 10.1109/ACCESS.2018.2853148.

[16] M. Simulink, "Electronics and electrical engineering Elektronika ir elektros inžinerija," *in Natural Gas*, vol. 55, pp. 1–6, 2018.

[17] F. Zenalden, S. Hassan, and A. Habbal, "Peer selection in device-to-device communication based on multi-attribute decision making," *2020 IEEE Int. Conf. Informatics, IoT, Enabling Technol. ICIoT 2020*, no. November, pp. 570–574, 2020, doi: 10.1109/ICIoT48696.2020.9089605.
[18] B. Sudharsan, J. G. Breslin, and M. I. Ali, "Adaptive strategy to improve the quality of communication for IoT edge devices," *IEEE World Forum Internet Things, WF-IoT 2020 - Symp. Proc.*, no. October, 2020, doi: 10.1109/WF-IoT48130.2020.9221276.
[19] Z. Belghazi, N. Benamar, A. Addaim, and C. A. Kerrache, "Secure Wifi-Direct using key exchange for IoT device-to-device communications in a smart environment," *Futur. Internet*, vol. 11, no. 12, 2019, doi: 10.3390/fi11120251.
[20] X. Xiang, W. Liu, N. N. Xiong, H. Song, A. Liu, and T. Wang, "Duty cycle adaptive adjustment based device to device (D2D) communication scheme for WSNs," *IEEE Access*, vol. 6, pp. 76339–76373, 2018, doi: 10.1109/ACCESS.2018.2882918.
[21] "No Title", Ye, Q., Al-Shalash, M., Caramanis, C., & Andrews, J. G. "Distributed resource allocation in device-to-device enhanced cellular networks." *IEEE Transactions on Communications*, *63*(2), 441–454, 2014, doi: 10.1109/TCOMM.2014.2386874.
[22] Y. Huang, A. A. Nasir, S. Durrani, and X. Zhou, "Mode selection, resource allocation, and power control for D2D-enabled two-tier cellular network," *IEEE Trans. Commun.*, vol. 64, pp. 3534–3547, doi: 10.1109/TCOMM.2016.2580153.
[23] P. Cheng, L. Deng, H. Yu, Y. Xu, and H. Wang, "Resource allocation for cognitive networks with D2D communication: An evolutionary approach," (pp. 2671–2676), 2012, doi: 10.1109/WCNC.2012.6214252.
[24] "Efficient resource allocation in device-to-device communication using cognitive radio technology." *In IEEE Transactions on Vehicular Technology*, 66(10024–10034), 2017, doi: 10.1109/TVT.2017.2743058.
[25] M. Belleschi, G. Fodor, and A. Abrardo, "Performance analysis of a distributed resource allocation scheme for D2D communications," *in IEEE globecom workshops (gc wkshps),* (pp. 358–362), 2011, doi: 10.1109/GLOCOMW.2011.6162471.
[26] Y. Zhang, E. Pan, L. Song, W. Saad, Z. Dawy, and Z. Han, "Social network aware device-to-device communication in wireless networks," *IEEE Trans. Wirel. Commun.*, vol. 14, no. 1, pp. 177–190, 2015, doi: 10.1109/TWC.2014.2334661.
[27] "No Title", Liu, J., Zhang, S., Nishiyama, H., Kato, N., & Guo, J. (2015, April). A stochastic geometry analysis of D2D overlaying multi-channel downlink cellular networks. In *2015 IEEE Conference on Computer Communications (INFOCOM)* (pp. 46–54). IEEE. doi: 10.1109/INFOCOM.2015.7218366.
[28] Q. Ye, M. Al-Shalash, C. Caramanis, and J. G. Andrews, "Distributed resource allocation in device-to-device enhanced cellular networks," IEEE," 2015, doi: 10.1109/TCOMM.2014.2386874.
[29] M. Tehrani, M. Uysal, and H. Yanikomeroglu, "Device-to-device communication in 5G cellular networks: Challenges, solutions, and future directions," *IEEE Commun. Mag.*, vol. 52, no. 5, pp. 86–92, 2014, doi: 10.1109/MCOM.2014.6815897.
[30] L. Wang, H. Wu, W. Wang, and K. C. Chen, "Socially enabled wireless networks: Resource allocation via bipartite graph matching," *IEEE Commun. Mag.*, vol. 53, no. 10, pp. 128–135, 2015, doi: 10.1109/MCOM.2015.7295474.
[31] M. Botsov, M. Klügel, W. Kellerer, and P. Fertl, "Location dependent resource allocation for mobile device-to-device communications," *IEEE Wirel. Commun. Netw. Conf. WCNC*, no. April 2014, pp. 1679–1684, 2016, doi: 10.1109/WCNC.2014.6952482.
[32] F. Wang, C. Xu, L. Song, Z. Han, and B. Zhang, "Energy-efficient radio resource and power allocation for device-to-device communication underlaying cellular networks," IEEE, 2013, doi: 10.1109/WCSP.2012.6543016.
[33] C. Xu, L. Song, Z. Han, Q. Zhaq, X. Wang, X. Cheng, and B. Jiao, "Efficiency Resource Allocation for Device-to-Device Underlay Communication Systems: A Reverse Iterative Combinatorial Auction Based Approach", *IEEE J. Sel. Areas Commun.*, vol. 31, pp. 348–58, doi: 10.1109/JSAC.2013.SUP.0513031.
[34] S. Sanyal, and P. Zhang, "Improving Quality of Data: IoT Data Aggregation Using Device to Device Communications," *IEEE*, 2018, doi: 10.1109/ACCESS.2018.2878640.

[35] W. Xia, S. Shao, and J. Sun, "Relay selection strategy for device-to-device communication," 2013, doi: 10.1049/cp.2013.0065.
[36] P. P. Ray, "A survey on Internet of Things architectures," *J. King Saud Univ. Comput. Inf. Sci.*, vol. 30, pp. 291–319, 2018, doi: 10.1016/j.jksuci.2016.10.003.
[37] J. Chen, C. Liu, H. Li, and X. Li, "A categorized resource sharing mechanism for device-to-device communications in cellular networks," *Mob. Inf. Syst.*, vol. 2016, pp. 5894752, 2016, doi: 10.1155/2016/5894752.

6 Design and Implementation of an On-Board System for a Spider Robot

Nasir Nauman, Hirra Mustafa, Abrar Ashraf, and Hisham Khalil

6.1 INTRODUCTION

In the modern era of technological development, spider robots have become a new approach that needs to be explored. Normally, robots have wheels to move around and they are limited to plain surfaces. In the search and rescue operation in deep forest, the spider-type robots are more likely to move freely as compared to wheel-based robots. To overcome this issue, researchers are considering the different approaches such as four- or six-leg based robots.

In [1], a four-leg spider was developed with the self-adaptive approach while the controlling scheme has been based on LabVIEW. One can observe from Figure 6.1 that the spider was made with metallic legs and abdomen which usually made the robot heavier. A work was found on a six-leg spider with walking algorithm [2] and feedback using ultra-sonic sensing. In this work, the detail of movements of an individual walking pattern has been listed in tabular format. Moreover, the presented spider still seemed too bulky – Figure 6.2 shows the robot. A spider design and implementation are also given in [3] wherein a Raspberry Pi is used to control remotely. In [4], a spider monkey robot is presented for tangled motions and compared to real climbing patterns for a real monkey while few details have been published related to this work. Another article presented a design with a six leg spider robot where it was controlled by the spider kit [5].

Furthermore, a comprehensive review of the spider robot for characterizing environment is discussed in [6]. Furthermore, the characteristics of real spider legs and comparison with the other insects are discussed in [7]. This paper presents a design and implementation of a novel spider robot. The abdomen of the spider is developed with a printed circuit board (PCB) wherein the main circuit is printed. This thing implied that no other structure is used to develop the main body or abdomen of a spider robot.

To control the motion of the robot, an algorithm is developed with two modes of operation such as walk forward or reverse while the other one is excised mode. Good control of servo motors is achieved with an in-house Android application which is connected with Bluetooth module. Moreover, Section 6.2 presents the hardware design of the abdomen while Section 6.3 describes the details of the algorithm. Furthermore, the results and discussion are given in the last section with a comparison of robots with the previously presented designs.

6.2 ABDOMEN AND LEGS DESIGN OF SPIDER ROBOT

In this section, the abdomen of the spider is presented wherein the PCB retrial named as RF4 is used for this whereas the PCB is designed in Auto-desk Eagle (version 9.4). To overcome the mechanical

 DOI: 10.1201/9781003497851-6

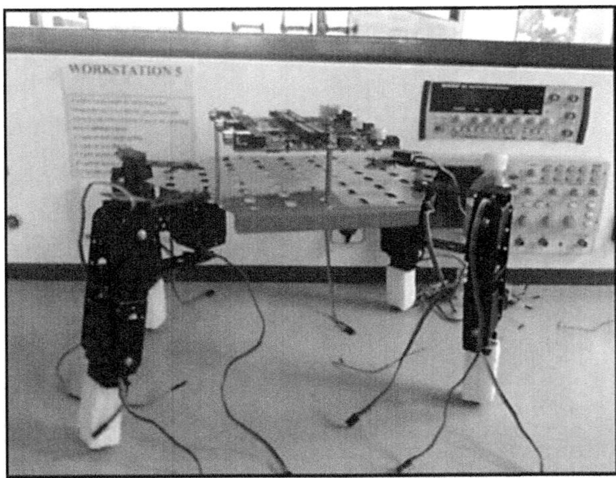

FIGURE 6.1 Self-adaptive four leg spider robot [1].

FIGURE 6.2 Six leg spider with sonar and camera [2].

stiffness of PCB a ring of acrylic sheet is conformed around the abdomen of the robot. An arc radius of 100 mm was subtracted from the main body as shown in Figures 6.3(a) and (b).

The basic purpose of this arc to create the range so that the servo motors have been used and the six legs can move easily without collision. Such as the spider has six legs therefore, total 18 servo motor (MG90s) metal gear with one bearing have been used [8]. Additionally, to control the robot Arduino Nano has been selected along the two 16 channel 12-bit PWM (PCA9685) with I2C bus protocol whereas both PWMs are in master and slave configuration receptively.

The legs of the spider are designed according to the real spider leg shown in Figure 6.4(a) wherein two basic part named as Femur and Tibia. Additionally, two joints such as Coxa and Patella are also shown in the same figure. Similarly, Figure 6.4(b) presents the 3D designed leg with a diameter of 10.35 mm for Coxa and Patella whereas the length of Femur is 60 mm. To create walking friction claws provide in a real spider leg, the silicon is used for this purpose in the fabricated leg.

However, the complete block diagram of the electronic hardware is shown in Figure 6.5 wherein 2-ampere buck DC-DC converter (LM2596S) is used to power up the PWM as well as Arduino

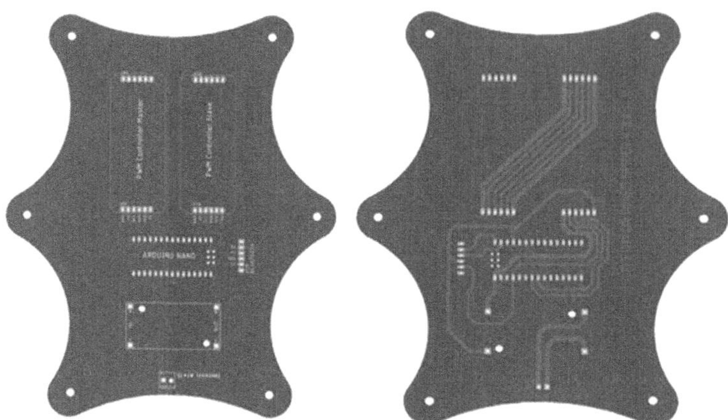

FIGURE 6.3 Abdomen design of spider (a) Top view of PCB based abdomen (b) Bottom view of PCB based abdomen design.

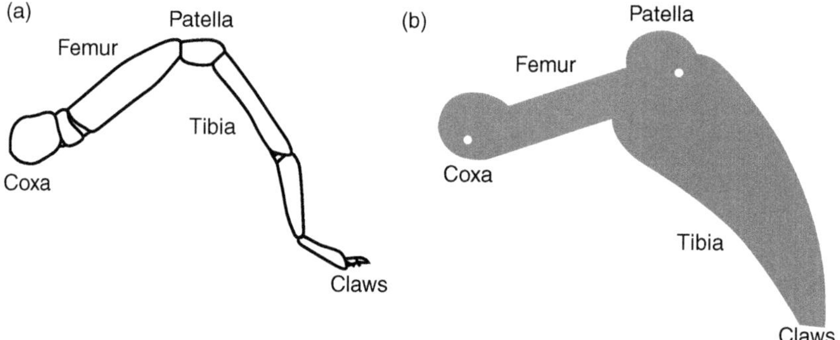

FIGURE 6.4 Spider robot 3D leg design (a) structure of real leg [7] (b) 3D model of spider leg assembled.

Nano (3.0). The controller has the clock speed 16 MHz while the connectivity of robot with Android handset is carried by the Bluetooth module (HM-10, 4.0). The detailed pin configuration of a servo motor to the micro-controller is also presented in Figure 6.5. As shown in Figure 6.3, all modules can be attached with the main abdomen of the spider by a female pin so that the further complexity of circuit can be reduced. The pins A4 and A4 is attached with PWM while Bluetooth module is connected with transmitting (Tx) and receiving (Rx) ports of Arduino.

6.3 ALGORITHM FOR SPIDER WALK AND EXCISE

The current requirements of servos and other modules are carefully designed, to avoid the unnecessary vibration or overshoots during the loaded condition.

In this section, the firmware design of the robot is discussed. Initially, the spider is programmed for two sets of operations. First, one is a walking pattern while the other one is excising. When the spider is powered up, all the servo motors are initialized at 90◦ so that the robot can stand up. It will remain in this state unless further instruction is received by the controller through the Android handset. Before the second step, the six legs of the spider need to be tuned in such a way that every leg is aligned with the corresponding diagonal leg as shown in Figure 6.8. Moreover, the walking pattern is defined such as the central servo of each leg moves +30° and after a specific delay is a move to 30°. The thing implies that if the first leg moves at +30° then the next leg in the same

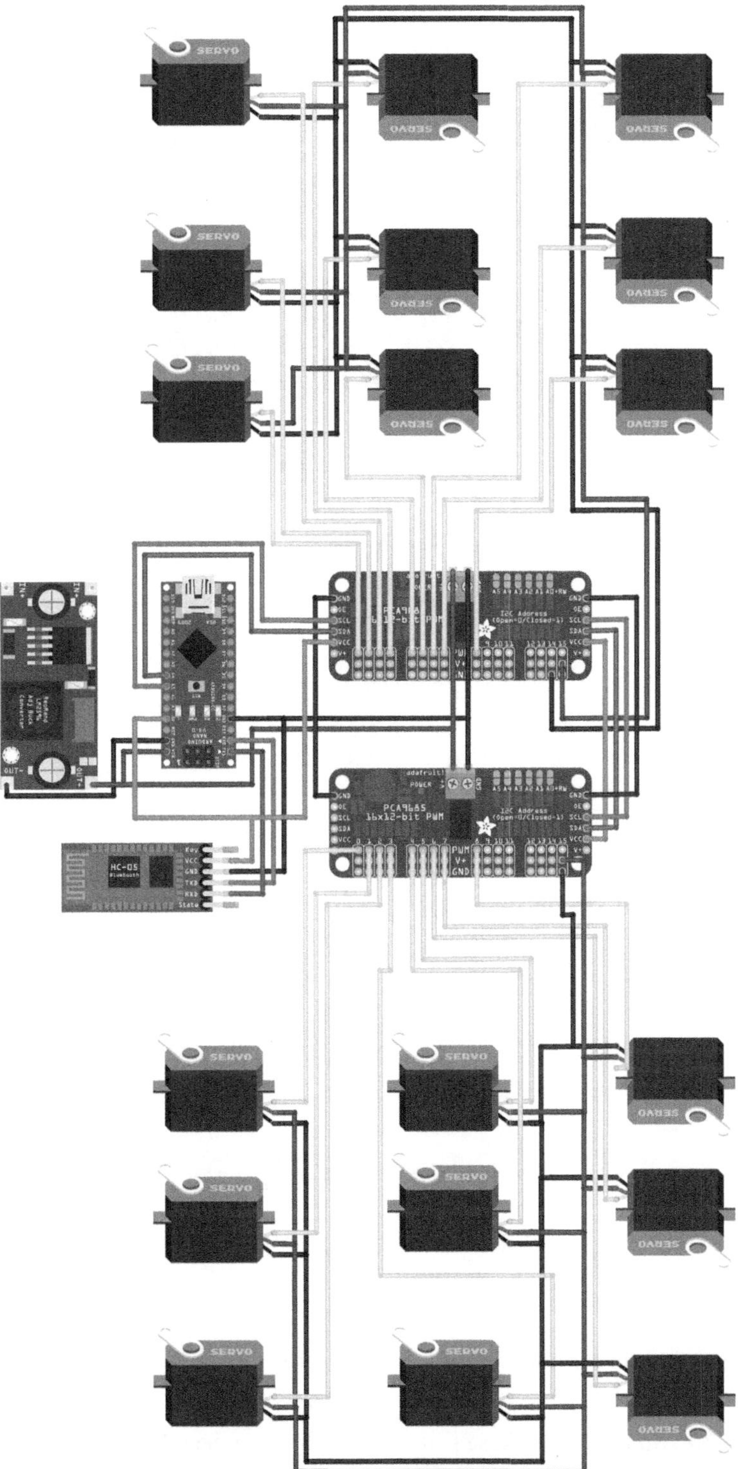

FIGURE 6.5 Complete block diagram of abdomen design with pin configuration of spider.

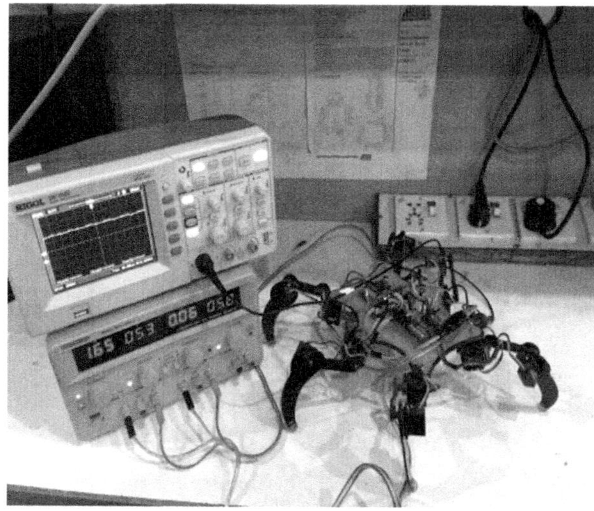

FIGURE 6.6 Operational setup of complete proposed system on board spider robot.

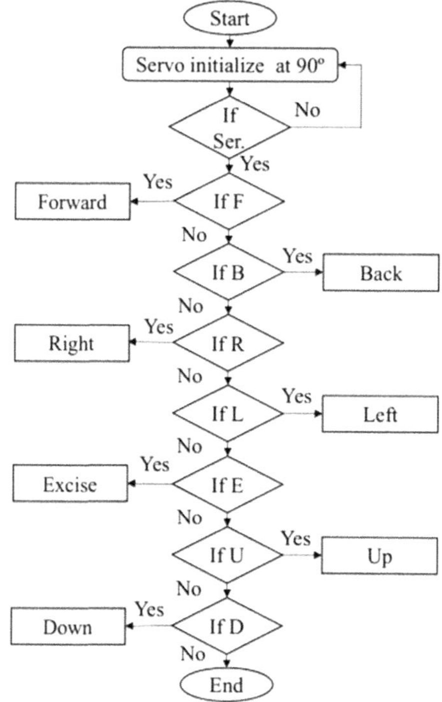

FIGURE 6.7 Flow-chart of algorithm for walk and excise mode.

side must be at 30°. Similarly, a reverse pattern is imposed on the other side of the legs in the loop fashion. Consequently, the spider moves forward with slipping the leg due to the silicon boot as shown in Figure 6.6. Figure 6.7 shows the controlling algorithm with necessary details of characters that have been programmed with required loops.

To evaluate the performance of servo motor the excise mode is programmed with two characters such as U for up and D for down.

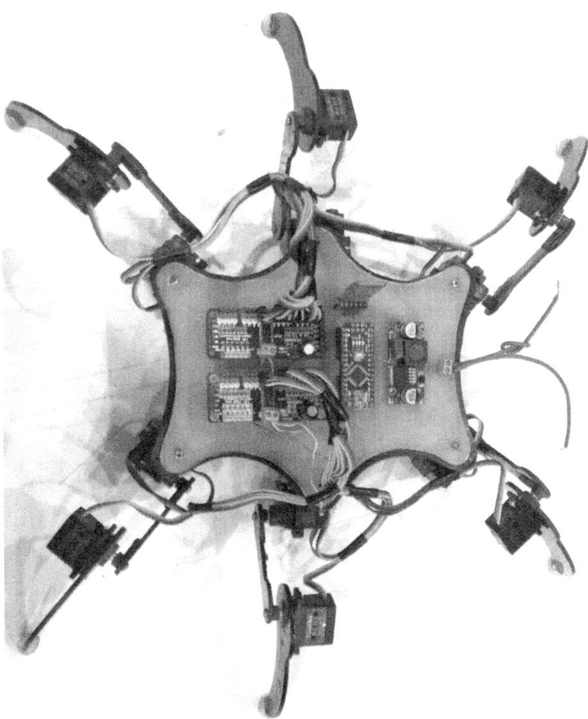

FIGURE 6.8 Top view of proposed six leg (18 servo) spider robot with the system on board structure.

6.4 RESULTS AND DISCUSSION

The testing of the spider robot is carried out in the lab environment in two domains; such as electronics testing as well as mechanical performance. The major object is to create a stable walking pattern. For this purpose, the output of PWM is tested though oscilloscope (RIGOL DS 1052E) to verify the unwanted ripple in the design system or reverse the current of servo motors. Figure 6.7 shows a stable output. Initially, a problem was identified such as the legs of the robot has vibrations. Therefore, the current supplies have been settled in such a way that each servo motor was drowning the required current. In Figure 6.9, the sequence of current signal for servo motor has been shown such as it starts from right to left-hand legs.

Then secondly, the mechanical testing and walking pattern are tested and the issue of slipping is noted during the walk. To overcome the dilemma, the Claws was redesigned with silicone boots and consequently, the slipping was successfully removed. In the current model of system onboard spider design, there is no batteries installation on the abdomen to reduce the payload of the robot. Table 6.1 lists the comparison with already reported spider robots. It has been observed that only the proposed spider has a system on board while all others were developed with metallic structures.

6.5 CONCLUSION AND FUTURE WORK

This paper presents the system onboard spider robot with abdomen designed with PCB and special purpose legs have been developed. Moreover, an algorithm is developed for multiple modes of operations such as walk and excise. The designed robot size and weight are compared with other robots in Table 6.1. It has also been tested in different operating environments. In conclusion, a spider can be developed with 3D printing technology so that the mechanical performance can be increased.

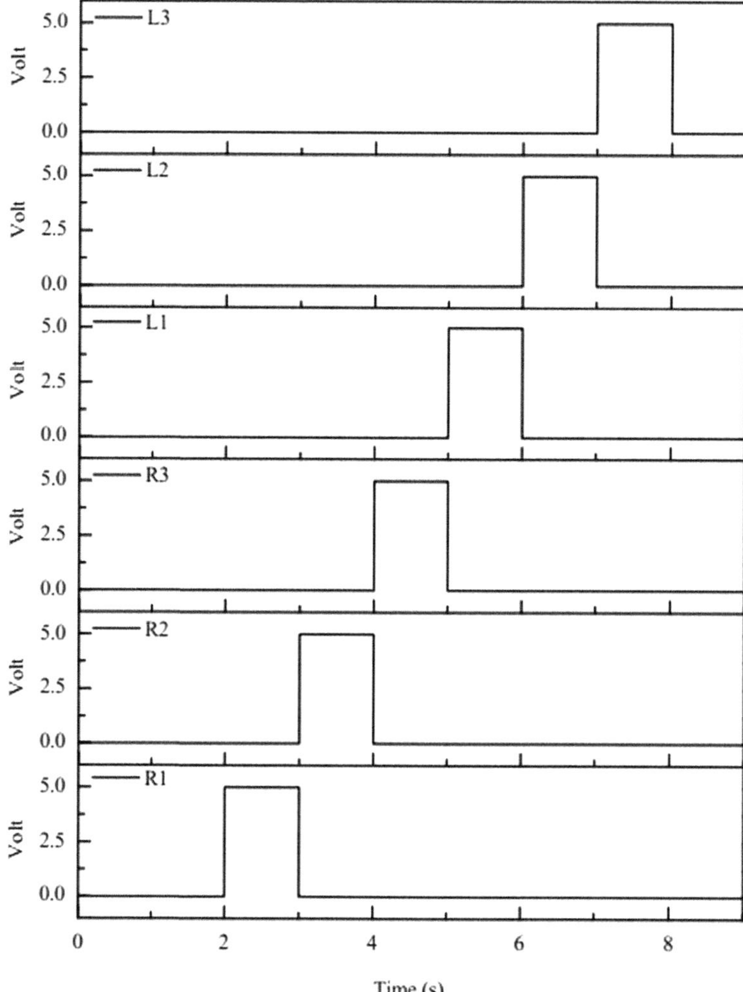

FIGURE 6.9 Sequential control of six legs right to left (bottom to top in the given figure).

TABLE 6.1
Comparison between Proposed and Previously Presented Spider Robots

Ref	Year	Spider type	Abdomen design	Application/Control
1	2013	4 leg	Metallic	Adaptive
2	2015	6 leg	Metallic	Search and Rescue
3	2015	6 leg	Metallic	Search
4	2017	4 leg	Metallic	Climbing monkey
5	2017	6 leg	Metallic	PC control
Proposed spider	2019	6 leg	PCB sys. on board	Android

REFERENCES

1. Teymourzadeh, R., Mahal, R. N., Shen, N. K., & Chan, K. W. (2013, December). Adaptive intelligent spider robot. In *2013 IEEE Conference on Systems, Process & Control (ICSPC)* (pp. 310-315). IEEE. Author, F., Author, S.: Title of a proceedings paper. In: Editor, F., Editor, S. (eds.) CONFERENCE 2016, LNCS, vol. 9999, pp. 1–13. Springer, Heidelberg (2016).
2. Karakurt, T., Durdu, A., & Yilmaz, N. (2015). Design of six legged spider robot and evolving walking algorithms. *International Journal of Machine Learning and Computing*, 5(2), 96.
3. Zak, M., & Rozman, J. (2015, November). Design, construction and control of hexapod walking robot. In *2015 IEEE 13th International Scientific Conference on Informatics* (pp. 302–307). IEEE.
4. Li, W. Y., Wang, Y. H., Kuo, C. T., & Lin, P. C. (2017, September). Design and implementation of a spider monkey robot. In *2017 International Conference on Advanced Robotics and Intelligent Systems (ARIS)* (pp. 62–62). IEEE.
5. Verma, G., Rai, P., Chauhan, B., Kumar, A., Pandey, P., & Karnail, V. (2017). Hardware implementation of autonomous hexapod spider robot. *International Journal of Information Technology*, 9, 395–398.
6. Tsitsimpelis, I., Taylor, C. J., Lennox, B., & Joyce, M. J. (2019). A review of ground-based robotic systems for the characterization of nuclear environments. *Progress in Nuclear Energy*, 111, 109–124.
7. Auzina, E. (2015, November). Exploring the possibility and need of making a spider inspired footpad for robotic legs using a 3D printer for use in a low energy system. In *2015 IEEE 3rd Workshop on Advances in Information, Electronic and Electrical Engineering (AIEEE)* (pp. 1–5). IEEE.
8. Lee, J. W., & Kim, T. W. (2011, November). Design and experimental analysis of embedded servo motor driver for robot finger joints. In *2011 8th International Conference on Ubiquitous Robots and Ambient Intelligence (URAI)* (pp. 548–551). IEEE.

7 Revolutionizing Underwater Sensor Networks

A Novel Routing Approach for Efficient Data Collection with Atom-Inspired Design

Muhammad Ahmad Pasha, Fatima Ijaz, Aleema Imran, and Ahmad Shaf

7.1 INTRODUCTION

Oceans cover 70% of the earth's surface. The ability to monitor underwater conditions has been strengthened thanks to recent technological advancements. Underwater wireless sensor networks (UWSNs) can be used for a variety of purposes, including mine detection, oil and gas pipeline monitoring, assisted navigation, disaster avoidance, oceanographic data collection, and ocean sampling [1, 2]. Acoustic communication is favored in UWSNs because it delivers higher data rates over longer distances, whereas radio frequency (RF) signals attenuate rapidly in water and have constrained transmission ranges [3]. However, there are numerous obstacles to acoustic communication as well, including multipath interference and bandwidth limitations, which can result in significant overhead from retransmissions [4].

Other long propagation delays, high error rates, long end-to-end delays, ineffective data collection, brief network lifespan owing to frequent topology changes, and high energy consumption from multi-hop data transmissions are further problems with UWSNs [5, 6]. Unlike terrestrial networks, the majority of UWSN applications demand reliable data gathering and dissemination, but this is challenging given the dynamic underwater environment [7, 8]. Energy holes may result from the traditional practice of sending sensed data in several hops directly to a sink node [9, 10].

To address this, we investigated the use of autonomous underwater vehicles (AUVs) for data collection. AUVs can collect data from sensor nodes, transmit it using the shortest path trees to sink nodes, and then forward it there, minimizing the number of hops and increasing energy efficiency [11, 12]. The AUV only stops at chosen nodes, known as gateway nodes (GNs), for data gathering to decrease the amount of transmission power required by sensor nodes [13, 14]. The values of the hello packet received signal strength indicator, the locations of the nodes along the AUV's route, and the amount of remaining energy are all considered for choosing GNs [15, 16]. Member nodes (MNs) associate with the GNs by flooding data packets to the GNs utilizing the shortest path tree technique. An entirely new GN with greater leftover energy is selected from its surrounding peers whenever a GN has lost all its energy below a certain threshold.

To maximize data collection while minimizing journey time, the AUV's trajectory is essential [17]. Unpredictable energy use and substantial packet loss can result from an erratic trajectory. Recent research indicates that an elliptical trajectory is ideal, although higher end-to-end latency is caused by the longer time needed for MNs farther away from GNs to associate with the GNs [18].

DOI: 10.1201/9781003497851-7

The high end-to-end latency in cooperative data collecting and transmission is addressed in this study by the proposal of an effective routing protocol termed ASEDG. To complete routing tasks, ASEDG employs a horizontal and vertical elliptical trajectory and an atomic-shaped delay model. This results in the creation of the most GNs while minimizing the time needed for MNs to associate vertically with GNs. To further reduce latency and balance energy usage, ASEDG additionally makes use of mobile sink nodes along the trajectory's edge. The final sink receives the data that the mobile sinks transport from the GNs.

7.2 LITERATURE REVIEW

The AEDG navigation protocol was suggested by the researchers of [1], which improves network speed and lifetime while reducing packet loss due to less energy use. This is accomplished with the aid of the AUV which gathers data from GNs and passes it to the collector. By associating the fewest number of mobile nodes (MNs) with the gateway nodes (GNs) and rotating the AUV depending on residual energy, the shortest path tree (SPT) method reduces packet loss. The path taken by the AUV is not exactly oval. The problem that still must be resolved is the lengthier end-to-end delay of the AEDG.

A distributed AUV data-gathering system was introduced by the authors of [4]. Data from a cluster-based underwater wireless sensor network (UWSN) are collected using an AUV in this manner. The AUV only visits the specified nodes, known as route nodes, as opposed to every node or cluster, to lower the transmission power of the nodes. Mobility, according to [9], expands the sensor coverage area.

The authors in [15] created Energy-efficient Routing Protocol with AUV Assistance to increase network lifetime by reducing energy usage. This method makes use of GNs that spin by their remaining energy. However, it does not restrict MNs' interaction with GNs; as a result, this becomes the justification for the GNs' high energy usage. The GNs also carry a high risk of data loss.

The authors presented the underwater routing protocol with AUV assistance (AURP) in [18] for efficient data collecting and network optimization. By transmitting the data to the sink via many AUVs, this technique lowers the overall quantity of data transmissions. It increases data transfer rates, guarantees accurate data collection, and lessens energy usage. Nonetheless, this method collects data from MNs using specified GNs. As a result, the majority of the GNs' energy is spent on transmitting data, which reduces the low data transmission rates caused by a lengthy network lifetime.

Data is moved from the point of origin to the sink using the [19] Relative distance-based forwarding (RDBF) routing protocol. The principle of fitness determines which node forwards data. The gap between UWSN nodes can be calculated mathematically, as shown in [20]. [21] describes a hierarchical UWSN data collection technique that employs numerous AUVs to scan large areas. These multiple AUVs combine to form a sporadic multi-jump from one network to another – AUV coordination is required for data transmission to reach the sink.

The contributors of [22] suggested the AREP, in which every node retains a table of routes to preserve information about nearby neighbors to ascertain if the link is active. This protocol aims to increase the network's lifespan and collect data from GNs. MNs were linked to GNs using the SPT algorithm to reduce association time and energy consumption. With the use of connected dominating sets (CDS), SEDG provides an elliptical trajectory for AUV movement [23].

A channel-aware routing protocol (CARP) that fully utilizes link-quality data for forwarding packets is described in [24]. It efficiently eliminates void and shadow regions while avoiding loop-free routing by leveraging fundamental topological information. For total latency in UWSNs, a network protocol for diagonally and vertically routing (DVRP) was proposed in [25]. Creating a flooding zone angle points packets in the direction of the sink. By the overflowing pattern and the level of energy, the sensors locally decide how to forward packets.

[26] proposed an enhanced hydro cast that dispatches an AUV to gather data. Routing to the washbasin is carried out greedily using the sensor nodes' pressure levels as a guide. The size of the packets and transmission power were two key factors considered by the developers of [27] while addressing the problem of network lifetime. Integer linear programming was used to offer a method for jointly optimizing these two parameters and extending the network lifetime.

A unique approach was presented in [28], where an AUV utilized Dubin curves to collect information about diverse objects dispersed around a 3D area. It creates a 2D route and converts a 3D target to a 2D plane in the initial stage. The Dubin curve's 2D global coordinates are translated into three dimensions in the following step using an Euler rotation transformation.

The route taken by an AUV has a significant impact on both the selection of data samples and the amount of energy used by sensor nodes. By minimizing node energy usage for several paths, an AUV cannot choose a path for gathering data activities. A better lawnmower pattern for AUVs was developed to fill this gap; routing and the energy consumption of sensor nodes were used [29]. The foundations, benefits, and limitations of modeling and path search methods for AUVs were discussed by the authors in [30]. They also improved the original methods and described how to fix certain technological shortcomings.

7.3 PROBLEM DEFINITION

Most data routing algorithms suggest a path between the source and destination nodes that makes use of several intermediary sensor nodes. However, there is occasionally a significant risk of data collecting gaps, node collapse, high energy utilization, and loss of packets. To deal with these possible problems, an AUV has been included in the routing processes. Currently, while considering effective data collecting, energy consumption, and dependable data delivery, the elliptical trajectory is the most advantageous option for the AUV movement. For data collection, the AUV travels along a predetermined path. However, there are some issues with both the trajectory itself and the technique used to gather data after it.

The following factors would be considerably improved by the adoption of a new trajectory design methodology:

A. First, to create linked dominating set (CDS) nodes, the sensor nodes first use excessive energy.
B. Secondly, it would likely take a lot of time and energy to manipulate the CDS to get an elliptical form.
C. Third, different network area sizes have been accommodated by using fixed-sized elliptical shapes.

The shortest path tree (SPT) is not used in a particular network region, which causes backtracking and a sizable delay when linking MNs and GNs. High end-to-end network delays are the result of nodes that are located further vertically from the horizontal elliptical trajectory.

7.4 SYSTEM MODEL

The AUV movement or trajectory design, overall network architecture, operational principles, and constraints are all covered. A few properties are used to arrange the nodes.

1. Constituent nodes, or MNs.
2. Gateway nodes, or GNs.
3. A vehicle that is autonomous underwater (AUV).

7.5 TRAJECTORY DESIGN TECHNIQUES

A revolutionary routing protocol we developed is called atomic-shaped effective delay and data gaining (ASEDG) to collect data quickly while minimizing unnecessary end-to-end delay. Our system employs the maximum number of GNs to maximize data collection, and delay models are applied to minimize latency.

7.5.1 Enhancing the Existing Trajectory Design

The elliptical course of action that is being seen is supported by the path for the AUV movement that has been constructed. The CDS, a group of network dominator nodes, provides several routes to the other network nodes. The MST of the CDS is built after the CDS has been formed. A Hamiltonian circuit (HC), which is an arbitrary course for the AUV's movement, is established once the MST is created, as shown in Figure 7.1. For creating an elliptical form, the CDS does not provide an image. The CDS nodes produce a random trajectory, which the HC then modifies into a circular trajectory and elliptically translates into a trajectory whose form is independent of the size of the network region. Previous research hasn't addressed the problem of a network's expansion or shrinkage, where the fraction of the ellipse's size changes.

7.5.2 Introduction to the Atomic Shape Trajectory Approach

A technique that uses elliptical horizontal and vertical trajectories has been proposed. The atomic structure served as inspiration for the concept of two paths. Both elliptical trajectories employ two AUVs. They maintain a fair amount of distance between one another to prevent collisions. An AEDG sensor node is positioned at the network's center, which contrasts with the scale of the network. By considering a main axis and two variables, b, the node in the center can generate elliptical horizontal trajectories of varied sizes (minor axis). Using the area's X-axis and Y-axis, one can determine the major and minor axes of a circular elliptical form, respectively.

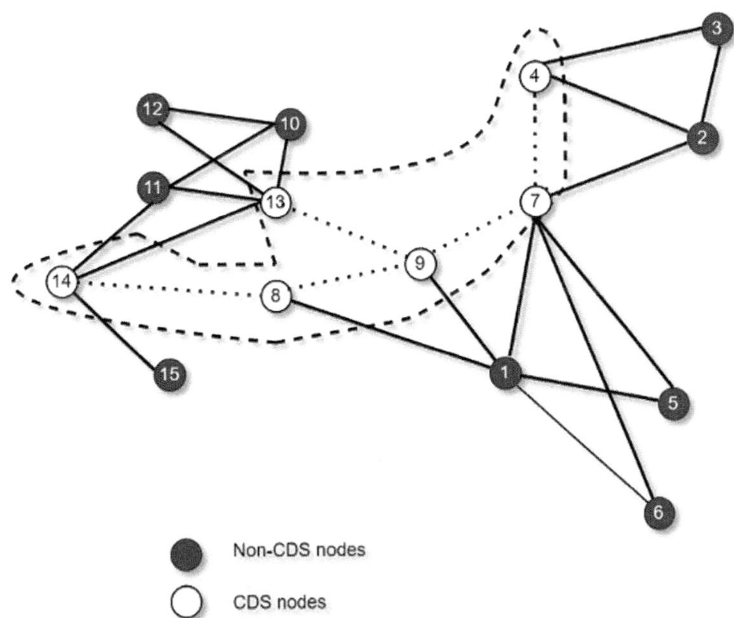

FIGURE 7.1 AUV's trajectory with HC.

7.5.3 Major and Minor Axes to Enhance Trajectories

We create numerous elliptical forms using the major as well as the minor axes of the shape of an elliptical in order to obtain the value of b static. The shapes are displayed in Figure 7.2.

The horizontal elliptical trajectory was produced using this method. Figure 7.3's central node produced a second elliptical trajectory using the initial trajectory's major and minor axes values as well as the same network region vertically.

The horizontal elliptical trajectory was produced using this method. As seen in Figure 7.3, the centrally located node produced a second elliptical trajectory that crossed the same network space vertically by leveraging the major and minor axes from the first.

7.5.4 Atomic Shape

The form turns into a circle when the main axis, b, is compared in size to the minor axis. There is insufficient coverage of the network's data at the point when an ellipse is created in vertical space if the major axis is intended to be 2b, which is double the value of the minor axis. To pick the farthest

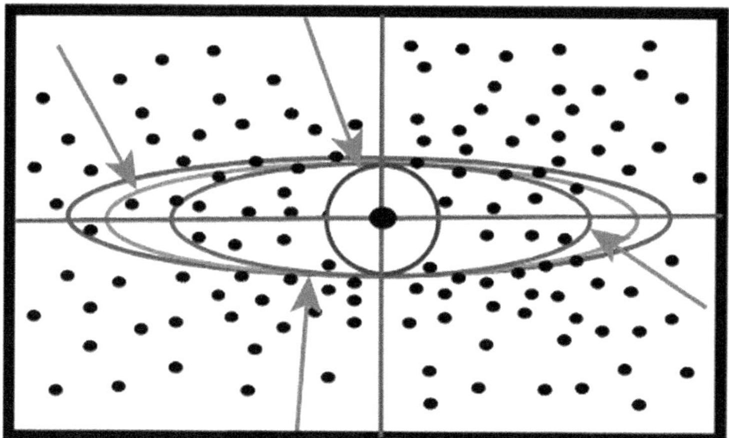

FIGURE 7.2 Multiple ellipses.

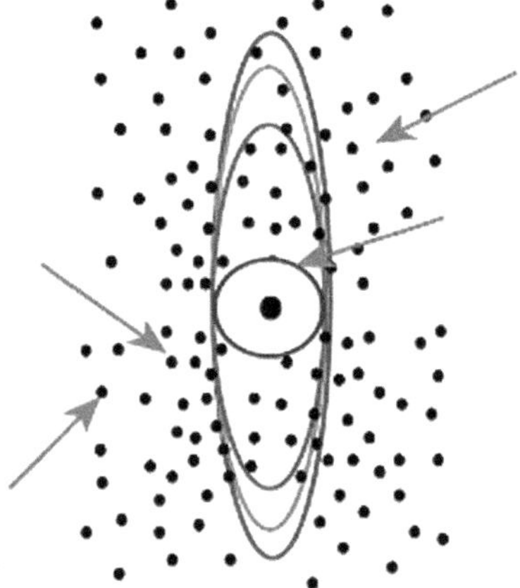

FIGURE 7.3 Atomic shape against multiple ellipses.

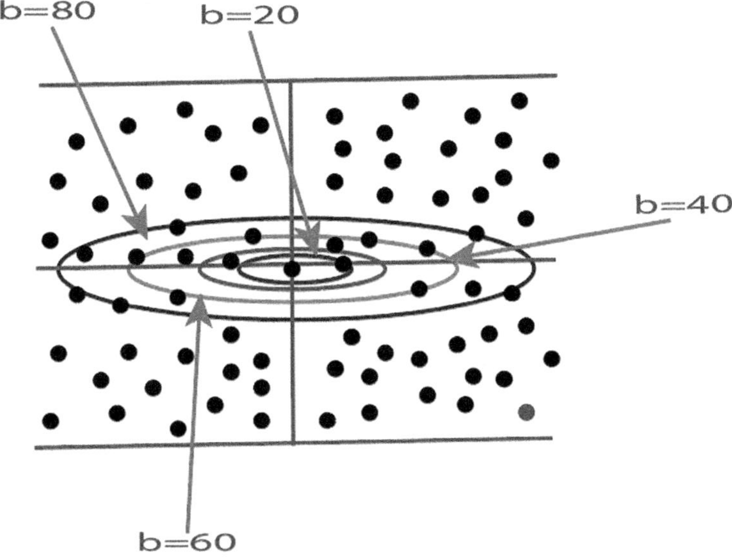

FIGURE 7.4 Selection of optimal ellipses.

nodes that are in the upward zone as GNs, the vertical zone either creates a vertical ellipse in step 3b that spans the network's important vertical region, or disqualifies remote nodes by setting the primary axis to be three times that of b. If the principal axis is taken to be 4b, data gathering would be hindered and an excess of GNs would be produced, increasing the network's use of energy. Using the major axis 3b, which is three times larger than the small axis, we build a straight-line trajectory which crosses the ideal region of the network.

When a = 3b, the only horizontal elliptical shape is optimal. Figure 7.4 shows a few elliptical geometries alongside various b values. The horizontal distance the ellipse covered was considered when choosing the best one. The horizontal distances of all the ellipses were compared to the network's Centre point because each ellipse had a horizontal distance of 2a. The distance with the closest approximation to the network's Centre was found to be ideal.

Since the Centre of the 500 500 m2 rectangle in Figure 7.4 is 250, the optimal values for b and an are 40 and 120.

At that point, three minor axes connect to one of the ellipse's primary axes. The values of b and a can be altered based within the limits of this ratio. The corresponding b and a values, for a 1000 x 1000 m2 area, for instance, are 80 and 240, and 20 and 60 for a 250 x 250 m2 area.

7.6 DELAY MODEL

Delay is a critical performance factor to take into account while developing a network. The two kinds of delays are considered, and their models are also provided. MNs first communicate with other MNs before communicating with GNs.

7.6.1 MNs to MNs

How long it takes a data packet to get from one node to another node in a network depends on the latency of the nodes in the network. The nodes may have significant forwarding delays if the following forwarder is located distant from the first node or farther away. The nodes' energy consumption and the likelihood of packet loss are both significantly increased by this prolonged delay.

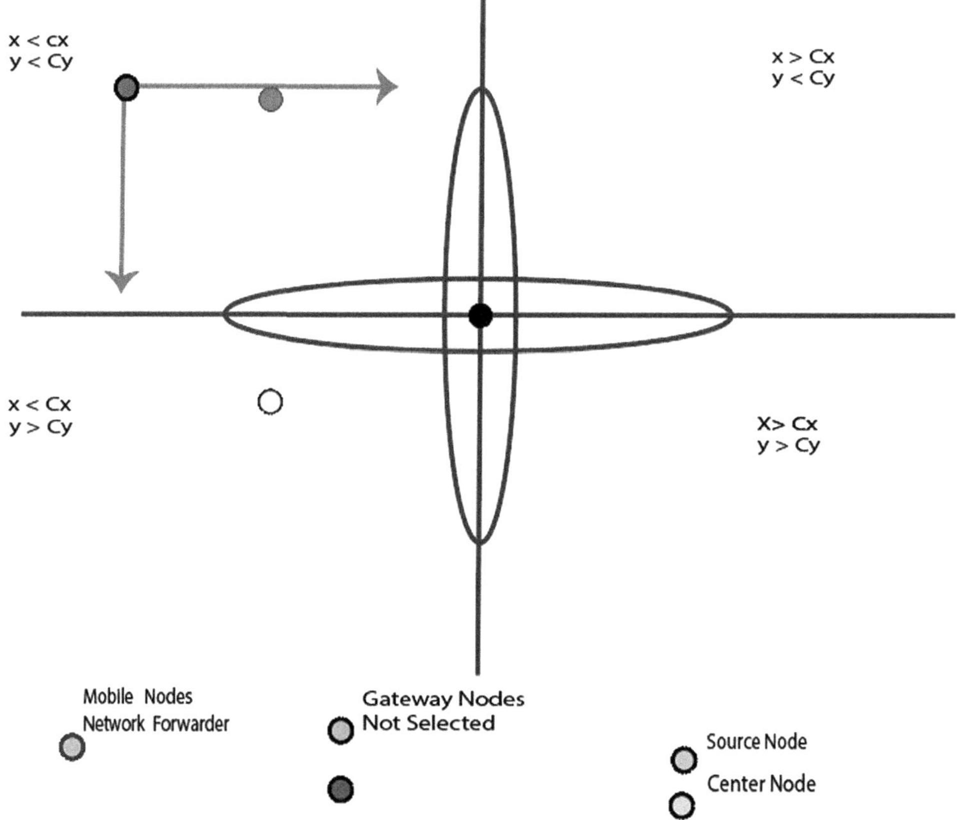

FIGURE 7.5 Next forward selection.

The algorithm for choosing the next forwarder in ASEDG has considered the node-to-node delay. Choosing the next forwarder is important since every node in the network is aware of its location and one node was strategically positioned in the network's middle. A source node can forecast the following carrier node's path by the subsequent forwarder's location or order can be calculated using an algorithm.

For this method, we divide the quadrants into equal parts so that each portion of the network becomes an origin node when it first establishes a region along the X and Y axes in that quadrant, both horizontally and vertically. The forwarders must reside on or very close to the line dividing the horizontal and vertical regions. Algorithm 7.1 explains a procedure for selecting the next forwarders. This will allow us to cut down on the delay brought on by the successive forwarders' poor selection. Additionally, the way the nodes are chosen is illustrated in Figure 7.5.

Algorithm 7.1 Identifying a Subsequent Forward Node

1: Set num as the number of network nodes.
2: Define an array s of size num to store the status of each node.
3: Create a dynamic array of integers with size num.
4: Set dx as the X-axis value of the central node.
5: Set dy as the Y-axis value of the central node.
6: Initialize a variable idx to 0.

```
 7: For i = 1 to num do
 8: If s[i] is "Alive" then
 9: Set arr[idx] as i.
10: Increment IDX by 1.
11: Else
12: Discard the node.
13: End if
14: End for
15: Set array size as the size of arr (IDX).
16: Set in size as the size of an integer.
17: Set length as arraySize divided by intSize.
18: For i = 0 to length - 1 do
19: Set an as the X-axis value of the ith node (arr[i]).
20: Set b as the Y-axis value of the ith node (arr[i]).
21: If g > dx and h > dy then
22: Call NextForwarder(g, h + 1), NextForwarder(g+ 1, h), NextForwarder(g + 1, h + 1).
23: Else if g > dx and h == dy then
24: Call NextForwarder(g, h + 1), NextForwarder(g, h), NextForwarder(g, h + 1).
25: Else if g > dx and g < dy then
26: Call NextForwarder(g, h), NextForwarder(g + 1, h), NextForwarder(g + 1, h).
27: Else if g == dx and h > dy then
28: Call NextForwarder(g, dy), NextForwarder(g, dy), NextForwarder(g, h).
29: End if
30: End for
```

The ensuing forwarders will be selected up until every forwarder has arrived at the GNs. The source node serves as the root node, while the first potential forwarders serve as child nodes, resembling a tree-like topology. The GNs play the part of leaves in this topology.

In the algorithm, cx and cy stand in for the position of the Centre node, where N is the total number of sensor nodes in the network. On the X and Y axes, respectively, g and h stand in for the source node's coordinates. The node in the Centre divides the network into four halves. The source node has four choices for building a path to the origin node thanks to the network region's division into four quadrants.

The node initially acknowledges it assesses its current location in relation to the middle aware node. It establishes its bounds in the second phase by looking at where the core node is located. A node communicates with its nearby nodes during the third phase, a hello packet (HP), and they reply with an acknowledgment that includes their location. By comparing its location to the locations of the recipients, the source node selects the following probable forwarders.

In each location, the following forwarder must adhere to two requirements: The first requirement is that they must be situated between the source node's G and H axes or at their intersection. The second requirement is that, depending on where the source node is located, it must satisfy one of four conditions. If a node appears in the initial area of the source node, the horizontal G and vertical H axes of the subsequent carrier must be larger than those.

The next forwarder in the second area must be on a different side from the source node. In the third zone, the horizontal G and vertical H axes of the subsequent carrier must, respectively, be larger and smaller than other types of the source node. The forwarder's subsequent G and H axes should be less than the origins in the fourth area.

These drawbacks will enable every area's node to carefully select the subsequent carrier. Once the packet has found its succeeding forwarder, the forwarder node repeats the process until it reaches its destination. Any two network nodes can be routed using this method.

Based on the coordinates of each node, the method starts by identifying the network's Centre node. With the central node at the intersection, it then divides the network into four quadrants. The source node establishes its quadrant and seeks the subsequent forwarder that satisfies the local positional restrictions. Up until the packet reaches the destination node, each forwarder will repeat this operation. Without requiring global network information at each node, this routing technique can route between any two nodes in the network.

7.6.2 MNs to GNs

The time it takes for a package to go from MNs to GNs is referred to as the delay from MNs to GNs. When nodes are deployed in a network in a region of high density, a delay is caused from MNs to GNs owing to distant communication, which results in a greater number of linkages between MNs and GNs. This collaboration is referred to as MNs to MNs. When the transmission distance and the associated duration of the MNs with the GNs are greater, multiple linkages can also form. In UWSN, there is an MNs-to-GNs delay.

There are several ways to go between nodes in UWSN. Since our suggested trajectory is grid-based, GNs are generated along its vertical trajectory. The GNs' horizontal and vertical trajectories are created by the nodes. It thus minimizes the lag brought on by long-distance communication between MNs and GNs. The MNs create pathways that connect numerous GNs. The SPT aids in determining which of the various routes leading to the GNs is the shortest in distance.

Based on the calculated distances of the MNs to all prospective GNs, the SPT algorithm creates the links between the MNs. When the length or weight of each edge is known, a common graph data structure challenge is determining the shortest path.

Assume that the vertices and edges of the graph G are (V, E). Give each V a distance estimate. Make a note of the source node, which serves as the origin, the present node, and a sign that none of the other nodes have been visited. Determine the rough distances between each of the current node's undiscovered neighbors. By comparing the estimated and actual distances, assign the value for the shortest distance to each V. Once the current node has considered every neighbor, mark the neighboring nodes as visited. If the destination node has already been marked, mark it as visited.

7.7 SIMULATION RESULTS

To evaluate ASEDG's performance, we tested it using Aquasim NS-2 and contrasted it to the performance of the already-existing AEDG routing protocol.

7.7.1 Durational Delay

The channel's length and transmission rate channel has an impact on the durational delay. The lengths between the nodes the acoustic signal was transferred over were the only thing that changed its constant velocity of 1500 m/s. The AEDG's dense deployment of nodes allowed for the establishment of additional linkages between the MNs and GNs, which caused distant communication to be delayed noticeably. The time spent connecting to the GNs, which may have been somewhat shortened, was used to determine the MNs' vertical positions about the horizontal elliptical trajectory.

The ASEDG's elliptical trajectories, both horizontal and vertical, encompassed the largest network area and the majority of GNs.

By developing GNs near an oval vertical trajectory, the vertically positioned MN connections were shortened and required less time to form. Compared to the AEDG, this led to an end-to-end delay reduction.

The elliptical trajectories in the ASEDG were designed to link to most GNs and offer network coverage across a wide area. Due to their larger separation from the GNs, the MNs in a vertical posture took longer to establish connections. The connection times for the MNs that were vertically positioned were shortened by generating GNs along the oval vertical trajectory. Compared to the AEDG, this resulted in a shorter end-to-end delay.

7.7.2 Delay without Node Rotation

Many times, the effects of dynamic network topologies with shifting connections have been disregarded or ignored. The routing paths between nodes in networks with a dynamic architecture are constantly changing. Higher data packet collision rates may occur from this, especially in constrained, compact portions of the network.

The same data packets must be retransmitted in the direction of the destination node when data packet collisions take place. As the packets are resent, these collisions cause delays, which lower network performance. The time it takes for packets to arrive at their destinations increases because of these persistent minor delays from resending collision packets.

The asymptotic equilibrium distribution gradient (AEDG) may be reached more quickly by the adaptive stochastic epidemic gradient descent (ASEDG) algorithm in some regions of the network. This may occur if the ASEDG moves outside of the network's core into an area with fewer connections. The AEDG can form more quickly than it does in the dense, highly linked areas of the network because there are fewer connections and fewer chances for packet collisions. In these sparser outer regions, the amount of time needed for the ASEDG to reach equilibrium is frequently shortened.

7.7.3 Network Throughput

Throughput is the rate at which packets are successfully delivered to the sink node. The number of packets transferred per second in the direction of one counts the sink node. The nodes for AEDG and ASEDG include always active to increase the network's dependability. The GNs spin. In the AESDG, two circular pathways that span the network's vertical and horizontal axes offer extra GNs. They transmit data coming from far-end nodes. This lessens expectations for data forwarding and streamlines the data transfer process. Shorter transmission lines experience fewer data collisions and losses.

For several protocols, the minimum and maximum packet speeds are displayed. In ASEDG, there are now more successfully delivered packets at the sink node. Nodes take longer to transmit packets. Losses of packets become less likely. Successful deliveries increase in ratio.

7.7.4 Energy Utilization

In AEDG, relationships between MNs and GNs is ideal. Few linkages between MNs and GNs exist, therefore the majority of nodes continue to be operational and relay data. This prolongs the life of the network. The shortest pathways are generated, together with the most GNs. As a result, the network duration is shortened and GNs use more energy. It is decreased by the relationships of MNs with GNs in ASEDG.

7.8 CONCLUSION

The collection of data in dangerous underwater conditions is the focus of this investigation. As a remedy, a routing protocol is suggested. We present data and an AUV-aided delay effective routing strategy (ASEDG) for UWSNs. The paths for data gathering and AUVs are modeled using a network intermediate node. Throughput-balanced end-to-end delay, energy use, and network lifetime goals are all met while this protocol addresses excessive end-to-end latency.

The proposed approach and the AEDG routing protocol are contrasted using NS2 simulations. Our protocol meets its objectives. Results indicate that the AEDG uses more energy because there are more GNs.

REFERENCES

[1] Javaid N, Ilyas N, Ahmad A, Alrajeh N, Qasim U, et al. An efficient data gathering routing protocol for underwater wireless sensor networks. *Sensors* 2015; 15 (11): 29149–29181. doi: 10.3390/s151129149.

[2] Kim HW, Cho HS. SOUNET: Self-organized underwater wireless sensor network. *Sensors* 2017; 17 (2): 283. doi: 10.3390/s17020283.

[3] Poncela J, Aguayo M, Otero P. Wireless underwater communications. *Wireless Personal Communications* 2012; 64 (3): 547–560. doi: 10.1007/s11277-012-0600-z

[4] Khan JU, Cho HS. A distributed data gathering protocol using AUV in underwater sensor networks. *Sensors* 2015; 15 (8): 19331–19350. doi: 10.3390/s150819331

[5] Chang SH, Shih KP. Tour planning for AUV data gathering in un- derwater wireless. In: *18th International Conference on Network-Based Information Systems*; Taipei, Taiwan; 2015. pp. 1–8. doi: 10.1109/NBiS.2015.120

[6] Khalid M, Ullah Z, Ahmad N, Arshad M, Jan B, et al. A survey of routing issues and associated protocols in underwater wireless sensor networks. *Journal of Sensors* 2017; 2017: 7539751. doi 10.1155/2017/7539751.

[7] Kao CC, Lin YS, Wu GD, Huang CJ. A comprehensive study on the internet of underwater things: applications, challenges, and channel models. *Sensors* 2017; 17 (7): 1477. doi: 10.3390/s17071477

[8] Ali T, Jung LT, Faye I. Three hops reliability model for underwater wireless sensor network. In: *International Conference on Computer and Information Sciences*; Kuala Lumpur, Malaysia; 2014. pp. 1–6. doi: 10.1109/IC- COINS.2014.6868378

[9] Cheng C, Li L. Data gathering problem with the data importance consideration in underwater wireless sensor networks. *Journal of Network and Computer Applications* 2017; 78: 300–312. doi 10.1016/j.jnca.2016.10.010

[10] Li N, Mart´ınez JF, Meneses Chaus JM, Eckert M. A survey on underwater acoustic sensor network routing protocols. *Sensors* 2016; 16 (3): 414. doi: 10.3390/s16030414

[11] Cai W, Zhang M, Zheng YR. Task assignment and path planning for multiple autonomous underwater vehicles using 3D Dubins curves. *Sensors* 2017; 17 (7): 1607. doi: 10.3390/s17071607

[12] Shaf A, Ali T, Farooq W, Draz U, Yasin S. Comparison of DBR and L2-ABF routing protocols in an underwater wireless sensor network. In: *15th International Bhurban Conference on Applied Sciences and Technology*; Islamabad, Pakistan; 2018. pp. 746–750. doi: 10.1109/IB-CAST.2018.8312305

[13] Jain SK, Mohammad S, Bora S, Singh M. A review paper on autonomous underwater vehicle. *International Journal of Scientific & Engineering Research* 2015; 6 (2): 38.

[14] Goyal N, Dave M, Verma AK. Data aggregation in underwater wireless sensor network: recent approaches and issues. *Journal of King Saud University – Computer and Information Sciences* 2019; 31: 275–286. doi: 10.1016/j.jksuci.2017.04.007

[15] Ahmad A, Wahid A, Kim D. AEERP: AUV aided energy-efficient underwater acoustic sensor network routing protocol. In: *Proceedings of the 8th ACM Workshop on Performance Monitoring and Measurement of Heterogeneous Wireless and Wired Networks*; Barcelona, Spain; 2013. pp. 53–60. doi: 10.1145/2512840.2512848

[16] Ilyas N, Alghamdi TA, Farooq MN, Mehboob B, Sadiq AH, et al. AEDG: AUV-aided efficient data gathering routing protocol for underwater wireless sensor networks. *Procedia Computer Science* 2015; 52: 568–575. doi: 10.1016/j.procs.2015.05.038

[17] Hollinger GA, Choudhary S, Qarabaqi P, Murphy C, Mitra U, et al. Underwater data collection using robotic sensor networks. *IEEE Journal on Selected Areas in Communications* 2012; 30 (5): 899–911. doi: 10.1109/JSAC.2012.120606

[18] Yoon S, Azad AK, Oh H, Kim S. AURP: an AUV-aided underwater routing protocol for underwater acoustic sensor networks. *Sensors* 2012; 12 (2): 1827–1845. doi: 10.3390/s120201827

[19] Li Z, Yao N, Gao Q. Relative distance-based forwarding protocol for underwater wireless networks. *International Journal of Distributed Sensor Networks* 2014; 2014: 173089. doi: 10.1155/2014/173089

[20] Hosseini M, Chizari H, Poston T, Salleh MB, Abdullah AH. Efficient underwater RSS value to distance inversion using the Lambert function. *Mathematical Problems in Engineering* 2014; 2014: 175275. doi 10.1155/2014/175275.

[21] Khan JU, Cho HS. Data gathering scheme using AUVs in large scale underwater sensor networks: a multihop approach. *Sensors* 2016; 16 (10): 1626. doi: 10.3390/s16101626

[22] Han G, Liu L, Bao N, Jiang J, Zhang W, et al. AREP: An asymmetric link-based reverse routing protocol for underwater acoustic sensor networks. *Journal of Network and Computer Applications* 2017; 92: 51–58. doi: 10.1016/j.jnca.2017.01.009

[23] Ilyas N, Akbar M, Ullah R, Khalid M, Arif A et al. SEDG: scalable and efficient data gathering routing protocol for underwater WSNs. *Procedia Computer Science* 2015; 52: 584–591. doi: 10.1016/j.procs.2015.05.043

[24] Basagni S, Petrioli C, Petroccia R, Spaccini D. CARP: A channel aware routing protocol for underwater acoustic wireless networks. *Ad Hoc Networks* 2015; 34: 92–104. doi: 10.1016/j.adhoc.2014.07.014

[25] Ali T, Jung LT, Faye I. Diagonal and vertical routing protocol for underwater wireless sensor network. *Procedia Social and Behavioral Sciences* 2014; 129: 372–379. doi: 10.1016/j.sbspro.2014.03.690

[26] Javaid MN, Ali B, Yahya A, Khan ZA, Qasim U. Improved hydrocast: a technique for reliable pressure-based routing for underwater WSNs. In: *10th International Conference on Innovative Mobile and Internet Services in Ubiquitous Computing*; Fukuoka, Japan; 2016. pp. 284–290. doi: 10.1109/IMIS.2016.139

[27] Yildiz HU, Gungor VC, Tavli B. Packet size optimization for lifetime maximization in underwater acoustic sensor networks. *IEEE Transactions on Industrial Informatics* 2019; 15 (2): 719–729. doi: 10.1109/TII.2018.2841830

[28] Wang Y, Zheng YR. 3-dimensional path planning for the autonomous underwater vehicle. In: *OCEANS 2018 MTS/IEEE*; Charleston, SC, USA; 2018. pp. 1–6. doi: 10.1109/OCEANS.2018.8604783

[29] Nam H. Data gathering protocol based AUV path planning for long duration cooperation in underwater acoustic sensor networks. *IEEE Sensors Journal* 2018; 18 (21): 8902–8912. doi: 10.1109/JSEN.2018.2866837

[30] Li D, Wang P, Du L. Path planning technologies for autonomous underwater vehicles - a review. *IEEE Access* 2019; 7: 9745–9768. doi: 10.1109/ACCESS.2018.2888617

8 Extending WSN Lifetime with Cluster Head Election

Noroze Fatima, Danish Shehzad, Ali Imran, Ahsan Imtiaz, and Muhammad Arif

8.1 INTRODUCTION

The sensors of a wireless sensor network are arranged in a sensor field to track and measure certain environmental traits and gather information about those events. The sensors are tiny devices with constrained battery life, memory, computational capacity, low data rates, limited processing bandwidth, and irregular network performance and other resources [1]. Despite their limitations, the sensors can provide us a fairly accurate picture of the field being detected when they are spread out widely.

1.1 Wireless Sensors Network

The wireless network sensors are wired and wireless networks were unable to cover the same geographic area as WSN. Once deployed in the field, they may be continuously watched after or may be left unattended in a variety of conditions [2]. The potential of the WSN will be used to offer effective and affordable solutions to a variety of issues. To deal with the sensor restrictions, it is required to put in place processes or practices. Using clustering techniques in WSN can help address some of those limitations by enabling the processes of the sensors, grouping them into clusters, and assigning specific responsibilities to the sensor in the clusters before transferring the information to higher levels.

Applications that need to scale the number of nodes to hundreds or thousands benefit most from clustering. In this context, scalability emphasizes the necessity for data aggregation, effective resource usage, and load balancing [1]. Additionally, when communicating with the base station, a number of routing protocols have the potential to use clustering to create a hierarchical structure and lower the path cost.

The three primary WSN components that can be divided when working with clusters are sensor nodes (SNs), base stations (BSs), and cluster heads (CH). To view the surroundings and gather data, the network's sensors, or SNs, are organized in a set. The primary responsibility of an SN in a sensor field is to recognize events, process nearby data right away, and then broadcast the outcomes. The three primary WSN components that can be divided when working with clusters are sensor nodes (SNs), base stations (BSs), and cluster heads (CH). To view the surroundings and gather data, the network's sensors, or SNs, are organized in a set. The primary responsibility of an SN in a sensor field is to recognize events, process nearby data right away, and then broadcast the outcomes. The BS is the location where data from the sensor nodes is processed before being made accessible to the end user. It is typically regarded as fixed and located a long way from the sensor nodes. Between the SNs and the BS, the CH serves as a gateway. The cluster head's responsibility is to complete tasks that are common to all cluster nodes, like consolidating data before sending it to the BS [3]. While

DOI: 10.1201/9781003497851-8

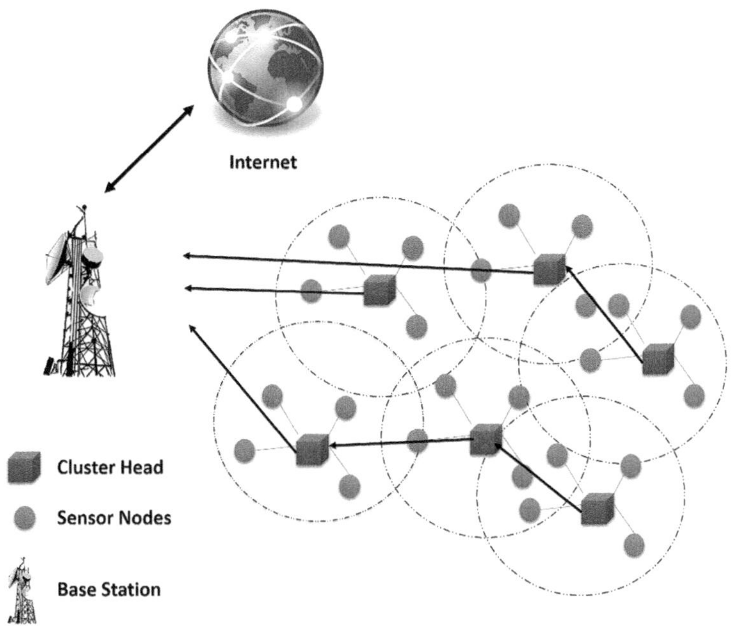

FIGURE 8.1 Network architecture for wireless sensors.

the BS acts as the sink for the cluster nodes, the CH acts as the sink for the cluster heads. By replicating the structure that was established between the sensor nodes, the sink, and the base station, the various layers of the hierarchical WSN are produced.

It is important to consider factors like cluster size and design, how to choose the cluster head, how to manage collisions between and within clusters, and energy-saving concerns while establishing a cluster. Due to the investigated efficiency of using a hierarchical communication strategy between the network piece show clustering works' design is one of the most crucial challenges for the proper functioning of the WSN. We can distinguish three key stages in the clustering establishment process across all cluster-based protocols:

- Cluster head election;
- Cluster formation (set-up phase);
- Phase of data transmission.

Each of these stages can be implemented using a variety of strategies. One option is to employ a predetermined distribution of the SN and CH, a dynamic strategy was used for sensor placement and CH election (Figure 8.2).

In various suggestions, the cluster heads are arranged in a hierarchy (Figure 8.3) as done in Bandyopadhyay & Coyle (2003) and Manjeshwar & Agarwal (2001) to facilitate inter-cluster communication Multi-level hierarchies enable better energy allocation and overall energy use. Maintaining the hierarchy could be costly in large, dynamic networks where nodes disappear as soon as their energy supply runs out.

One of the primary objectives of hierarchical organization in sensor networks is energy efficiency, which enables a longer network lifetime. A cluster head can combine and aggregate the data it receives before transmitting it to the base station. In extremely dense networks, it is possible to select a subset of nodes for low-power sleep mode operation without compromising connectivity or network coverage. An effective cluster head plan the states of its member nodes in this situation.

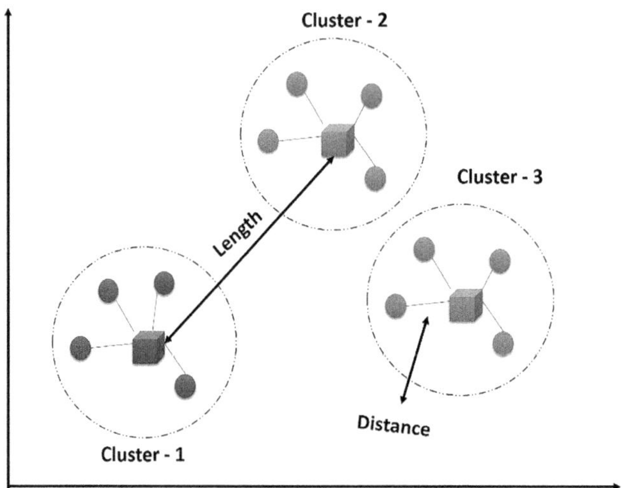

FIGURE 8.2 Distance and length of cluster selections.

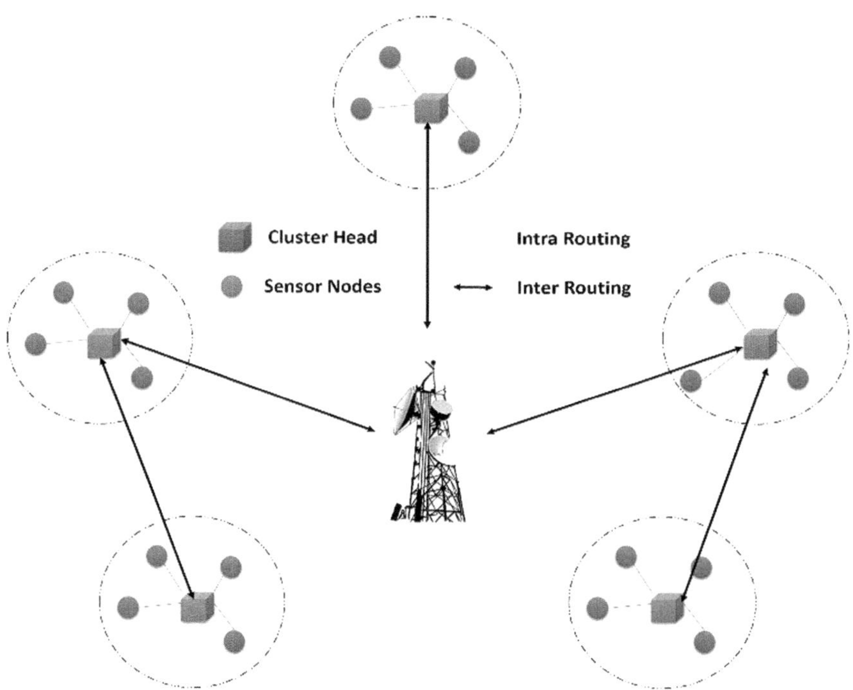

FIGURE 8.3 Multi-hop inter-cluster communication.

Furthermore, if a round-robin technique is a practice among the member nodes of the cluster, medium access collision can be avoided. Collisions may necessitate the retransmission of data by nodes, using up additional energy.

Balanced intra-cluster traffic leads to a severely skewed load distribution on cluster heads in terms of inter-cluster communication (Figure 8.3). The farther the cluster head is from the base station, the more energy it uses, and the sooner it will expire in singlehop communication where cluster heads use direct connection to connect to the base station. Even if multi-hop inter-cluster

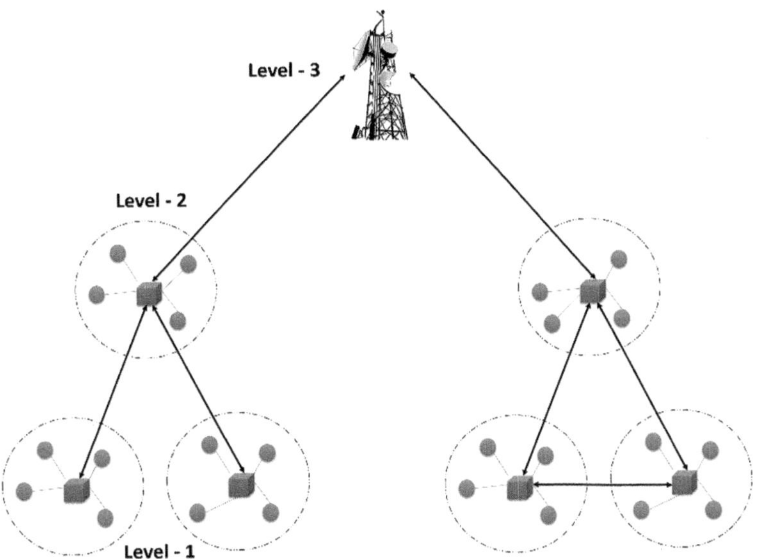

FIGURE 8.4 Three-level hierarchy of cluster head with base station.

communication is used, the so-called hot spot problem still exists since nodes adjacent to the base station have to deal with larger traffic loads. Many-to-one traffic theory that distinguishes WSN is to blame for this. Nodes close to a hot spot lose energy more quickly and die significantly more quickly than cluster heads farther away. This might cause significant connection (network partition).

Since LEACH is totally distributed, substantial network understanding is not required. Large area networks cannot use it since it creates a one-hop intra-cluster and inter-cluster topology. It is expected that cluster heads have a long communication range and can directly communicate with the sink. Due to the fact that the cluster heads are ordinary sensors and the sink is frequently positioned far away, this assumption is not always accurate. Additionally, the additional cost created by the ads phase at the start of each round by dynamic clustering may reduce the energy gain. LEACH cluster head rotation assumes a homogeneous network and is unable to guarantee actual loadbalancing in the case of nodes with varied initial amounts of energy since the choice to elect a cluster head is probabilistic and does not take energy into consideration. A node with extremely low energy will prematurely perish after serving as the cluster head nodes with higher energy for the same number of rounds. This can have an impact on connectivity and network coverage (Figure 8.4).

8.2 LITERATURE REVIEW

Various clustering procedures were compared in a survey by Arboleda et al. [1]. The survey's authors briefly studied WSNs' proactive and reactive algorithms, LEACH-based protocols, as well as some fundamental clustering concepts like cluster structure, cluster kinds, and clustering advantages. The key differences between these protocols were evaluated, and examples of contemporary applications for them were provided.

The currently employed clustering methods for WSNs were reviewed by Kumarawadu et al. [4] And categorized according to the cluster formation characteristics and CH election criteria. The authors of the survey looked at the major design challenges and discussed performance issues related to clustering protocols in addition to classifying identity-based clustering algorithms, neighborhood information-based clustering algorithms, probabilistic clustering algorithms, and biologically inspired clustering algorithms.

Deosarkar et al. [5] focuses on analyzing different clustering schemes' CH selection processes using the categories of deterministic scheme, adaptive scheme, and mixed metric scheme. The costs of CH selection were examined for cluster formation, CH distribution, and cluster creation. The need for a clustering technique that is more reliable, expandable, and energy-efficient for data collection in WSNs was also raised.

Jiang et al. [6] presented a classification of the three primary benefits of clustering approaches for WSNs, including greater scalability, lower overheads, and easier maintenance. Using a total of eight clustering attributes, WSN clustering techniques. Additionally, the authors examined six well-known WSN clustering methods, including LEACH, PEGASIS, HEED, EEUC, etc., as well as numerous features, and compared these WSN clustering techniques.

In the survey described in [7], the workings of different clustering protocols were discussed, and the advantages and disadvantages of each of these approaches were summarized. There are only seven popular clustering algorithms for WSNs, such as LEACH, TL-LEACH, EECS, TEEN, and APTEEN [8], that were chosen by the survey's authors. The survey also contrasted the energy usage and network longevity of these clustering techniques.

TABLE 8.1
Cluster Head Selection Techniques and Limitations

Protocols	Features	Limitations
MEACM	Superior performance by reducing the sensor node's energy usage increases the number of dead sensor nodes, throughput, and network lifetime	Simulation is preferred to real-time experiments. Requires that the scalability of sensor nodes within each cluster be taken into account.
ACOPSO	A smaller packet size increases the network's lifespan effective energy conservation	Defining the basic design parameters is crucial cannot specify the scattering difficulties.
DCT- CDA	Increases network lifetime boosts the stability of the network	Security must be taken into account.
GWO	Improved network throughput and lifespan energy usage made simple	Uncertain acquisition procedure MAC and network layer consideration must be combined.
WOA-C	Increases network longevity Improving scalability	Network performance should be enhanced. There is no consensus on the properties of Cluster Heads.
GECR	Improved network lifecycle optimum use of energy	The focus of upcoming research is on validating and utilizing the proper metaheuristic algorithms. It is unclear which option is best.
Load balanceing protocol	Enhanced network throughput, stability, and packet drop rate	Consideration should be given to energy harvesting schemes, and network longevity needs to be further improved.
RE attempt	Improved network throughput and lifespan energy usage made simple	Plans to concentrate on crosslayer design procedures MAC and network layer consideration must be combined.
K-means	Maximum output reduction in energy consumption	Finding the ideal solution in a fair amount of time is challenging.
FCM	Maximizes energy	Information loss is a possibility.
DBSCAN	According to belong-to, do not belong to, and ambiguity, it classified the cluster. Become more energy efficient	Less limitations are taken into account.
LEACH	It gauges how long the sensor nodes will last. Reduce inter-cluster distance.	The importance of packet routing

Boyinbode et al. [7] published a survey on WSN clustering methods. After exploring the major challenges for clustering algorithms, nine well-known clustering algorithms for WSNs, including LEACH, TL-LEACH, EECS, HEED, and EEUC, were briefly summarized in the survey. The authors further contrasted various clustering techniques based on variables such as residual energy, uniformity of CH distribution, cluster size, delay, hop distance, and cluster generation methods.

In [9], the parameters utilized for CH selection, the presence of a centralized control during clustering, and hops between nodes and CH in intra-cluster communication are simply three simple categorization criteria that the authors used to define the clustering architecture in WSNs.

The survey [10] also emphasized the difficulties associated with clustering WSNs and briefly described a few clustering routing strategies.

Reduced frequent remote transmission is important to increase network lifetime due to the limited energy of the sensor nodes [11]. Therefore, direct contact between nodes and base stations is discouraged. The network nodes can be organised into a number of clusters, and each cluster can choose a cluster head who is responsible for transmitting data from the entire cluster to the base station. In order to save energy, just a few number of nodes are needed to transmit data over long distances, and other nodes will handle short-distance transmission.

We provide an enhanced LEACH-based inter-cluster multi-hop routing mechanism in [12]. Remaining energy and the distance between the node and the base station are assumed to have an impact on the threshold function of cluster head election, and the multi-hop routing transmission technique is employed for inter-cluster communication. [15,16] proposes a distributed unbalanced clustering routing method for WSNs. The time broadcast method chooses the cluster heads, and the cluster head can transmit data to the base station in line with social welfare to reduce the quantity of control message exchange function. In [15] in order to limit the amount of data transmitted by the node, the compressed sensing method is created. The redundant data is then reduced using data gathered from the cluster head. A multi-level heterogeneous network topology control algorithm is created using a technique for creating a reverse connected dominating set tree in [16] to improve the network topology. Two different types of cluster head election probability are proposed to build bio-level heterogeneous clusters in wireless sensor networks (WSNs) in [17], which presents a stable election procedure for clustered heterogeneous wireless sensor networks (SEP) [5]. It can delay the initial node's demise; however, it is not applicable in a heterogeneous WSN system. For heterogeneous networks, a distributed clustering algorithm is put forth in [18]. Since each sensor's initial energy varies, the competition among nodes for the position of cluster head is diverse. Additionally, the likelihood will change when a node's residual energy changes over time.

8.3 PROPOSED SOLUTION

Data collection requires the use of a routing system that is energy-efficient. The significance and capacities of each sensor node are identical. This highlights the necessity to increase the sensor nodes' and the sensor network's lifetime. The PE-LEACH Protocol's stated goals are to decrease energy usage and improve network lifespans. The Gaussian distribution model in this case effectively covers the sensor network region. In addition, node aggregation makes use of the conditional probability theorem. Figure 8.2 illustrates the PE-LEACH Protocol's flow. Assume there is a base station and N nodes in a sensor network. Over a region, BS is split up. Both the base station's location and the sensor nodes' positions are known in advance. Let's have a look at a 2D network

Cluster formation in sensor networks is strongly influenced by the latency to receive a neighbor node message and the remaining energy. The process is divided into rounds, with each round beginning with the search for the optimum CH. How the cluster is formed is determined by the following processes.

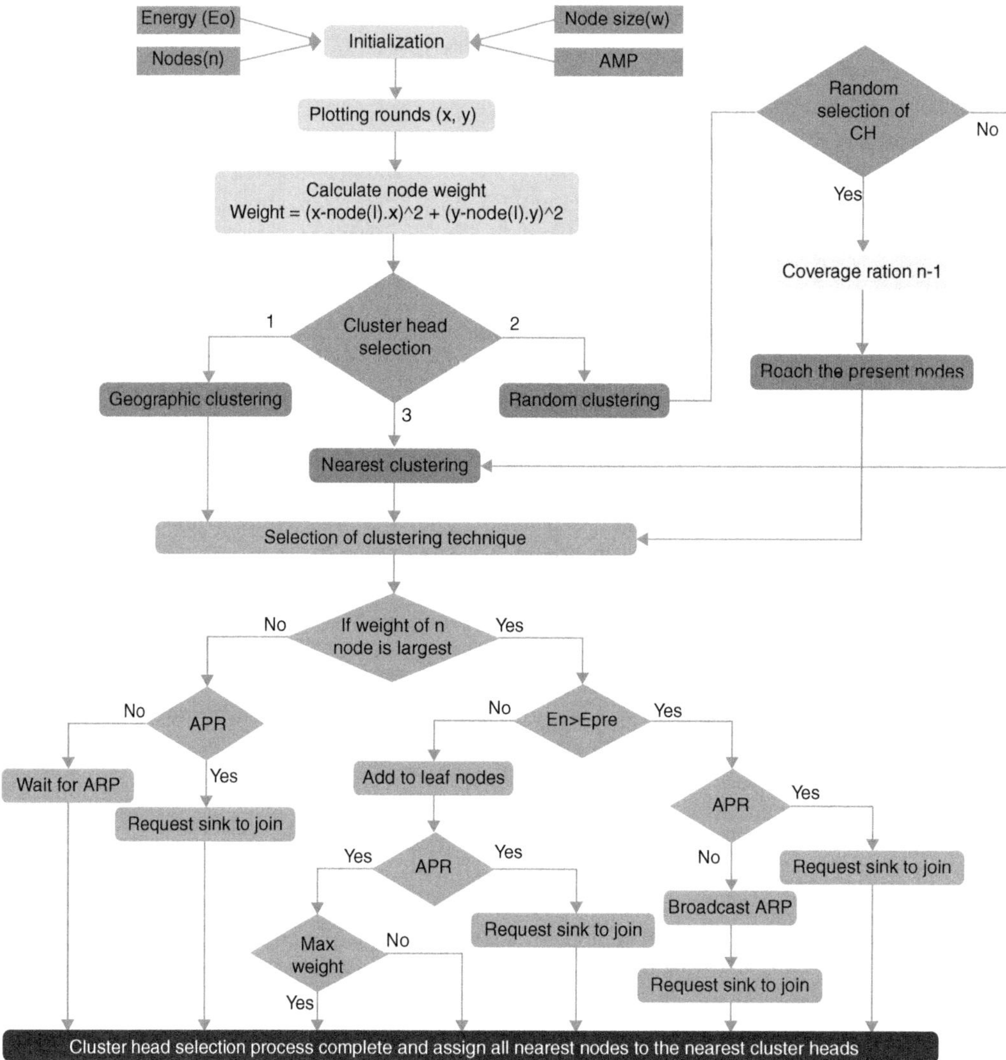

FIGURE 8.5 Proposed technique for cluster head selection.

8.3.1 Step 1: Initialization and Information Retrieval about Neighbors

The neighbor node information is found by broadcasting beacon signals throughout the network [19, 20]. First, we must specify the region we want to include in our clustering. Define two numbers, xa and xb, for the area determination, where xa and xb are $ 0, 1, 2, 3, etc. and xa and xb initially equal 0. In the finite range, there are n nodes total, but there is initially no lovely node. Every node has an energy of Eo, and a data package has a size of k. The data will be sent for sorting following the initialization process [21].

8.3.2 Step 2: Implement a Sorting Algorithm

The sorting approach is applied to get a list of all data package (k) within a certain hop distance. Ordered in decreasing priority is the list [22]. The weight (w) is calculated of nodes in the finite filed and bases on the plot graph and their positions. The weight will be calculated as

$$Weight = (x - node(I).\, x)\string^2 + (y - node(I).\, y)\string^2$$

The coordinates of that node are represented by x and y in the above equation, and (I) represents the energy of the system with many sensor nodes [23].

8.3.3 Step 3: Candidate for Cluster

When a node's two-hop neighbor is not enclosed, examine each stage 2 member individually and select any one of them as a candidate for the cluster [24, 25]. These simplifications form the basis of the computations: Assume that there is a lengthy intra cluster transmission stage. Since inter cluster transmission is long enough and all data nodes can send data to their CH, all CH with data can transfer their data to the BS [26]. The suggested PE-LEACH Protocol's Figure 2 Flow and data aggregation must be carried out by the CH. There we have three cluster head selection techniques. The weight (w) determines the cluster head based on the Light(l) and the distance(d) in the finite field [27]. The random clustering performs in the first stage and determine the cluster head randomly then every other node wait for the approval to join the sink.

$$E_o > E_{present}$$

The system waits for the ever node to join and then the cluster selection process complete. Finally, the sorting procedure is carried out using the adjacent nodes' remaining energy [28].

8.4 RESULTS AND DISCUSSION

The proposed method has been implemented using MATLAB, and the outcomes have been contrasted with those of the protocols CRPD, LEACH, and MODLEACH. In comparison to CRPD, the adoption of EEMCS has improved energy utilization by 20–30%, and it is more stable than LEACH and MODLEACH in large-scale networks. Furthermore, the network lifetime of CRPD [29], MODLEACH, and LEACH is 62%, 58%, and 46%, respectively, compared to 77% for EEMCS on 150 nodes in a 50 50 m2 region? Due to EEMCS's dynamic nature and very low energy consumption, nodes can remain in the network for a longer time [30]. In every situation, EEMCS has a higher throughput than the current protocols.

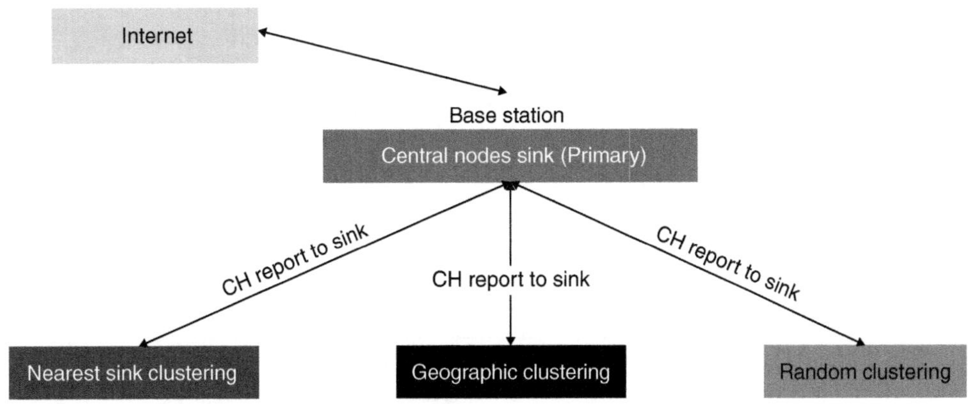

FIGURE 8.6 Cluster head selection techniques.

8.4.1 Advantages of a Strong CH Election

The selection of CH nodes in sensor networks can provide the following three benefits: Increasing the networks' longevity Heterogeneous networks are used [31].

1. **Data transmission.** Typically consumes a lot less energy than transmission in homogenous networks from the sensor node to the BS.
2. **Enhancing data.** It is commonly known that links typically have low forwarding reliability. With each hop, the packet delivery rate is significantly decreased.
3. **Decreasing latency.** Computational heterogeneity can reduce the latency in nearby nodes for data transmission.

8.4.2 Data Gathering using Data Ensemble

When a cluster of nodes is replaced by a single node, the joint deployment of the network is not altered [32]. During node aggregation, the data ensemble process also takes place. It's essential to locate the macro node for data aggregation. The process now consists of two steps:

- Path definition
- Pair of combinable nodes

8.4.3 Power Efficient Data Routing

After gathering the data from various clusters, the CHs must give it to the BS [8]. The forwarding nodes are thus selected based on the nodes with the highest residual energy.

8.5 CONCLUSION

An energy-efficient routing mechanism must be used for data collection; each sensor node serves the same function and has the same skills. This highlights the necessity to increase the sensor nodes' and the sensor network's lifespan. The PE-LEACH Protocol's goals are to cut down on energy use and increase network lifespans. Here, the sensing network region is effectively covered by the Gaussian distribution model.. In addition, the conditional probability theorem is used in node aggregation.

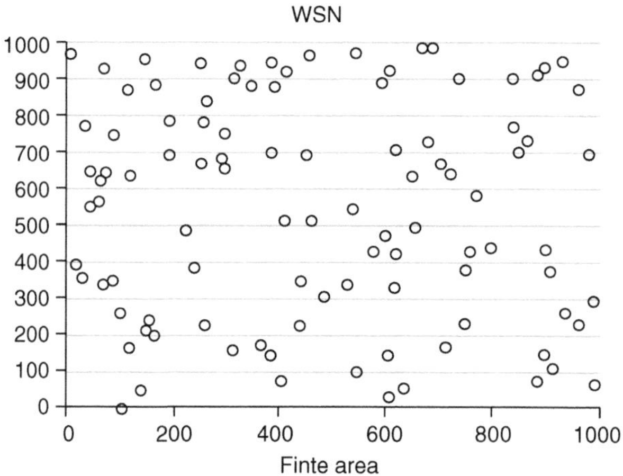

FIGURE 8.7 Clusters in wireless sensor network.

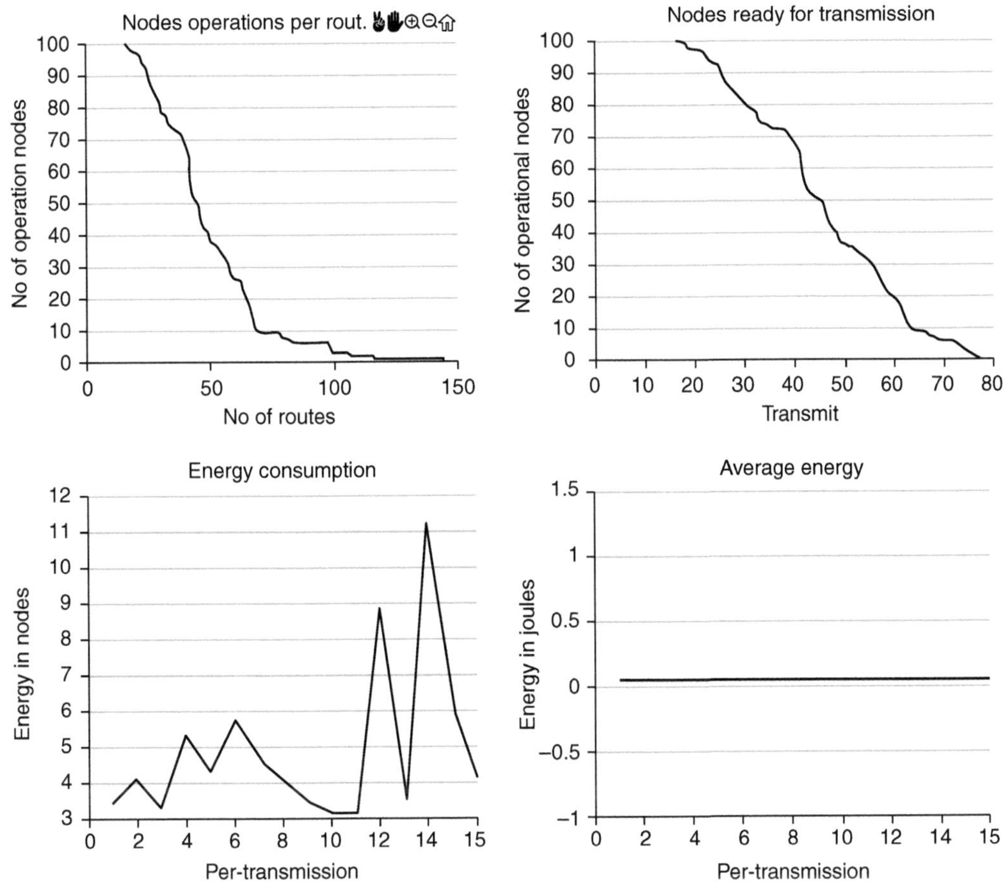

FIGURE 8.8 Operations to perform rounds for transmission and energy.

REFERENCES

[1] N. Nasser, C. Arboleda, M. Liliana, "Comparison of clustering algorithms and protocols for wireless sensor networks," *IEEE,* p. 6, 2006.

[2] D. J. Dechene, A. El Jardali, M. Luccini, A. Sauer, "A Survey of Clustering Algorithms for Wireless," https://uwo.ca/, p. 10, 2005.

[3] P. Goel, E. A. Goyal, "Clustering in Wireless Sensor Networks: Survey," *International Journal of Advance Research and Innovative Ideas in Education,* vol. 3, p. 361–365, 2017.

[4]T. M. Behera, S. K. Mohapatra, U. C. Samal, M. S. Khan, M. Daneshmand, A. H. Gandomi, "Residual Energy Based Cluster-head Selection," *IEEE,* p. 8, 2019.

[5] D. S. K. S. Deosarkar, "IOT Based WSN," *ICCIP,* p. 11, 2020.

[6] C. Jiang, D. Yuan, Y. Zhao, "Towards Clustering Algorithms in Wireless Sensor," *in IEEE wireless communications and networking conference,* p. 1–6, 2009.

[7] U. Sajjanha, P. Mitra, "Distributive Energy Efficient Adaptive Clustering Protocol for Wireless Sensor," *IEEE,* p. 5, 2007.

[8] S. Su, S. Zhao, "An Optimal Clustering Mechanism based on Fuzzy-C means for Wireless Sensor Networks," *Sustainable Computing: Informatics and Systems,* Elsevier, vol. 18, p. 127–134, 2018.

[9] D.Xu, J. Gao, "Comparison Study to Hierarchical Routing Protocols in Wireless Sensor Networks," *Procedia Environmental Sciences*, vol. 10, pp. 595–600, 2011.

[10] W. Wu, N. Xiong, C. Wu, "Improved clustering algorithm WSN," *IET,* vol. 6, p. 47–53, 2017.

[11] X. Liu, "A Survey on Clustering Routing Protocols," *Mdpi,* p. 41, 2012.

[12] H. Junping, J. Yuhui, D. Liang, "A Time-based Cluster-Head Selection Algorithm for LEACH," *IEEE,* p. 5, 2008.
[13] H. Yang, B. Sikdar, "Optimal Cluster Head Selection in the LEACH Architecture," *IEEE,* p. 8, 2007.
[14] A. E. J. i. M. L. A. S. D. J. Dechene, "A Survey of Clustering Algorithms for WSN," *UWO.ca,* p. 9, 2015.
[15] M. Haque, T. Ahmad, M. Imran, "Review of Hierarchical Routing Protocols," *Springer,* p. 10, 2018.
[16] H. Z. a . F. L. Moufida Maimour, "Cluster-based Routing Protocols for Energy," *Intechopen,* p. 24, 2009.
[17] V. Katiyar, N. Chand, S. Soni, "Clustering Algorithms for Heterogeneous Wireless Sensor Network: A Survey," *International Journal of Applied Engineering Research,* 1(p. 273), 2010.
[18] N. Saini, J. Singh, "A Survey: Hierarchal Routing Protocol in Wireless Sensor," *ISSN,* vol.14, p. 8, 2014.
[19] D. Jia, H. Zhu, S. Zou, P. Hu, "Dynamic Cluster Head Selection Method for Wireless Sensor," *IEEE, 16(8), 2746–2754* , 2015.
[20] B. Singh, D. K. Lobiyal, "Energy-Aware Cluster Head Selection Using Packet Retransmission in WSN," *Elsevier,* 4(171–176), 2012.
[21] S. M. Sang, H. Kang, "Distance Based Thresholds for CH in WSN," *IEEE,* p. 4, 2012.
[22] G. S. Arumugam, T. Ponnucham, "EE-LEACH: Development of Energy-Efficient," *EURASIP Journal on Wireless Communication and Networking,* vol. 2015, p. 9, 2015.
[23] Y. U. Xiu-Wua, Y. U. Hao, L. Yong, X. Ren-Rong, "A Clustering Routing Algorithm Based on Heterogeneous Algorithm," *ELSEVIER,* p. 10, 2020.
[24] Y. Liang, H. Yu, "Energy Adaptive Cluster-Head Selection for Wireless Sensor Networks1," *IEEE,* p. 5, 2005.
[25] N. A. Latiff, C. C. Tsimenidis, B. S. Sharif, "Energy-Aware Clustering for Wireless Sensor Networks," *IEEE,* p. 5, 2018.
[26] F.Xiangning, S. Yulin, "Improvement on LEACH Protocol of Wireless Sensor Network," *IEEE,* p. 5, 2007.
[27] X. Liu, J. Wu, "A Method for Energy Balance and Data Transmission Optimal Routing in Wireless Sensor Networks," *19(13), 3017,* 2019.
[28] Z.Yonga, Q. Pei, "A Energy-Efficient Clustering Routing Algorithm," *Elsevier,* p. 7, 2012.
[29] Y.Yin, J. Shin, Y. Li, P. Zhang, "Cluster Head Selection Using Analytical Hierarchy Process," *IEEE,* p. 5, 2017.
[30] M. C.Thein, T. Thein, "An Energy Efficient Cluster-Head Selection for Wireless Sensor Networks," *IEEE,* p. 5, 2010.
[31] M. Zeng, X. Huang, B. Zheng, X. Fan, "A Heterogeneous Energy Wireless Sensor Network," *Hindawi,* p. 12, 2019.
[32] T. M.Behera, U. C. Samal, S. K. Mohapatra, "Energy-efficient modified LEACH protocol for IoT application," *IET Wireless Sensor Systems, 8*(5), 223–228, 2018.
[33] O. Boyinbode, H. Le, M. Takizawa, "A Survey on Clustering Algorithms for Wireless Sensor Networks," *International Journal of Space-Based and Situated Computing, 1*(2-3), 130–136, 2010.

9 A Parametric Analysis of Routing Protocols in Wireless Sensor Networks

Sajjad Abbas, Danish Irfan, M. Arfan Jaffar, Syed Asad Ali Naqvi, Naveed Ali Khan Kaim Khani, and Muhammad Yousaf

9.1 INTRODUCTION

In today's expanding industrial life, sensor technology is crucial. Machines that operate automatically, regulate and control the several operations via sensors. Safety systems, environmental sensors, and monitoring structural reliability of building components are also highly important. Typically, a processing unit attached to the central conductor via some cables processes and interprets the sensor signals from these sensors. The cost of complicated sensor wiring limits the adaptability and reusability of sensors [1]. It is now possible to build a compact gadget using data processing, wireless connection, and microchip sensing techniques. Links are not essential because they are transmitted over radio, but it is not necessary to transfer data through wires. The Mott filter may combine or process data exchange at its own. The range of wireless communication is con- strained. To increase the range of all motes until it reaches the intended device, rules that are partly based upon the multi-hop routing protocol signal may be transmitted [2]. The wireless sensor network (WSN) in this outlying region is designed to communicate data among all linked devices. To monitor humidity, pressure, motion, and other envi- ronmental parameters like the health of pollutants or sound, many wireless sensor nodes are gathered in a designated work area with movement sensor nodes in WSNs [3]. The main reason is because whereas stone on the battlefield served as the main battleground when defense-related uses of WSNs were initially created, they are now being ex- panded to include field management, traffic monitoring, and industry monitoring. The storage capacity and processing speed of the sensor nodes determine how much band- width is available, thus it is crucial to identify additional boundaries in terms of energy cost and physical restrictions [4]. Wireless sensor nodes are made up of radio trans- ceiver; micro-controller and power batteries. WSNs with a variety of distinctive quali- ties such as Scalability, Geographical dispersion over a vast region and Node failures can be handled by WSNs. The capacity to live in a range of conditions and having a dynamic network architecture are two more features of the mobile node [5]. Sensor nodes are compact, inexpensive, multi-purpose with wireless communication capabili- ties across short distances [6]. They communicate via radio waves. Network nodes for WSNs can be placed anywhere they are needed to observe the physical environment.

DOI: 10.1201/9781003497851-9

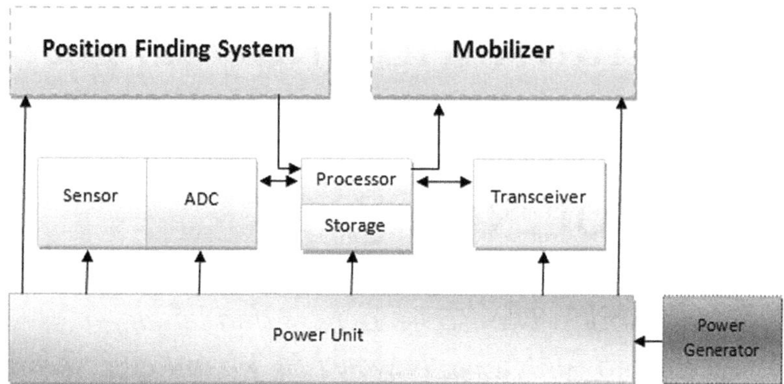

FIGURE 9.1 Sensor node architecture.

9.2 DISCUSSION ON ROUTING PROTOCOLS

9.2.1 Wireless Sensor Routing Protocols

The five categories of WSN protocols including Topology, Cluster, Position, Geo-cast and broadcast-based routing protocols. Among the mentioned categories, our target protocols fall under topology-based routing protocols, which are pictorially depicted in Figure 9.2.

9.2.2 Topology-Based Routing Protocols

Its primary objective is to connect information in data transmission across the current network.

Proactive Routing Protocols. WSNs, which seek to offer dependable and effective communication between sensor nodes, heavily rely on proactive routing methods. These protocols are intended to create and maintain a routing table for each network node, which provides details on the routes to other network nodes. In spite of node failures or other network disturbances, this helps to ensure that data packets are delivered effectively and dependably to their destination [7].

Reactive Routing Protocols. Reactive protocols are another important type of protocols used in Wireless Sensor Networks (WSNs). Unlike proactive routing protocols, which maintain a table having route information about each node, reactive protocols are designed to establish a route only when it is needed. This helps to conserve energy and reduce network overhead, making reactive protocols particularly well-suited for large-scale WSNs [8].

Ad-hoc On Demand Distance Vector (AODV): Wireless ad-hoc Networks used AODV protocol which is a Reactive in nature [9]. In order to save energy and lower network overhead, it is intended to build a route only when it is required. When used in large-scale wireless networks, where nodes may move continually and the network architecture may change often, AODV is especially well-suited [10].

For wireless ad hoc networks, AODV is an effective and dependable routing technology. Data packets are consistently delivered even in the face of shifting network circumstances because to its route finding and maintenance methods, which also assist to preserve energy and lower network overhead [11].

Dynamic Source Routing (DSR): DSR is a routing protocol used in Wireless Ad-hoc Networks. It is a reactive protocol that only creates a route when it is required, reducing network overhead

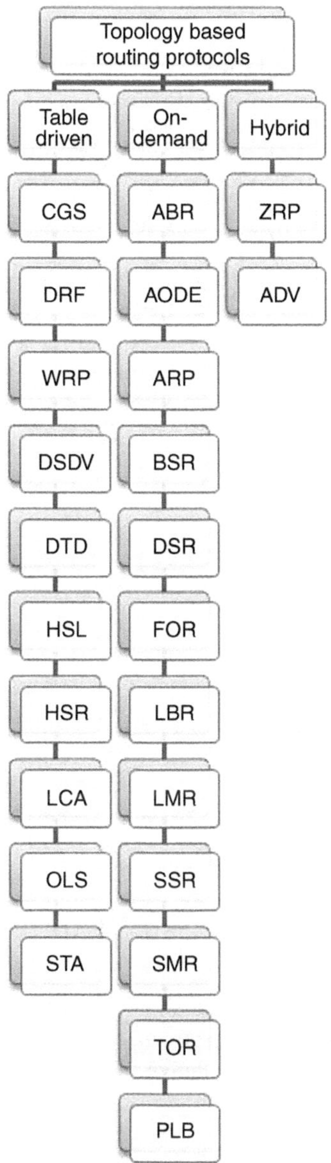

FIGURE 9.2 Topology-based routing protocols.

and preserving energy. DSR works especially effectively in mobile ad hoc networks, where nodes may move around a lot and the network topology may change quickly [12]. DSR is a trustworthy and effective routing technology for wireless ad hoc networks overall. Despite shifting network circumstances, data packets are consistently delivered thanks to its source routing technique and route discovery and maintenance algorithms [13] .

Destination-Sequenced Distance Vector (DSDV): DSDV is a routing protocol based on proactive routing technology used in Wireless ad hoc networks [14]. It is a distance-vector protocol that keeps track of the optimal path to each destination node in the network. In static ad hoc networks, where nodes don't move around a lot and the network topology is generally stable, DSDV is especially

TABLE 9.1
Parameters for Simulation

Parameter	Value
Protocol used	DSDV, AODV, DSR
Time of Simulation	400 s
Nodes (in Numbers)	5, 10, 15, 20, 25, 30, 35
Area of Simulation	600 m x 600 m
Pause-Time	0, 40, 80, 120, 160, 200, 240, 280, 320 s
Type of Connection	TCP, UDP
Max Speed	5, 10, 15, 20, 25 m/s
Mobility Model	Manhattan Grid and Random Waypoint
Simulator	NS 2.0

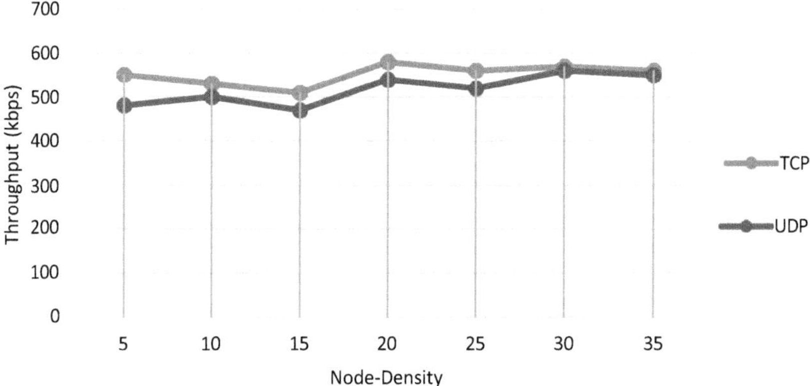

FIGURE 9.3 Comparison of AODV throughput over TCP and UDP connections w.r.t. node density.

well suited for usage [15]. Overall, the wireless ad hoc network routing system DSDV is dependable and effective. The net- work can swiftly adjust to changes in the network architecture because to the proactive approach it takes and the route poisoning method it uses.

9.3 SIMULATION RESULTS

9.3.1 Parameters

Table 9.1 provides details of the parameters.

9.3.2 Traffic Analysis of AODV w.r.t. Node Density

The impact of Node Density on the rate at which UDP and TCP pass via AODV is covered in this section. Figure 9.3 indicates the throughput of UDP and TCP over AODV. According to the findings, adding more nodes boosts the UDP and TCP throughput. The relationship between throughput and node density is linear. TCP has a better throughput per unit time than UDP. TCP's high throughput is a result of the default TCP window size and 0.1 Mbps data rate.

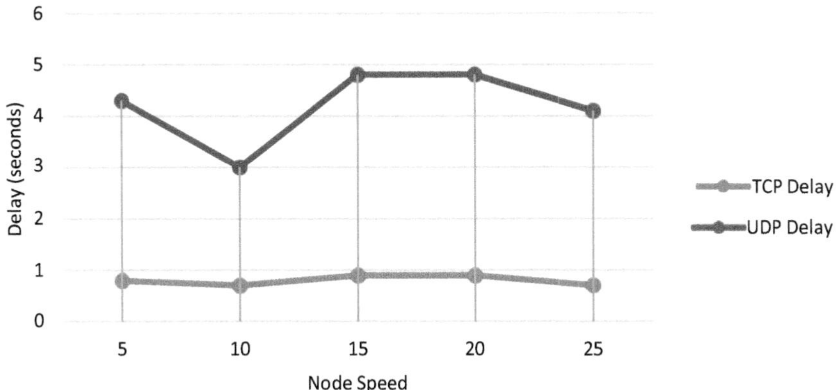

FIGURE 9.4 TCP & UDP connection delay comparison for node speed over AODV.

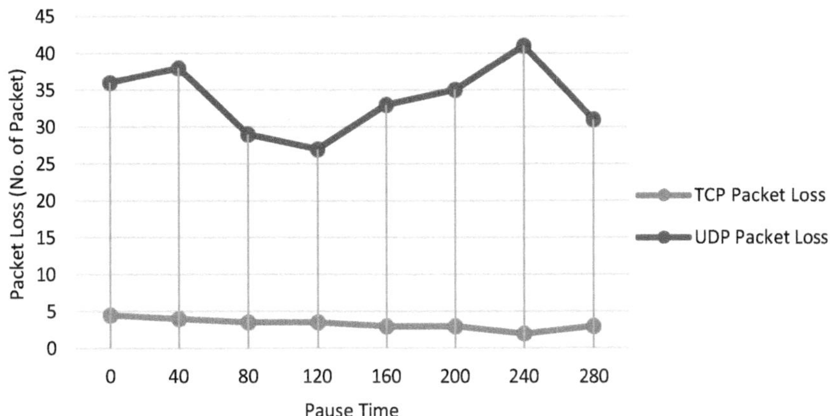

FIGURE 9.5 Loss of packets, comparison of the AODV connection via TCP & UDP with pause time.

9.3.3 Traffic Analysis of AODV w.r.t. Node Speed

The advanced characteristic of nodes in MANET is node speed. In this article, we go through the impact of node speed on delay. Figure 9.4 shows that at speeds between 5 m/s and 25 m/s, node speed has a about similar influence on TCP performance. When compared to higher speeds, UDP works better at lesser speeds. There is less delay be- tween 5 and 10 but more over 15 m/s.

9.3.4 Traffic Analysis of AODV w.r.t. Pause Time

Variable pause times are used to analyze packet loss. Figure 9.5 illustrating packet loss analysis in terms of pause time that the Pause Time has less impact on AODV over TCP than it does on UDP. UDP has a higher packet loss ratio than TCP as a result of the aforementioned factor. A greater AODV over UDP packet loss ratio is the result of more pauses.

9.3.5 Traffic Analysis of DSDV w.r.t. Node Density

The quantity of nodes in a network is known as node density. Using node density, we assess the throughput. For the analysis of throughput in terms of node density, the simulation results shown

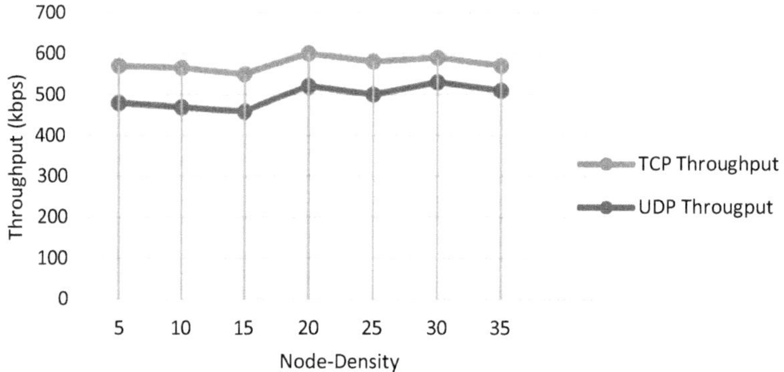

FIGURE 9.6 Throughput analysis for DSDV using TCP and UDP with node density.

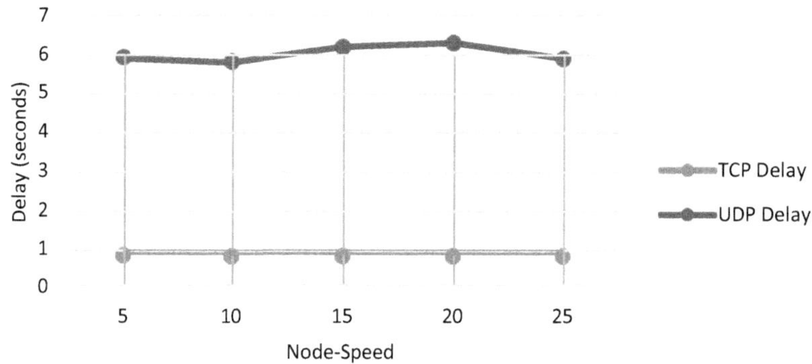

FIGURE 9.7 Delay analysis for DSDV using TCP and UDP with node speed.

in Figure 9.6 indicate that the node density has a direct impact on the network throughput in both UDP and TCP. By adding more nodes, DSDV speed is improved over UDP and TCP. TCP generally has a better throughput than UDP. This is as a result of the provided data rate's smaller size when compared to the TCP window's size. Because of this, data may transit across the TCP window without worrying about packet loss.

9.3.6 Traffic Analysis of DSDV w.r.t. Node Speed

For the purpose of studying the metric delay's performance, we measured the node speed. The analysis of delay in terms of node speed as shown in Figure 9.7 demonstrates that UDP degrades performance for high node speeds, i.e., by increasing speed, UDP latency rises, but expanding nodes has no impact on TCP performance. We definitely saw this and came to the conclusion that TCP should be utilized for high-speed nodes.

9.3.7 Traffic Analysis of DSDV w.r.t. Pause Time

In MANET, Pause Time is a factor that is frequently employed. It speaks of the moment when a mobile node stops communicating. On different pause times, we examine the same metrics. The analysis of packet loss in terms of pause time, the simulation findings demonstrate that while pause

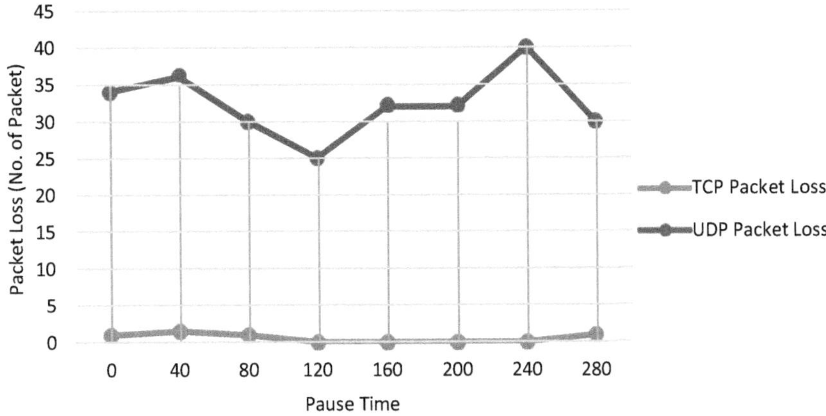

FIGURE 9.8 DSDV across TCP and UDP compared for packet loss to determine pause time.

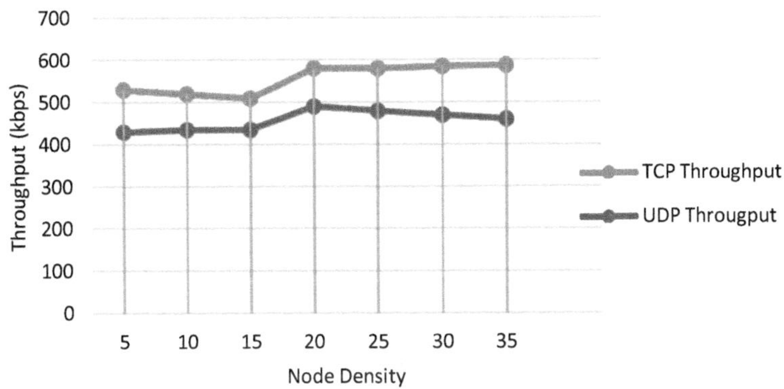

FIGURE 9.9 Node density for throughput comparison of DSR across TCP and UDP connections.

time has no effect on DSDV over TCP, it has a significant influence on DSDV over UDP. According to Figure 9.8, UDP packet loss is greater at shorter pause times than at larger pause times.

9.3.8 Traffic Analysis of DSR w.r.t. Node Density

We determine the network's throughput by adjusting node density. Figure 9.9 demonstrates the influence of node density over DSR on TCP and UDP. TCP and UDP throughput both rose with the addition of nodes. TCP has a better throughput than UDP does on an individual basis. UDP has a lesser capacity than TCP because it lacks a good flow mechanism, which results in significant packet loss.

9.3.9 Traffic Analysis of DSR w.r.t. Node Speed

In this section, we assess Delay in terms of Node Speed. Figure 9.10 demonstrates that UDP performs worse than TCP, and it also demonstrates that UDP latency grows after 25 m/s. In the instance of TCP, the differences between the values of 5 and 20 m/s are incredibly minor. So, based on the results of the trials, I came to the conclusion that Node Speed has no real impact on TCP's functionality.

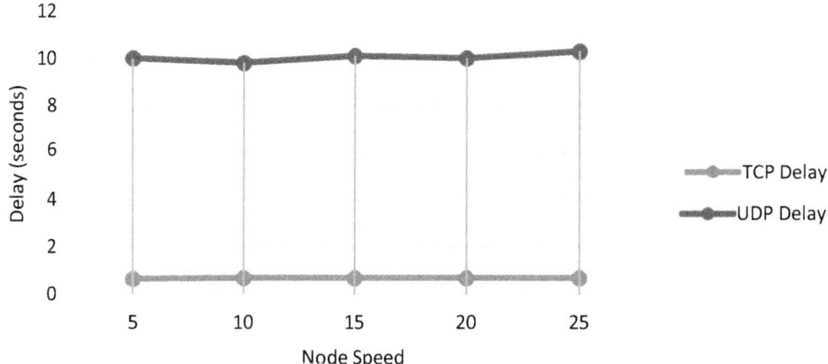

FIGURE 9.10 Delay analysis for AODV using TCP and UDP connection with node speed.

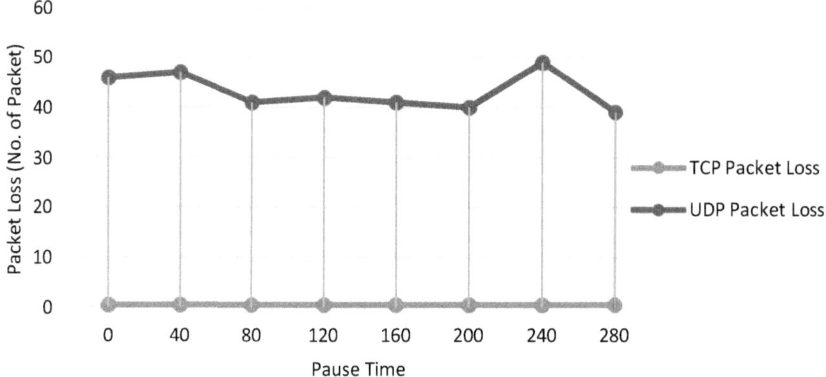

FIGURE 9.11 Packet loss analysis for DSR using TCP and UDP connection with pause time.

9.3.10 Traffic Analysis of DSR w.r.t. Pause Time

Here we assess Packet Loss in terms of Pause Time and it is shown in Figure 9.11. The line graphs make it obvious that the packet loss relationship of TCP is consistent and close to zero but in contrast, UDP has a much greater Packet Loss Ratio than TCP. The Figure also makes it evident that raising Pause Time improves UDP performance.

9.4 CONCLUSION

Ad hoc routing protocol analysis is carried out using a genuine WSN situation. AODV, DSR, and DSDV were analyzed using simulation in a genuine WSN environment. Given that the AODV protocol's packet delivery ratio is higher than that of the other two protocols, we can say that it is suitable for transferring sensitive information in WSN. However, because the AODV protocol has the largest end-to-end delay, it falls short in situations when transmission time should be extremely short. DSR works well for rapid transmission but is unsuitable for carrying information due to significant packet loss. We can see that using TCP over DSDV will result in higher throughput. UDP should not be utilized in high delay situations because node density has a greater impact on UDP latency than TCP does. Our simulation approach suggests TCP in many mobility scenarios since simulation results demonstrate it has a higher throughput than UDP. TCP performs better than UDP in environments with large levels of mobility as well. Compared to UDP, Pause Time has less of an impact on TCP's throughput, latency, and packet loss.

REFERENCES

1. Lilhore, U.K. , Khalaf, O.I. , Simaiya, S. , Tavera Romero, C.A., Abdulsahib, G.M. , M, P. , Kumar, D.: A Depth-Controlled and Energy-Efficient Routing Protocol for Underwater Wireless Sensor Networks. *Int. J. Distrib. Sens. Netw.* 18, 155013292211171 (2022). https://doi.org/10.1177/15501329221117118.
2. Sampoornam, K.P., Saranya, S., Vigneshwaran, S., Sofiarani, P., Sarmitha, S., Sar- umathi, N.: A Comparative Study on Reactive Routing Protocols in VANET. In: *2020 4th International Conference on Electronics, Communication and Aerospace Technology (ICECA)*. pp. 726–731. IEEE, Coimbatore, India (2020). https://doi.org/10.1109/ICECA49313.2020.9297550.
3. Pedditi, R.B., Debasis, K.: Energy Efficient Routing Protocol for an IoT-Based WSN System to Detect Forest Fires. *Appl. Sci.* 13, 3026 (2023). https://doi.org/10.3390/app13053026.
4. Naim, Z., Hossain, Md.I.: Performance Analysis of AODV, DSDV And DSR in Vehicular Adhoc Network (VANET). In: *2019 International Conference on Ro- botics,Electrical and Signal Processing Techniques (ICREST)*. pp. 17–22. IEEE, Dhaka, Bangladesh (2019). https://doi.org/10.1109/ICREST.2019.8644313.
5. Sharad Institute of Technology College of Engineering, Ichalkaranji, Maharashtra, India, Sarao, P., Sharma, M.: Reactive and Proactive Route Evaluation in MANET. *J. Commun.* 143–149 (2022). https://doi.org/10.12720/jcm.17.2.143-149.
6. Daud, S., Gilani, S.M.M., Riaz, M.S., Kabir, A.: DSDV and AODV Protocols Per- formance in Internet of Things Environment. In: *2019 IEEE 11th International Conference on Communication Software and Networks (ICCSN)*. pp. 466–470. IEEE, Chongqing, China (2019). https://doi.org/10.1109/ICCSN.2019.8905256.
7. Manuel, A.J., Deverajan, G.G., Patan, R., Gandomi, A.H.: Optimization of Routing-Based Clustering Approaches in Wireless Sensor Network: Review and Open Research Issues. *Electronics*. 9, 1630 (2020). https://doi.org/10.3390/electron-ics9101630.
8. Sehrawat, P., Chawla, M.: Interpretation and Investigations of Topology Based Routing Protocols Applied in Dynamic System of VANET. In Review (2022). https://doi.org/10.21203/rs.3.rs-1833768/v1.
9. Singh, R., Singh, N.: Performance Assessment of DSDV and AODV Routing Pro- tocols in Mobile Adhoc Networks with Focus on Node Density and Routing Over- head. In: *2020 International Conference on Emerging Smart Computing and In- formatics (ESCI)*. pp. 298–303. IEEE, Pune, India (2020). https://doi.org/10.1109/ESCI48226.2020.9167627.
10. Daanoune, I., Abdennaceur, B., Ballouk, A.: A Comprehensive Survey on LEACH- Based Clustering Routing Protocols in Wireless Sensor Networks. *Ad Hoc Netw.* 114, 102409 (2021). https://doi.org/10.1016/j.adhoc.2020.102409.
11. Basil, N.: Wireless Subscribers with AODV Routing Protocol Simulated on Mo- bile Ad-hoc Networks utilized Ad-hoc on-Demand Distance Vector. In Review (2023). https://doi.org/10.21203/rs.3.rs-2478764/v1.
12. Mahdi, H.F., Abood, M.S., Hamdi, M.M.: Performance Evaluation for Vehicular Ad-hoc Networks Based Routing Protocols. *Bull. Electr. Eng. Inform.* 10, 1080–1091 (2021). https://doi.org/10.11591/eei.v10i2.2943.
13. Mhmood, A., Zengi, N.A.: Performance Evaluation of MANET Routing Protocols AODV and DSDV Using NS2 Simulator. *Sak. Univ. J. Comput. Inf. Sci.* 1–10 (2021). https://doi.org/10.35377/saucis.04.01.780465.

14. Sarao, P.: TCP and UDP Traffic Based Data Transmission for Ad-hoc on Demand Routing. *J. Gujarat Res. Soc.* 21, 57–65 (2019).
15. Shaban, A.M., Kurnaz, S., Shantaf, A.M.: Evaluation DSDV, AODV and OLSR routing protocols in real live by using SUMO with NS3 simulation in VANET. In: *2020 International Congress on Human-Computer Interaction, Optimization and Robotic Applications (HORA)*. pp. 1–5. IEEE, Ankara, Turkey (2020). https://doi.org/10.1109/HORA49412.2020.9152903.

10 Implementation of IoT and WSN Technologies for Health Monitoring

Tehseen Mazhar, Muhammad Iqbal, Muhammad Iqbal, Yazeed Yasin Ghadi, Ateeq Ur Rehman, and Waseem Abbasi

10.1 INTRODUCTION

Everything is interconnected in the current culture. Any gadget with an internet connection is a smart device. The Internet of Things is a vast network of connected, locally aware things that share and exchange data, to put it simply. Thus, the Internet of Things is a network of interconnected electrical devices. The Internet of Things is crucial for the remote control of digital devices. The potential to foster commerce, professional development, and social interaction exists for tangible and digital goods. Nano chips can communicate with routers thanks to sensors, actuators, and software. The adoption of IoT is rising. Due to its many benefits, including cheaper cost, less infrastructure, a wider variety of network topologies, and less maintenance, wireless sensor networks (WSNs) are increasingly being employed in practically all monitoring applications in the digital age. Climate change, weather, natural catastrophes, traffic, forests, healthcare, location, and other things are all monitored by wireless sensors and WSN. Healthcare sensors control, watch over, regulate, alert, and track people.

Sensors quickly evaluate and identify this behavior, completely atomizing the healthcare system and removing the need for human involvement. Soon, healthcare will undergo rapid change. For patients, medical practitioners, and healthcare system managers, many valuable Internet of Things apps [1] have been created. These technologies improve healthcare by making it possible to treat severe diseases in real time and manage medical emergencies, among other things. Several aspects of health, including blood pressure, insulin levels, cardiac issues, and fitness, may be tracked and reported to a doctor or medical facility using Internet of Things apps. The development of wireless networking has touched every element of our life. The Internet of Things is growing quickly. The Internet of Things has the potential to alter our way of life by tying together all of our many technology devices.

Consequently, continuous access to reliable communications is necessary, especially in busy enterprises. When connected to the Internet, smart devices may share information (things). The Internet of Things enables advancements in technology. Many modern home appliances, security cameras, and environmental monitoring devices include transceivers, microcontrollers, and protocols for transferring control and sensor data [2] (Figure 10.1).

The medical sector was an early adopter of sensor technology and the telemetry made possible by the Internet of Things to perform remote monitoring. With the possibility of remote monitoring, patients no longer need to make frequent journeys to the doctor's office. Before recent technological advancements, this was not conceivable. This may be helpful for patients who are physically unable

DOI: 10.1201/9781003497851-10

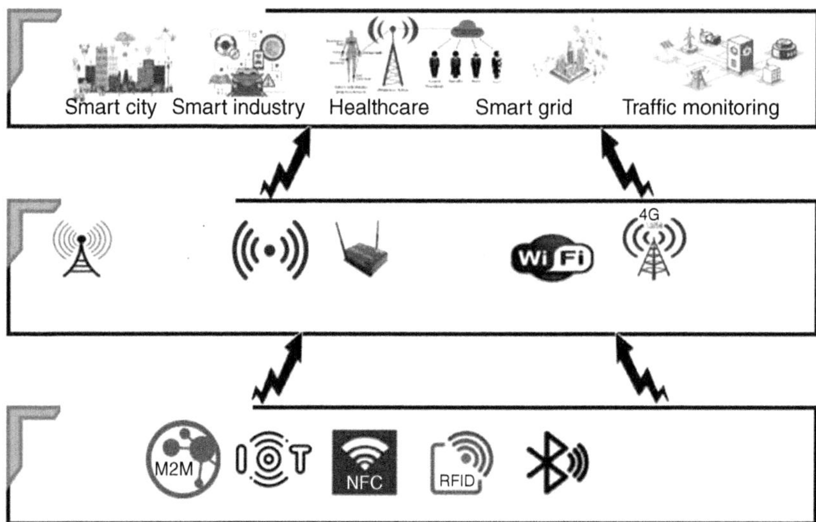

FIGURE 10.1 Generic three layer architecture in IoT [2].

to visit a clinic but still need routine medical monitoring for illnesses like diabetes, cardiovascular disease, or chronic obstructive pulmonary disease (COPD). Thanks to the possibilities provided by telemedicine, patients undergoing psychiatric or surgical operations may have the opportunity to have their status examined remotely (Figure 10.2).

Figure 10.3 demonstrates how the concepts of incidental, accessible, and universal computing form the basis of the Internet of Things. We analyze the relationships between IoT, M2M, CPS, and WSN (WSN). M2M and WSN technologies are essential for the IoT to function. Utilizing CPS and its continuous coordination between physical items and computational components may make it easier to develop dynamic M2M applications. For instance, computing, sensing, and actuation may be carried out using technologies like M2M, CPS, and WSN. A trustworthy M2M Wireless Sensor Network (WSN) that might provide CPS is the Internet of Things.

The fast spread of portable and wearable devices in this age of ubiquitous computing, when computing is nearly everywhere, has enabled improvements in healthcare made possible by the Internet of Things (IoT). With the development of cloud computing, data exchange between previously inaccessible devices improves ubiquitous computing. IoT acts as a basis for linking numerous systems, including smart homes, smart cities, and smart healthcare. These interconnected gadgets and sensors gather data for difficult activities like planning, prediction, and recommendation [3].

The structure of the paper is as follows: a general description-related work of IoT and WSN technologies is presented in Section 10.2. The Keyword Search, Inclusion and Exclusion, HealthCare IoT Application etc. is presented in Section 10.3. Section 10.4 describes the results regarding applications of IoT-based WSN. In Section 10.5 the detail of Challenges, Limitations, and Future Scope is presented. Finally, open issues of credit card fraud detection are presented in Section 10.6. Figure 10.4 shows Organization of Paper.

10.2 LITERATURE REVIEW

Medical, smart city, greenhouse monitoring, environmental, and air and water pollution applications use IoT-based WSN.

In healthcare, IoT and sensors have been utilized for drug administration, hospital operations, and remote patient monitoring. IoT's major utility in healthcare is remote patient monitoring. Healthcare practitioners may monitor patients' vital signs and symptoms in real time, spotting any

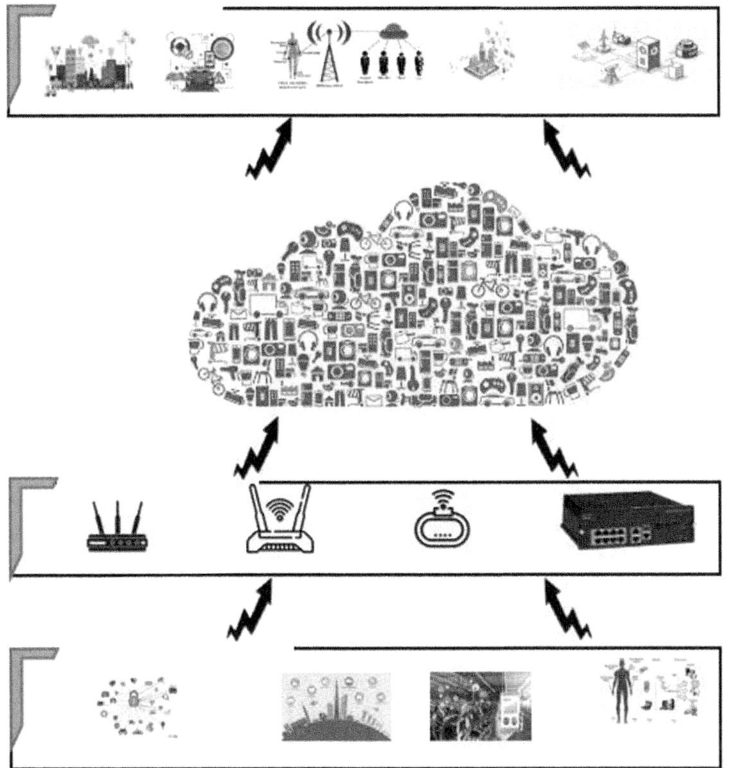

FIGURE 10.2 IoT network.

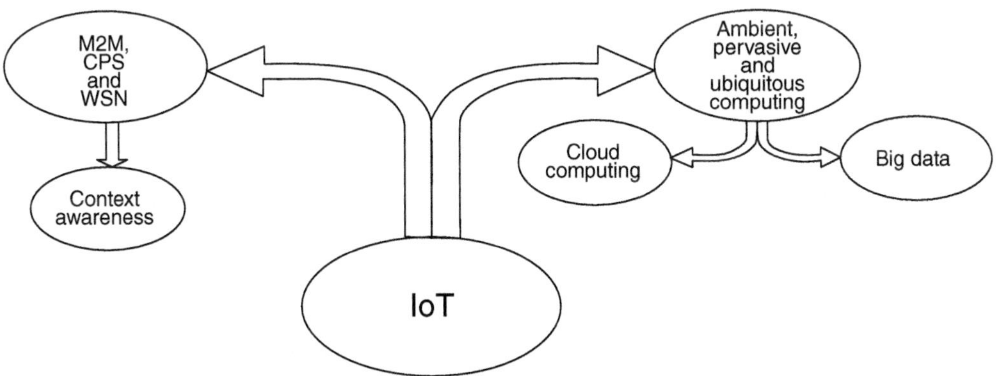

FIGURE 10.3 Relationship between IoT and related technologies.

health risks. [4] found that IoT-enabled remote monitoring devices lowered hospital readmissions in COPD patients. IoT healthcare incorporates sensors in hospitals. Sensors can monitor patients, personnel, surfaces, and medical equipment. This technology may boost hospital efficiency, patient safety, and disease prevention proved that RFID technology in hospitals boosted asset management and minimized medical equipment failure. IoT and sensors are being exploited in medication management. IoT-enabled items can remind and monitor medicine usage. [5] Revealed that IoT-enabled medication management systems improved patient outcomes and minimized medication mishaps.

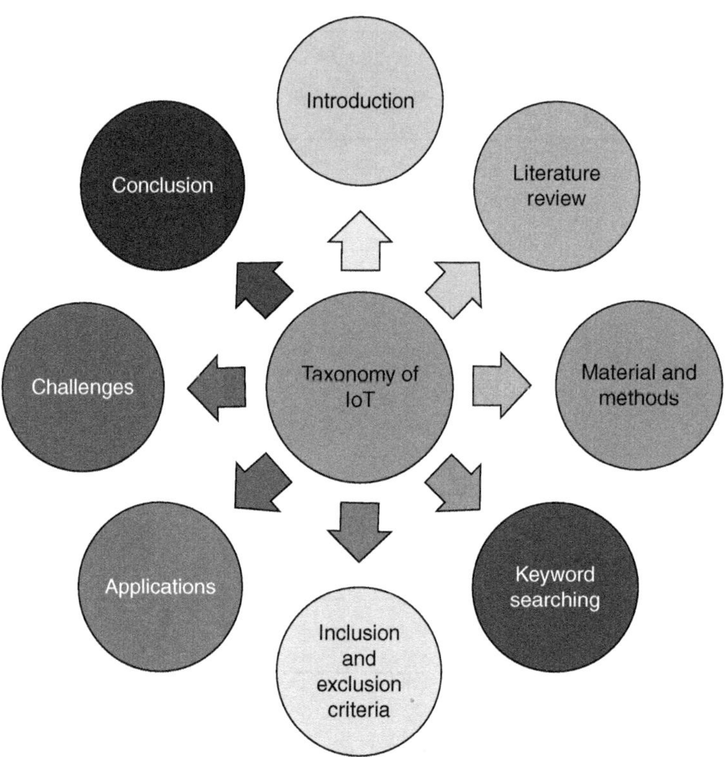

FIGURE 10.4 Organization of paper.

Sensors and IoT in healthcare are expanding swiftly and might boost patient outcomes and reduce medical expenses. In remote monitoring, wearable sensors, and medicine delivery, it works. A complete research of the IoT in healthcare: current state and future perspectives [6].

10.2.1 Healthcare Utilizes IoT and Cloud Computing

IoT applications employ cloud-based services to identify patients from home and handle several difficulties. Android applications, cloud IoT, and computers run healthcare. Due to IoT needs in academia, business, and society, [22] devised a new healthcare approach. These two studies suggest a technique for sending medical information to friends and family while monitoring heart rate using a smart health band.

The author of [7] studied H-IoT communication technology aspects in 2018. Four applications were cardiovascular, musculoskeletal, neuromuscular, and infectious disorders. They also investigated new communication technology difficulties. This research misses AI's usefulness in HIoT applications. E-health, telehealth, home monitoring, and RFID-based monitoring systems were assessed by [8]. They examined smart healthcare system interoperability, low latency tolerance, and reliability, missing HIoT system fundamentals. In another investigation, the author of [9] explored healthcare IoT applications. They focused on important components, jobs, and problems. They overlooked IoT security and privacy threats. The author of [10] evaluated H-IoT therapy. Sensing, communication, and data analytics were considered. They discussed existing challenges and provided solutions. This poll does not mention cloud computing and AI's participation in this research. Another study by [11] evaluated IoT-based health care service delivery. Survey advantages and pitfalls were considered. The IoT healthcare system's fundamental components are not

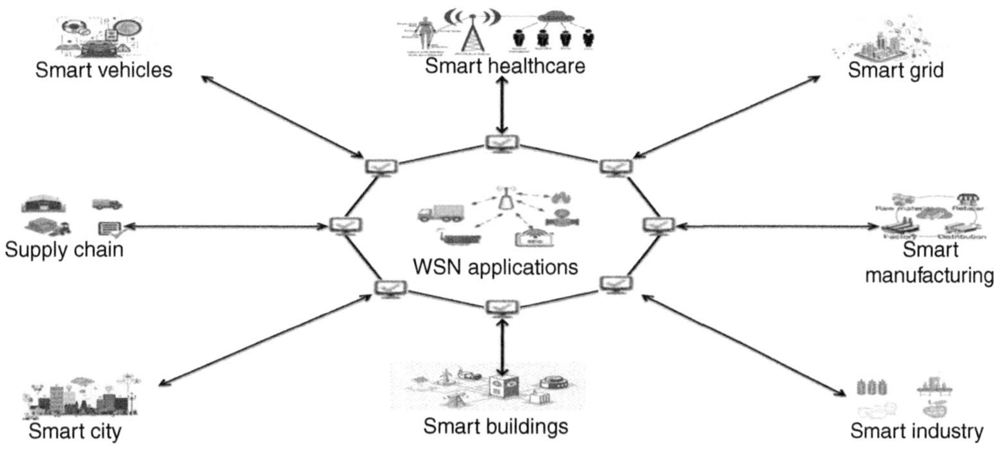

FIGURE 10.5 Application of IoT-based wireless sensor network (reproduced from [47].

investigated. This doesn't help readers comprehend IoT healthcare. The author of [12] researched H-IoT technology and applications. Like earlier surveys, important issues and unsolved subjects are covered.

Taking a patient's temperature is only one of several diagnostic tools used to track homeostasis. Changes in core body temperature may also be an early warning sign of injury, illness, or other conditions. To rule out several diseases, a doctor may take a patient's temperature. A person's temperature may be taken in a variety of ways, including by placing a thermometer in their mouth, ear, or rectal area. When using these methods, there is often a concern with patient comfort and a higher risk of infection. Recent developments in IoT-based technology, however, have made a variety of viable options easily available. A 3D-printed earpiece with an infrared sensor measures internal body temperature by contacting the tympanic membrane [13]. It included a CPU for processing data and a wireless sensor unit. Environment and exercise have little effect on the temperature around here. The core body temperatures of newborns were monitored in real time using lightweight, wearable sensors [14]. It may also alert parents if the child's temperature climbs over a potentially harmful level. Figure 10.5 shows the applications of IoT Based Wireless Sensor Network

10.3 MATERIAL AND METHODS

10.3.1 Keyword Searching

This research examined the wireless networking aspects of IoT and healthcare systems simultaneously. In the first stage, a plethora of documents, including crucial healthcare data, were collected. Second, we gathered literature on IoT healthcare systems and related sensors. To this end, we looked for terms associated with healthcare IT, the IoT, and sensors. Figure 10.6 displays the results of a search conducted across many databases using the keywords shown there.

10.3.2 Inclusion and Exclusion Criteria

Several scholarly articles were discovered on the mentioned Healthcare, Wireless Sensor Networks, Internet of Things, and Healthcare Applications in IoR. Applying the criteria presented in Table 10.1, the most relevant and concise material was collected from these publications written in English for general consumption. They correspond to common queries and study objectives in the areas of WSNs, IoT, healthcare, and healthcare-related IoT applications.

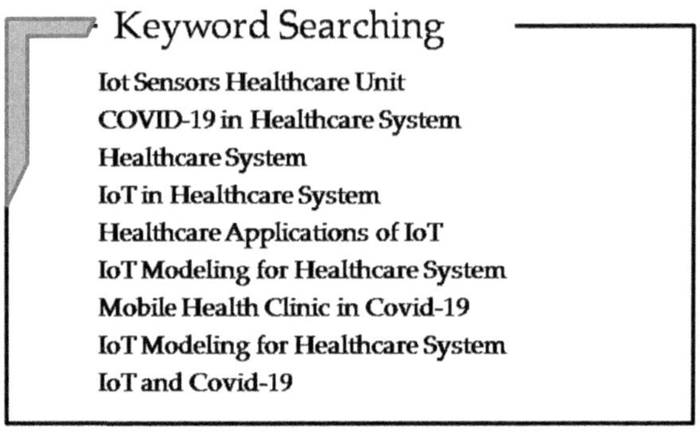

Keyword Searching

Iot Sensors Healthcare Unit
COVID-19 in Healthcare System
Healthcare System
IoT in Healthcare System
Healthcare Applications of IoT
IoT Modeling for Healthcare System
Mobile Health Clinic in Covid-19
IoT Modeling for Healthcare System
IoT and Covid-19

FIGURE 10.6 Research methods and keyword searching.

TABLE 10.1
Inclusion and Exclusion Criteria

Criteria for Inclusion
Only those papers that were authored in English should be included in the review.
Include only works published between 2016 and 2022
Particularly include articles with sufficient information on WSN, IoT, and Healthcare in IoT in the titles, keywords, abstracts, and conclusions.
Include papers where the main objective of the content is on WSN, IoT, or Healthcare in IoT.
Exclusion Criteria
Do not take into consideration any papers that have been written in a language other than English.
Gray papers Exclude.
Exclude papers that were not published between 2014–2021.
Less than three pages of research papers containing exclude
Exclude papers that failed to meet the inclusion criteria

Research Question

The following research questions are this study

1. What is the application of IoT that is based on WSN?
2. What is the role of Smart Technologies in IoT-Based Healthcare Systems for COVID-19?
3. What is the role of IoT Based Healthcare Challenges/limitations for COVID-19?

The most prevalent use of IoT in healthcare is the monitoring of physical activity via consumer electronics. Wearable technology, such as smart clothes and wristbands that track vital signs and activity levels, falls under this category. The fitness of an athlete is calculated by looking at all of the available information [15]. Several different types of sensors are advocated for use in [15] to measure health and fitness. In a three-layer architecture, the user interface to the central processing unit represents the device layer. The database server receives data that has been processed locally and stores it. A database may be accessed remotely to keep tabs on vital signs. The researchers in this study monitored sweat production with the use of an accelerometer, temperature and pulse sensors, and a Grove GSR sensor. A material similar to "smart fabric" connects all of the sensors. It is impossible to gauge the efficacy of gym training without first developing a personalized workout program [16]. This technique uses the Apple Watch and the Health app to obtain the necessary information. Evaluating cyclists' physical readiness is recommended in [17]. And it can even sniff

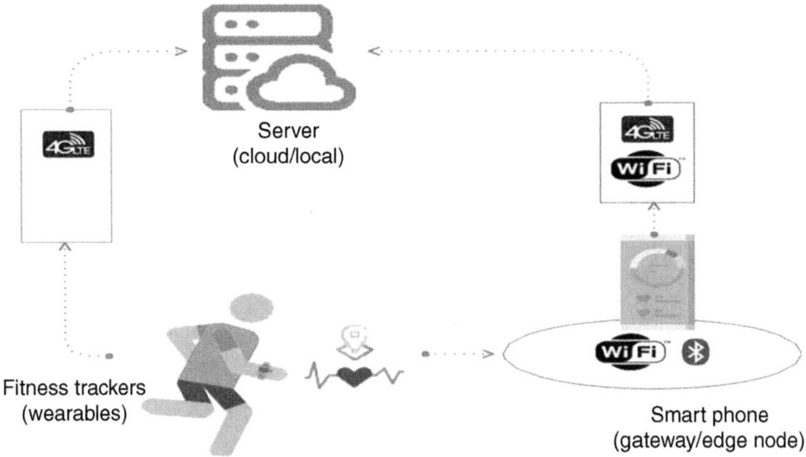

FIGURE 10.7 An overview of fitness tracking system.

out those who would steal your bike. The software features a two-pronged system architecture, with the backend and the communication mechanism being shared. There are sensors and bicycle safety devices at the user end of this three-stage method. While accelerometer data is used by the bike's safety system, heart rate is used to gauge the severity of any health issues. In tandem, they provide data to an Android app that monitors bike locks and health indicators. Fitness monitoring plans [18] often include tracking activities like walking, jogging, and relaxing. The structure is composed of three layers, one of which is a BSN sensor layer. Figure 10.7 shows an Overview of Fitness Tracking System.

10.4 RESULT

10.4.1 Applications of IoT-Based WSN

WSNs based on the Internet of Things is used in fields as varied as medicine, smart cities, greenhouse monitoring, and environmental assessment of air and water pollution. Below are examples of many different utilizations.

The IoT has completely altered the ways in which we live, work, and communicate. Technology advancements in real-time communication between devices and systems have enabled "smart" dwellings, urban infrastructure, and transportation systems. The medical field has seen significant IoT growth. Potentially, the Internet of Things and sensors will revolutionize the administration and delivery of medical care. Here, the benefits and drawbacks of using sensors and the Internet of Things in healthcare are outlined. Sensors and the Internet of Things might enhance clinical operations and encourage the development of novel diagnostic and treatment processes, as well as allow remote patient monitoring and management. The medical field is a big user of IoT devices. Indeed, H-IoT is used in healthcare applications. The H-IoT is a subset of the IoT. Internet of Things and Enhanced Internet of Things relies on WSNs and BSNs. IoT systems and H-IoT systems are technologically distinct. Data from the fitness tracker and wearables market indicates further expansion [19]. IoT Net is a healthcare centric system because of the proliferation of smart health monitoring devices and the development of the IoT connectivity infrastructure. Health monitoring is only one use of the Internet of Things. Patients' biometric data and vital signs might be monitored across the network to aid in accurate diagnosis and treatment [15]. Data sharing across the many participating institutions is impossible without a standard infrastructure. To advance H-IoT, a standardized or reference architecture is required. Multiple groups and businesses are working together to provide standardized

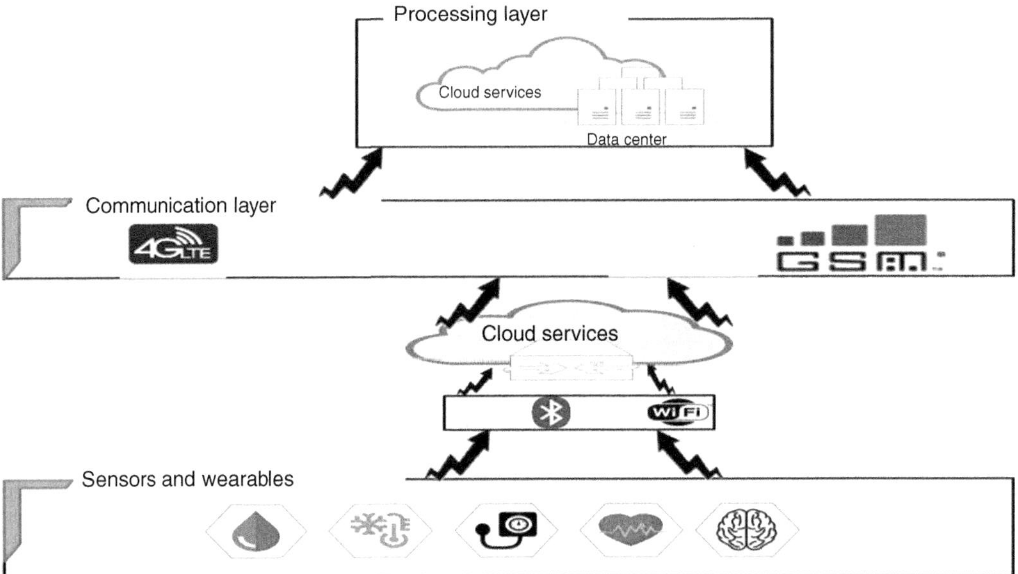

FIGURE 10.8 The three-tier architecture of the h-IoT systems.

designs for a wide range of Internet of Things uses [20]. A distributed, service-oriented architecture, The IEEE Standards Association establishes norms for Point-of-Service medical devices. Setup, node discovery, and node communication, as well as QoS components, are all outlined for the H-IoT scenario [21] (Figure 10.8).

Recent years have seen an increase in the use of sensors and the IoT in healthcare. Potentially enabling real-time monitoring, data collecting, and more efficient and effective treatment, these technologies have the potential to significantly alter the current state of healthcare delivery. The numerous applications of the Internet of Things and sensors in healthcare, such as telemedicine, remote monitoring, and device integration, will be discussed in this article.

We will also discuss the potential benefits and drawbacks of these technologies and provide recommendations for further study and development. When one is seeing something from a distance, Remote monitoring is a critical use of IoT and sensors in the medical industry. Patients may be monitored remotely, reducing the frequency with which they must visit the hospital. Patients with chronic conditions, including diabetes, cardiovascular disease, and chronic obstructive pulmonary disease, who require frequent monitoring but are unable to make regular hospital visits may benefit the most from this. Remote monitoring might potentially be used to check up on people's mental health or recovery from surgery or other medical treatments.

10.4.2 Smart Technologies of IoT-Based Healthcare Systems for COVID-19

Combining many Internet-of-Things tools with current medical infrastructures. Several businesses have increased their use of IoT tools as a direct consequence of COVID-19. As more and more devices are connected to the internet, more and more cutting-edge technologies are made available to individuals and businesses alike. To name a few, the following technologies may be of great help to future IoT healthcare systems:

10.4.2.1 Ambient Intelligence Communications Technologies

Ambient intelligence is crucial for users and patients. Its use in the healthcare industry benefits both patients and medical professionals [22]. Information about patients is gathered and analyzed with the use of wearable technology and sensors linked to the Internet of Things and a healthcare facility. COVID-19's embedded system of human–computer interaction, autonomous control, Internet of Things, and ambient intelligence may thereby help governments, healthcare systems, and patients.

10.4.2.2 Augmented Reality

In the age of Industry 4.0, augmented reality is a key component of cutting-edge innovation, education, and growth. One area where IT has made a significant difference is in healthcare. Augmented reality is used in the medical industry for a wide range of applications, including remote monitoring, surgery, instruction, and training. AR promotes and stresses regular hand washing as a means of combating COVID-19 and other diseases [23].

10.4.2.3 Wearable Devices

Wearable technology offers a low-cost, high-value answer to a variety of health problems. A wide range of sensors and wearable accessories may be used with the devices to gather data on the patient and their surroundings. Because it is stored digitally, it may be accessed from anywhere. Connecting Internet of Things wearables to mobile apps boosts their processing capability. Wearable technology has become an increasingly important aspect of IoT adoption in healthcare, as seen in Figure 10.9, which illustrates its use in the fight against pandemics such as COVID-19.

10.4.2.4 Smart Technology

As a result of COVID-19, businesses have adjusted to the new conditions. Some of the suggestions made in the MIT report include: increasing the distance between workstations; removing chairs from conference rooms; installing antimicrobial surfaces; installing thermal scanners; adjusting the building's temperature control; painting the floors; and enforcing strict cleaning regulations. Returning employees may be offered or required to undergo coronavirus testing by certain companies. No business was ready for COVID-19, not even the most innovative ones. These industries may better protect their workers and increase output without sacrificing quality by adopting a "Command Center" [24], which converts all cutting-edge technology into business intelligence technology. During pivotal junctures, the command center gathers all pertinent industry data for use in making strategic and operational decisions.

10.4.2.5 Ambient Assisted Living (AAL)

We're developing an AAL intelligent support system to improve the older population's wellbeing, security, comfort, and self-respect. Facilitators, physicians, and the healthcare system all stand to benefit from taking a more simplified approach to caring for the aged. COVID-19 integrates the Internet of Things (IoT), smartphones, and wearables. Using this technology, loving ones may alert their elderly relatives in the home or at a care facility of any impending threats or emergencies [25]

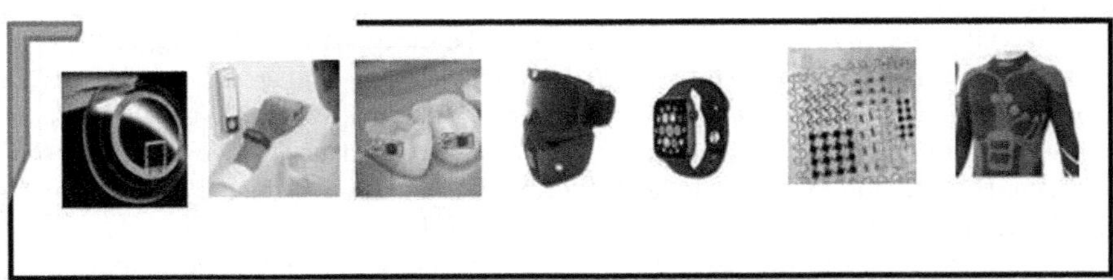

FIGURE 10.9 Wearable devices.

10.5 CHALLENGES, LIMITATIONS, AND FUTURE SCOPE

Healthcare problems have lately been addressed through technological solutions. Superior medical care is more accessible now than ever before. Networking, cloud computing, and Internet of Things enabled intelligent sensors have all contributed to the transformation of the healthcare sector. The Internet of Things has advantages and disadvantages much like any other technology [26].

Cost updates are essential for IoT gadgets to ensure they always function at the cutting edge. Several sensors and health care gadgets are linked in each IoT network. The costs associated with upkeep, maintenance, and repairs are detrimental to the company's bottom line and clientele. To save money, we need low-maintenance sensors.

It's common for IoT gadgets to rely on batteries for power. Changing batteries in networked sensors may be difficult. A massive battery supplied the power. Scientists from all across the globe are working together to develop medical devices that can maintain themselves. One option is to combine IoT with alternative power sources. Using these techniques may be useful in addressing the present energy crisis. Several medical groups exist to serve the public. Almost all of these products advertise themselves as benchmarks for their respective fields uncertainty.

An authoritative body is needed to standardize HIoT device communication protocols, data aggregation, and gateway interfaces. HIoT gadgets are necessary for the verification and standardization of EMRs. Researchers may work toward device standardization by joining working groups with the IETF, ETSI, IPSO, and other bodies.

Because of cloud computing, real-time monitoring has evolved. Cyberattacks on the healthcare network are a real concern. Information about patients or their treatments may be at risk. To prevent this kind of attack, HIoT systems must be well-protected.

HIoT medical and sensor equipment requires evaluation and implementation of identity authentication, secure booting, fault tolerance, authorization management, white-listing, password encryption, and secure pairing protocols to prevent attacks. Secure routing and message integrity are essential for wireless technologies like Wi-Fi, Bluetooth, ZigBee, and others. Since everyone using the system is also linked to the server, security flaws in the IoT might compromise sensitive patient data. Security may be bolstered via the use of algorithms and cryptographies.

Because medical devices are adaptable to many settings, they may be scaled easily. Gains in speed and output may be expected as a result of scaling. Being adaptable is crucial. Future and present system efficiency are improved. Medical equipment, sensors, and actuators drive an IoT system for the healthcare sector. The variety of IoT devices necessitates careful management of scalability in HIoT systems.

REFERENCES

[1] Ma, Y. ; Wang, Y. ; Yang, J. ; Miao, Y. ; Li, W. Big Health Application System Based on Health IoTand Big Data. *IEEE Access* 2016, 5, 7885–7897.

[2] Rehman, A.U.; Naqvi, R.A.; Rehman, A.; Paul, A.; Sadiq, M.T.; Hussain, D. A Trustworthy SIoT Aware Mechanism as an Enabler for Citizen Services in Smart Cities. *Electronics* **2020**, 9, 918. https://doi.org/10.3390/electronics9060918

[3] Isravel, D.P.; Silas, S. A Comprehensive Review on the Emerging IoT-Cloud Based Technologies for Smart Healthcare. In *Proceedings of the 2020 6th International Conference on Advanced Computing and Communication Systems (ICACCS)*, Coimbatore, India, 6–7 March 2020; pp. 606–611

[4] Raza, A.; Ayub, H.; Khan, J.A.; Ahmad, I.; S. Salama, A.; Daradkeh, Y.I.; Javeed, D.; Ur Rehman, A.; Hamam, H. A Hybrid Deep Learning-Based Approach for Brain Tumor Classification. *Electronics* 2022, 11, 1146. https://doi.org/10.3390/electronics11071146

[5] Kao, A.K., et al. IoT-Enabled Fall Detection and Monitoring System for Older Adults. *J. Med. Syst.* 2015, 39(12), 1–8.

[6] Ahmad, I.; Ullah, I.; Khan, W.U.; Rehman, A.U.; Adress, M.S.; Saleem, M.Q.; Cheikhrouhou, O.; Hamam, H.; Shafiq, M. Efficient Algorithms for E-healthcare to Solve Multiobject Fuse Detection Problem. *J. Healthcare Eng.* 2021, 2021, 1–16. https://doi.org/10.1155/2021/9500304

[7] Mazhar, M.S.; Saleem, Y.; Almogren, A.; Arshad, J.; Jaffery, M.H.; Rehman, A.U.; Shafiq, M.; Hamam, H. Forensic Analysis on Internet of Things (IoT) Device Using Machine-to-Machine (M2M) Framework. *Electronics* 2022, 11, 1126. https://doi.org/10.3390/electronics11071126

[8] Alam, M.M.; Malik, H.; Khan, M.I.; Pardy, T.; Kuusik, A.; le Moullec, Y. A Survey on the Roles of Communication Technologies in IoT-Based Personalized Healthcare Applications. *IEEE Access* 2018, 6, 36611–36631. [CrossRef]

[9] Shaikh, Y.; Parvati, V.K.; Biradar, S.R. Survey of Smart Healthcare Systems Using IoT(IoT). In *Proceedings of the 2018 International Conference on Communication, Computing and IoT(IC3IoT)*, Chennai, India, 15–17 February 2018; pp. 508–513.

[10] Ahmadi, H.; Arji, G.; Shahmoradi, L.; Safdari, R.; Nilashi, M.; Alizadeh, M. The Application of IoTin Healthcare: A Systematic Literature Review and Classification. *Univers. Access Inf. Soc.* 2019, 18, 837–869.

[11] Haider, S.K.; Jiang, A.; Almogren, A.; Rehman, A.U.; Ahmed, A.; Khan, W.U.; Hamam, H. Energy Efficient UAV Flight Path Model for Cluster Head Selection in Next-Generation Wireless Sensor Networks. *Sensors* 2021, 21, 8445. https://doi.org/10.3390/s21248445

[12] Asif, M.; Khan, W.U.; Afzal, H.R.; Nebhen, J.; Ullah, I.; Rehman, A.U.; Kaabar, M.K. Reduced-Complexity LDPC Decoding for Next-Generation IoT Networks. *Wirel. Commun. Mob. Comput.* 2021, 2021, 1–10. https://doi.org/10.1155/2021/2029560

[13] Sadiq, M.T.; Aziz, M.Z.; Almogren, A.; Yousaf, A.; Siuly, S.; Rehman, A.U. Exploiting Pretrained CNN Models for the Development of an EEG-based Robust BCI Framework. *Comput. Biol. Med.* 2022, 143. https://doi.org/10.1016/j.compbiomed.2022.105242

[14] Shahwar, T.; Zafar, J.; Almogren, A.; Zafar, H.; Rehman, A.U.; Shafiq, M.; Hamam, H. Automated Detection of Alzheimer's via Hybrid Classical Quantum Neural Networks. *Electronics* 2022, 11, 721. https://doi.org/10.3390/electronics11050721

[15] Arif, S.A.; Niaz, M.H.; Shabbir, N.; Zafar, M.H.; Hassan, S.R.; Rehman, A.U.. RSSI Based Trilatertion for Outdoor Localization in Zigbee based Wireless Sensor Networks (WSNs). *2018 10th International Conference on Computational Intelligence and Communication Networks (CICN)*, Esbjerg, Denmark, 2018, pp. 1–5. https://doi.org/10.1109/CICN.2018.8864943

[16] Muhammad Hussain, N.; Rehman, A.U.; Othman, M.T.B.; Zafar, J.; Zafar, H.; Hamam, H. Accessing Artificial Intelligence for Fetus Health Status Using Hybrid Deep Learning Algorithm (AlexNet-SVM) on Cardiotocographic Data. *Sensors* 2022, 22, 5103. https://doi.org/10.3390/s22145103

[17] Kanwal, S.; Tao, F.; Almogren, A.; Rehman, A.U.; Taj, R.; Radwan, A. A Robust Data Hiding Reversible Technique for Improving the Security in e-Health Care System. *Comput. Model. Eng. Sci.* 2022, 134(01), 201–219. https://doi.org/10.32604/cmes.2022.020255

[18] Saeed, N.; Bader, A.; Al-Naffouri, T.Y.; Alouini, M.S. When Wireless Communication Responds to COVID-19: Combating the Pandemic and Saving the Economy. *Front. Commun. Netw.* 2020, 1, 566853–566867.

[19] WAS. Making Vehicles Special. Available online: www.wasvehicles.com/en/home.html/ (accessed on 22 June 2022).

[20] Hassan, S.R.; Ahmad, I.; Nebhen, J.; Rehman, A.U.; Shafiq, M.; Choi, J.G. Design of Latency-Aware IoT Modules in Heterogeneous Fog-Cloud Computing Networks. *Comput. Mater. Contin.* 2022, 70(3), 6057–6072. https://doi.org/10.32604/cmc.2022.020428

[21] Whelan, K. Smart Ambulances: The Future of Emergency Healthcare. November 2018. Available online: https://emag.medicalexpo.com/smart-ambulances-the-future-of-emergency-healthcare/ (accessed on 22 June 2022).

[22] Mike, T. How IoT Helps Improve Healthcare. Available online: https://builtin.com/internet-things/iot-in-healthcare (accessed on 22 June 2022). Available online: www.quio.com/ (accessed on 22 June 2022).

[23] Arshad, J.; Rehman, A.U.; Othman, M.T.B.; Ahmad, M.; Tariq, H.B.; Khalid, M.A.; Moosa, M.A.R.; Shafiq, M.; Hamam, H. Deployment of Wireless Sensor Network and IoT Platform to Implement an Intelligent Animal Monitoring System. *Sustainability* 2022, 14, 6249. https://doi.org/10.3390/su14106249

[24] Zhang, C.; Härenstam, K.P.; Meijer, S.; Darwich, A.S. Serious Gaming of Logistics Management in Pediatric Emergency Medicine. *Int. J. Serious Games* 2020, 7, 47–77. [CrossRef]
[25] Sundas, A.; Badotra, S.; Bharany, S.; Almogren, A.; Tag-ElDin, E.M.; Rehman, A.U. HealthGuard: An Intelligent Healthcare System Security Framework Based on Machine Learning. *Sustainability* 2022, 14, 11934. https://doi.org/10.3390/su141911934
[26] Shanmugasundaram, G.; Sankarikaarguzhali, G. An Investigation on IoT Healthcare Analytics. *Int. J. Inf. Eng. Electron. Bus.* 2017, 9, 11–19.

11 A Survey on Approximate Hardware Accelerator for Error-Tolerant Applications

Sahibzada M. Waqas, Muhammad Zakwan, Muhammad Ashraf, Ghadah Naif AlWakid, and Mamoona Humayun

11.1 INTRODUCTION

Power consumption and data storage capacity are significant considerations throughout the design phase of new computer systems with dense processing. These devices are getting closer and closer to the essential energy constraints necessary for completely trustworthy computing [1]. It has been a driving force behind the development of innovative design strategies that aim to lower the amount of electricity and energy required. A proficient computational method called approximate computing (AxC) trades off computation results' precision for a more efficient use of system resources. It has evolved as a new paradigm that many applications, in which erroneous results are acceptable, prefer over conventional computing methods. In recent decades, approximate computing has developed as a unique strategy that is relevant to systems that have a fair resistance to faults [2]. This pattern may be used in computing that approximates the actual value of data. However, one of the most fundamental drawbacks of approximation computing is that it relies strongly on the data.

It is because the resulting approximate architecture is dependent on the data that was used for training. Therefore, the inaccuracy of the approximate circuit may still approach these design limits even if the final assignment is different from the training data that was utilized during the assumption made for the approximation process. The computational quality (precision of results) is compromised in order to lower the necessary computational effort (area/speed/energy/power) [3]. It is done by accepting acceptable outcomes that are created by inexact computations, mainly in areas in which human judgment plays a considerable role.

As a result of the fact that many applications do not need very accurate answers, human sensory limits can allow for the occasional acceptance of approximations. At the hard-ware level, approximation has been investigated to improve the energy usage of these applications that run on FPGA [4]. One example of this would be the use of approximate multipliers. Nevertheless, the level of accuracy that may be achieved using approximation computing is highly dependent on the particular application, inputs and preferences of the users. The key contributions of this study are as follows:

1. It examines all of the most recent innovative techniques that have been offered by researchers for approximations, including the use of approximate hardware accelerators for error-tolerant applications with emphasis on their practicality in the field of Deep Learning.
2. It highlights the findings of these researchers, along with a statistical analysis of their gains and results.
3. It provides a comparison of all of the proposed designs and provides future research areas of approximation for the Deep Learning architectures.

DOI: 10.1201/9781003497851-11

This paper is structured as follows: Section 11.2 elaborates on the approximate computing and error-tolerant applications, Section 11.3 presents the literature review of the recent researches with their results, Section 11.4 provides a comparative analysis table, Section 11.5 presents the future areas, while Section 11.6 provides the conclusion of this study.

11.2 APPROXIMATE HARDWARE ACCELERATOR (HWA)

Approximate Computing. Unfortunately, Moore's law is reaching its limit, and traditional computing methods are ineffective at enhancing performance any further, while still adhering to physical constraints like power consumption. These limits have given rise to a variety of intriguing problems as well as new challenges. Focusing on an acceptable lowering of computing accuracy without losing practicality is one of the most promising topics to investigate. This strategy is referred to as "approximate computing" that enables a controlled approximation in computation with acceptable error rates. By trading accuracy for other system parameters, this method seeks to enhance the performance, energy efficiency, and resource utilization of computing systems. In machine learning tasks, AxC can cut energy usage by up to 62% while keeping the accuracy levels to acceptable limits [1]. Additionally, AxC can significantly boost efficiency and resource utilization in many application fields, making it a practical methodology. Figure 11.1 shows if we increase the computational accuracy, the system becomes slower, and the more area / power it requires.

Accelerators that can conduct approximate calculations with tolerable error limitations are known as approximate hardware accelerators. Studies found that while keeping acceptable accuracy levels, approximation hardware accelerators can enhance energy efficiency by up to 16 times compared to traditional hardware units [2].

Error Tolerant applications. Applications that can handle some degree of computational error without affecting the system's overall functionality or performance are said to be error-tolerant. This category includes many applications in domains like machine learning, image processing, and signal processing. Approximate hardware accelerators are among other approximation techniques that may execute imprecise computations while retaining acceptable error boundaries, which is advantageous for specific applications. These accelerators can dramatically boost the energy efficiency and performance of computing systems by making use of the error tolerance of these applications. Another example is the Deep Learning (DL) chip from Google, where the Tensor Processing Unit (TPU) significantly improves the performance by utilizing established approximate computing methods.

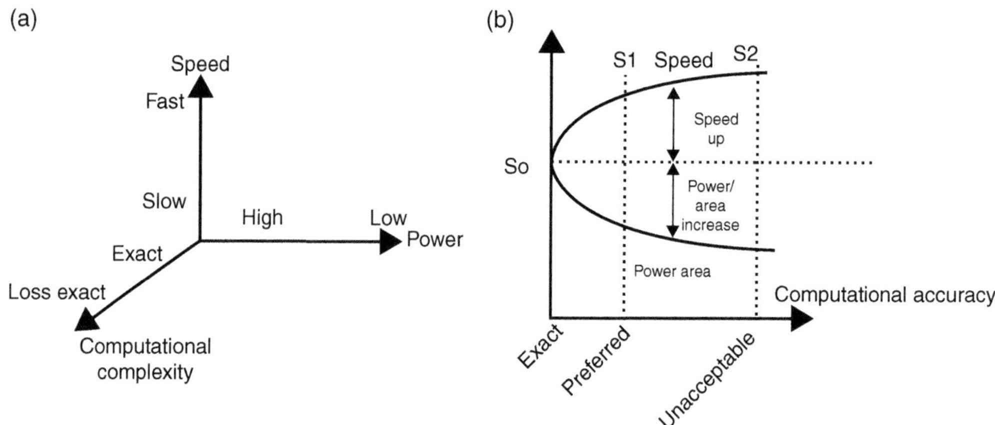

FIGURE 11.1 Approximate computing design space: (a) Power, performance vs. accuracy. (b) Compromise among area, speed and error.

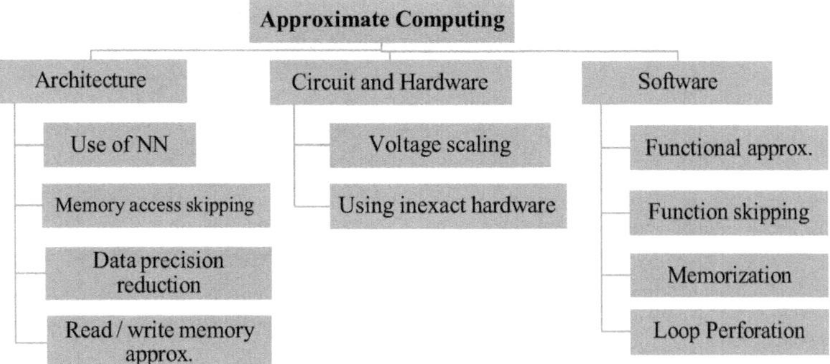

FIGURE 11.2 Taxonomy for approximate computing (AxC) including the commonly used methods.

Additionally, IBM research has started a project to develop on-chip AI accelerators using AxC [3] (Figure 11.2).

11.3 LITERATURE REVIEW

11.3.1 Reconfigurable Approx. HWA

Systems of computing that use the least amount of energy are often referred to as energy-efficient. On the other hand, the systems that were analyzed for this study may be classified as error-efficient [1] since they only produce the same number of mistakes that each of their applications can accept. In their paper, the researchers proposed an approximate Micro-Architecture Manager (MAM) that is also reconfigurable during runtime [5]. This runtime MAM can monitor the distributions relating to the workload of each approximate accelerator. It then re-configures the accelerators following the approximate micro-architecture, trained with input data distributions most similar to the current workload, thereby keeping the error under control. Because approximation computing may result in exceptionally high errors in output owing to unstable or dynamic workloads, the construction of a collection of approximate micro-architectures that are trained with a variety of input data distributions (IDDs) was also suggested in [5].

Findings: The approximation accelerator's random setting resulted in a 54% increase in the average inaccuracy, while SW and HW MAM versions had lower than 10%.

11.3.2 Fault Analysis for Designing Approximate HWA

The researchers suggested a technique of compiler-driven error analysis in the paper [3] to analyze the errors that are created by approximation adders in the layout of approximate accelerators. It was done in order to determine how accurate the design would be. It is feasible to compute the error behavior as the result of any accelerator layout by first determining the source of faults and then tracing their progression through the system. It was suggested that a methodology relying on Depth-First Search (DFS) analyze the nodes in the Data Flow Graph (DFG), moving backwards from the output to the inputs. It was done by presenting the accelerator as a Data Flow Graph (DFG). Figure 11.3 displays a DFG having a size 3 x 3 Gaussian kernel. The error at the output is calculated based on the amount the value shifts to the right at the output. It makes a request for the mistake that was made by its ancestors, which in this instance is the eighth adder (S8). This adder requires the error that was generated by the seventh adder (S7) in order to function properly up to the point when the inputs are reached. For example, the sum of the errors introduced by each of the approximation adders S1 and S2 at nodes S1 and S2 is determined by the configuration of the adders.

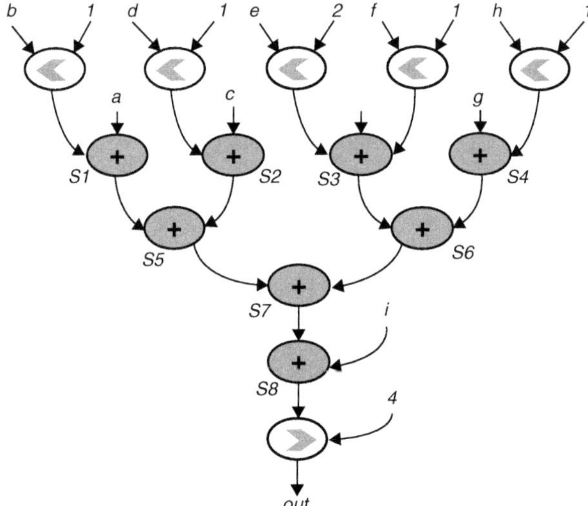

FIGURE 11.3 DFG to represent a size 3 x 3 Gaussian kernel. Approximate adders potentially substitute the additions.

Findings. Regarding Monte Carlo simulations and picture-based input data sets, research on various image-processing kernels has proven astonishingly accurate estimates. It has made it possible for speedier and more in-depth explorations of the error propagation that is produced by approximation calculations.

11.3.3 A Machine Learning-Based Approximate Hardware Accelerator Assuring Quality and Dynamic Partial Reconfiguration (DPR)

The role of partial reconfiguration is somewhat like a microprocessor that can switch between jobs in the sense that PR allows the hardware resources on any device to be time-multiplexed. It enables the FPGA's utility to be significantly increased since there is no longer a requirement to completely re-configure and re-establish links. Moreover, it allows self-healing and adaptive systems that ultimately require less space to work. In [4], the researchers proposed approximate accelerators with hardware efficiency and sound quality that are suitable for implementation in FPGA-based ML algorithms along with other error-resilient applications. Deep Neural Networks (DNNs), which are a sub-domain of ML algorithms, are appropriate for error tolerance; however, these complex models have relatively colossal storage and computational requirements. Thus, an energy-efficient implementation of DNNs will have unlimited practical advantages. The suggested methodology [6] incorporated the advantages of approximation computing, DPR, and ML, which had a good influence on the area and power consumption. Moreover, researchers [6] introduced a novel design selector for an ML algorithm, i.e., a decision tree that continuously monitors the input data and readily calculates the most appropriate approx. design; in response, the selected approximate design is then partially re-configured into the FPGA while maintaining the entire error-tolerant application.

Figure 11.4 depicts the internal design for the approx. accelerator with 16 multipliers. The inputs, i.e., Aj & Bj while $j \geq 0$ & ≤ 16, have 8-bit width. For the parallel execution, the proposed model utilizes the prevailing block RAM in Xilinx 7 series FPGAs, comprising 1,030 blocks of 36 Kbits.

Findings. The proposed design adaptation method was implemented around two thousand times while the target output quality (TOQ) was reached more than fifteen hundred times (PSNR used was greater than 30dB). An overall accuracy of 81.82% was achieved.

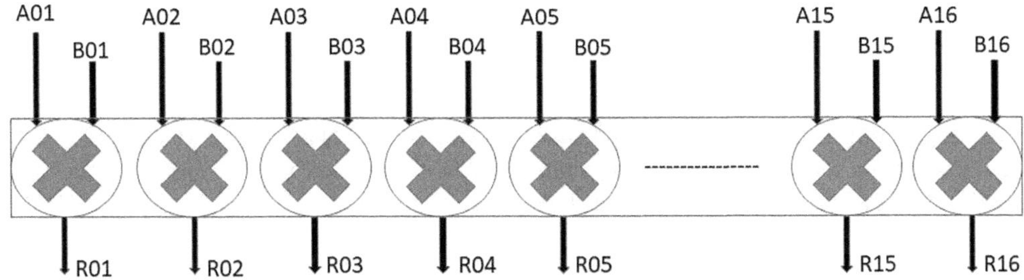

FIGURE 11.4 An accelerator comprising 16 similar approx. multipliers.

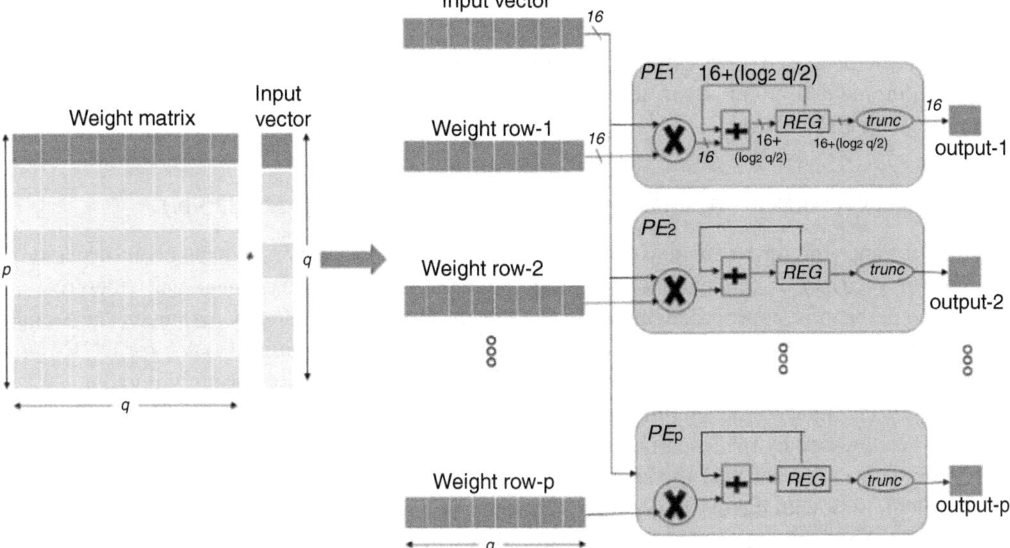

FIGURE 11.5 Proposed accelerator architecture for the matrix multiplication operation.

11.3.4 Approx. Multiplier and HWA for Mobile Computing of DL Applications

The complexity and depth of a DNN rise exponentially as the number of perceptrons increases, thereby requiring a more significant number of arithmetic operations to train the network. Moreover, massive computing power and significant memory are also essential for saving the network settings, which include weights and corresponding losses. Research [5] offers an approximate radix-4 Booth multiplier and HWA for deploying DL applications on power-constrained mobile or edge computing devices. The workloads are accelerated by the proposed accelerator using approximate multiplier-based parallel processing components.

Multiplications consume 96% of the power used by MACs in a standard DNN, so reducing the power multipliers' use can considerably improve the HWA's energy efficiency. As seen in Figure 11.5, the register adder, multiplier and truncation block are collectively called PE. The proposed accelerator mimics a large number of these PEs on the hardware built on the resources available on the examined FPGA fabric.

Findings. The total time needed for MVM operation was cut by more than 85% for the weights of sizes [64 64] and [256 256], compared to the traditional architecture. The total number of clock cycles needed for MMM operation was decreased by 70%.

11.3.5 An Area-Efficient Consolidated Configurable Error Correction for Approximate Hardware Accelerators

Researchers presented effective Error Detection and Correction (EDC) implementations for approximative hardware accelerators in the cited study [6]. It was shown how to rectify errors that have accumulated from several additions using consolidated error correction (CEC) for cascaded adders. Only an AND gate is needed for error detection.

Error Correction for Approximate HWA. The majority of real-world applications use a network of adders as hardware accelerators to provide more intricate arithmetic data routes. A Consolidated Error Correction (CEC) circuit can be created by combining two approximate adders instead of separately for each one through simple changes to the proposed EC circuit. The circuit's area-on-chip is thereby decreased.

Findings. The proposed architecture was demonstrated as efficient in terms of the clock cycle and space requirement in comparison to the latest available designs (space requirement reduced by around 30% in CEC-8).

11.3.6 Fault-Tolerant Accelerators for Deep Neural Networks (DNN)

The performance of conventional Machine Learning (ML) methods has perhaps been surpassed by DNNs. For a variety of applications, including language translation and image and video recognition, they are the most appropriate option. These DNNs, however, have many parameters and require a lot of processing power to train and test. Therefore, designing specialized HWA for DNN implementation to enhance power and performance efficiency is becoming increasingly popular.

In [7], the challenge of developing error-tolerant HWA for DNN execution in high fault rate technologies is addressed by the researchers. The study put forth Fault- Aware Pruning (FAP), a method that demonstrates that a sizable percentage of a DNN's nodes can be trimmed without any effect on results. Beginning with the discovery that each weight in the DNN translates to precisely one MAC unit, as described in the section on "DNN Acceleration on TPU," comes the suggested fault-tolerant TPU design.

Findings. The novel approaches enable TPUs to function even with errors as high as 50% while experiencing a tolerable reduction in the accuracy of results/map (ranging from 0.1% for TIMIT to 1.7% for AlexNet). Both methods have no runtime performance overhead.

11.3.7 Multi-Level Approximate Accelerator Synthesis under Voltage Island Constraints

The circuit's latency and the quantity of critical paths are reduced via logical and algorithmic approximations [8]. Due to the fact that fewer routes are impacted by the voltage fall, VOS further reduces energy usage at the cost of a minor increase in error. The approximate accelerator synthesis idea proposed in [9] is built upon a collection of approximate arithmetic components. The approximation is applied in different layers. Figure 11.6 shows the proposed architecture for approx. accelerator synthesis. The proposed design of multi-layer approx. the accelerator is depicted in Figure 11.6.

Findings. The solutions offered by this architecture outperform single-level approaches for available benchmark and error bound. For instance, 5% error bound. For the Sobel, matrix multiplication, and DCT benchmarks, respectively, the proposed designs consume 61%, 40%, and 49% less power than the actual (non-approximate) designs when the island restriction is not followed.

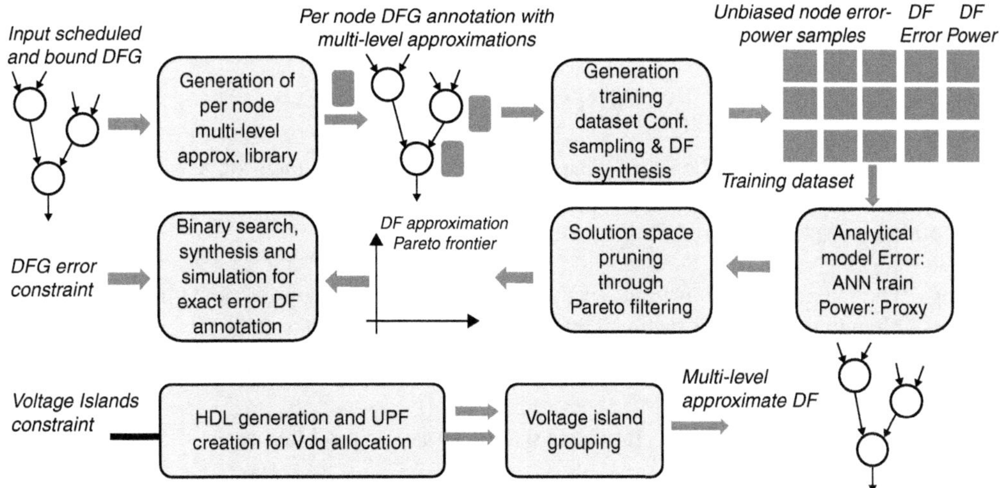

FIGURE 11.6 The presented architecture for multi-level approx. accelerators.

11.3.8 Complexity Reduction of Fault-Tolerant System via Approximate Computing

The disciplines of DSP and image processing are where many applications utilizing hardware accelerators originate. These programs have a rare capacity to tolerate specific output error levels. [10] suggests an automated process that will provide approx. Duplication With Compare (DWC) redundancy systems that are assured of producing accuracy within a specified maximum limits of error.

Findings. The proposed methodology was 28 times faster, on average than the exhaustive search in terms of required time, with the time increasing with the increase in FUs in the design.

11.3.9 Improving Power of DSP and CNN Hardware Accelerators Using Approximate Floating-Point Multipliers

Floating point (FP) arithmetic offers a more excellent range of values and higher accuracy for similar word length with a simplified programming approach, but at the same time, suffers from added hardware expense [11]. The work [12] discussed how the rising expense of floating-point multiplicators' hardware renders them unsuitable for embedded computing. An area/power efficient multiplier family called AFMU (Approximate Floating-point Multiplier) was introduced by researchers that employ two approximation methods in the resource-demanding mantissa multiplication and may be easily expanded to accommodate the setting of the approximation levels through gating signals.

Findings. According to the evaluation, AFMU produces energy gains between 3.6% and 53.5% for half-precision and 37.2% to 82% for single-precision, for mean relative errors of 0.5% to 3.33% and 0.1% to 2.20, respectively. This level of accuracy is accomplished while offering around 60–67% area and 57-6w% gains in power.

11.3.10 Approx. the Sum of Absolute Differences HW for FPGAs

A standard metric for block matching in many fault-tolerant systems is the sum of absolute differences (SAD) operation. In these applications, including stereo vision and video coding, the

SAD operation is a commonly employed metric for block matching [13]; also, it is the operation that uses the greatest time and energy (1).

$$S = \{(-1)^{Xn} X\} + \{(-1)^{Yn} Y - Yn\} \tag{11.1}$$

Where Xn and Yn can either be 0 or 1, keeping $Xn = 1$, 2's complement of X is calculated, while keeping $Yn = 1$, 1's complement of Y is calculated. The proposed adder/subtractor calculates the values of the two subtraction results for further addition.

Findings. The maximum and average errors of the demonstrated approx. SAD hardware is lower than those already described in the available literature. It utilizes up to 20% fewer LUTs and up to 38% less energy than the smallest approximation SAD.

11.3.11 MACISH: MAC Accelerators with Internal-Self-Healing

In comparison to conventional approximate computation approaches that limit errors, self-healing algorithms have shown to be a good compromise between quality and efficiency. Internal Self-Healing (ISH) is a ground-breaking methodology that researchers suggest in [14]. By enabling self-repairing within a system internally without needing a paired/parallel module, ISH opens the use of uneven data channels. For creating an approximative multiply-accumulate (xMAC), this ISH mechanism is used. According to Figure 11.7 below, the multiplier is thought of as an approx. stage and the accumulator as a repairing stage.

Findings. The presented ISH framework results in 55% better output for an iMac.

11.3.12 ENAP: An Efficient Number-Aware Pruning Technique of Approximate Configurations

[15] suggested using an effective number-aware pruning (ENAP) technique to choose the right configuration from several approximation units with distinct error characteristics in order to reduce resources while maintaining user-defined error tolerance.

Findings. Though not very significant, a compression rate of 0.0008% of search space, i.e., pruning, was demonstrated.

11.3.13 Approximate Computing Using Code Design in DNNs

In addition, coordinated software and hardware designs have been presented to produce effective, high-performing, and dedicated outputs using approximations. [16] proposed to integrate

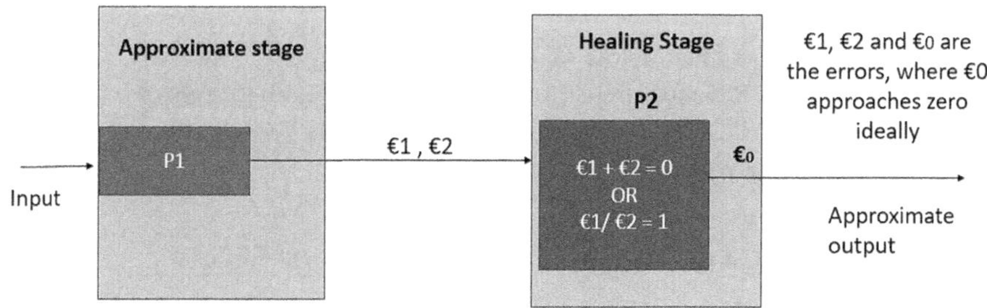

FIGURE 11.7 Proposed ISH computing methodology.

approximate circuits with DNN algorithms using the incremental network approximation (INA) technique with acceptable loss of accuracy. DNNs are encouraged to be fault tolerant by INA, which leads to greater trade-offs between accuracy and setup costs.

Findings. The approximate inference models that INA retrained caused reduction of up to 80% at different hardware design levels, while mere decrease of 2% accuracy.

11.3.14 Increase in Approximation Requirement for DL Architectures

As elaborated in the reviewed works [15,16], it becomes extremely challenging to de- ploy deep CNNs on resource-constrained edge devices with computational restraints. So, present researches have focused on approximating existing DL models, as much as possible. These approximations have reduced the FLOPS (floating-point operations per second) of existing DL architectures like VGG-16 and Resnet-50 by 77% and 47% respectively; with only a drop of 2% in accuracy [17]. Further researches in this domain have rendered significant and encouraging results.

11.4 COMPARISON SUMMARY

TABLE 11.1
Comparison Summary of Research Reviewed

Ref	Description of Research	Year	Results
[18]	Approximate Reconfigurable Hardware Accelerator (HWA)	2017	The average error was reduced from 54% to 10%
[19]	Approximate Accelerators Com- piler-Driven	2018	Accurate results and Monte Carlo simulations for image data
[4]	Approximate HWA–based on ML and DPR	2021	The accuracy of obtained results was 81.82%
[5]	Approx. multiplier and HW for edge computing of DL applications	2021	Timing of MVM & MMM re- duced by 88% & 70%, respectively.
[6]	Configurable Error Correction for Approx. HWA	2016	Space reduction of around 30% was shown
[7]	Fault-Tolerant Accelerators for DNN Execution	2019	Enables TPU operation even with error of 50% (AlexNet used)
[9]	Multi-Level Approx. Accelerator Synthesis Under Voltage Island Constraints	2019	Less power consumption of 61%, 40%, and 49%
[10]	Reducing the Complexity of Fault-Tolerant Systems Amenable to Approx. Computing	2021	In terms of speed, the proposed method was 28 times faster
[12]	Improving Power of DSP and CNN HW Using Approx. Floating- point Multipliers	2021	Power reduction of 57.4%, area up to 60.4% and power gains of 57.4% attained.
[13]	Approx. The sum of Absolute Differences HW for FPGAs	2021	A reduction of 38% in power us- age was demonstrated
[14]	Approx. MAC Accelerators de- sign with Internal Self-Healing	2019	Around 55% improved output for equivalent-efficiency designs
[15]	Pruning Methodology for Design and Space Exploration	2023	A compression rate of 0.0008% was achieved
[16]	Approx. computing in DNN us- ing code design	2019	Up to 80% reduction in hardware design level achieved

11.5 FUTURE CHALLENGES

For future researches, various Deep Learning algorithms need to be examined for sound output and lower implementation costs. Approximate CNNs for already existing models like AlexNet, VGG-16 and Incetion-V3 may further be pruned and validated by layer-wise pruning of approximable neurons [20], similarly an extensive error-immune analysis of approximate CNNs [21] must also be looked into. Further, research can be focused on its applications like lightweight and low-cost aerial imaging solutions for small drones [22].

11.6 CONCLUSION

Design approx. implemented at the hardware level, such as approximate multipliers, is a feasible way to improve the power efficiency of FPGA-based fault-tolerant systems based on the studied proposed models. However, it calls for caution because, when applied to static approximate systems, a few approximate computing techniques can result in significant output inaccuracies, particularly for non-static inputs. In the re- viewed research [3–17], different optimizations for data gathering, transporting, and storing have been proposed to enhance the data performance of systems. [19] and [4] provide accurate estimation results with regards to specific experimental set-ups; [6] provide significant area overhead reduction; [9], [12], and [13] provide noteworthy power requirement reduction; while [10] through the concept of asymmetric module redundancy significantly simplifies the resultant circuits. A novel design [7] enables TPUs to function even with errors around 50%, with minimum drops in results accuracy/ mAP. Leveraging the efficiency of multi-level approximation [9] produces high-speed solutions that outperform single-level state-of-the-art approximate techniques.

REFERENCES

[1] G. Rodrigues, F. Lima Kastensmidt, and A. Bosio, 'Survey on Approximate Computing and Its Intrinsic Fault Tolerance', *Electronics*, vol. 9, no. 4, p. 557, Mar. 2020, doi: 10.3390/electronics9040557.

[2] Q. Xu, T. Mytkowicz, and N. S. Kim, 'Approximate Computing: A Survey', *IEEE Des. Test*, vol. 33, no. 1, pp. 8–22, Feb. 2016, doi: 10.1109/MDAT.2015.2505723.

[3] W. Liu, C. Gu, M. O'Neill, G. Qu, P. Montuschi, and F. Lombardi, 'Security in Approx- imate Computing and Approximate Computing for Security: Challenges and Opportuni- ties', *Proc. IEEE*, vol. 108, no. 12, pp. 2214–2231, Dec. 2020, doi: 10.1109/JPROC.2020.3030121.

[4] M. Masadeh, Y. Elderhalli, O. Hasan, and S. Tahar, 'A Quality-assured Approximate Hardware Accelerators–based on Machine Learning and Dynamic Partial Reconfigura- tion', *ACM J. Emerg. Technol. Comput. Syst.*, vol. 17, no. 4, pp. 1–19, Oct. 2021, doi: 10.1145/3462329.

[5] K. Manikantta Reddy, M. H. Vasantha, Y. B. Nithin Kumar, C. K. Gopal, and D. Dwivedi, 'Quantization Aware Approximate Multiplier and Hardware Accelerator for Edge Computing of Deep Learning Applications', *Integration*, vol. 81, pp. 268–279, Nov. 2021, doi: 10.1016/j.vlsi.2021.08.001.

[6] S. Mazahir, O. Hasan, R. Hafiz, M. Shafique, and J. Henkel, 'An Area-Efficient Consoli- dated Configurable Error Correction for Approximate Hardware Accelerators', in *Proceed- ings of the 53rd Annual Design Automation Conference*, Austin Texas: ACM, Jun. 2016, pp. 1–6. doi: 10.1145/2897937.2897981.

[7] J. J. Zhang, K. Basu, and S. Garg, 'Fault-Tolerant Systolic Array Based Accelerators for Deep Neural Network Execution', *IEEE Des. Test*, vol. 36, no. 5, pp. 44–53, Oct. 2019, doi: 10.1109/MDAT.2019.2915656.

[8] S. Lee, L. K. John, and A. Gerstlauer, 'High-Level Synthesis of Approximate Hardware Under Joint Precision and Voltage Scaling', in *Design, Automation & Test in Europe Con- ference & Exhibition (DATE), 2017*, Lausanne, Switzerland: IEEE, Mar. 2017, pp. 187–192. doi: 10.23919/DATE.2017.7926980.

[9] G. Zervakis, S. Xydis, D. Soudris, and K. Pekmestzi, 'Multi-Level Approximate Acceler- ator Synthesis Under Voltage Island Constraints', *IEEE Trans. Circuits Syst. II Express Briefs*, vol. 66, no. 4, pp. 607–611, Apr. 2019, doi: 10.1109/TCSII.2018.2869025.

[10] Z. Zhu and B. C. Schafer, 'Reducing the Complexity of Fault-Tolerant System Amenable to Approximate Computing', in *2021 IEEE International Symposium on Circuits and Sys- tems (ISCAS)*, Daegu, Korea: IEEE, May 2021, pp. 1–5. doi: 10.1109/ISCAS51556.2021.9401605.

[11] P. Yin, C. Wang, W. Liu, E. E. Swartzlander, and F. Lombardi, 'Designs of Approximate Floating-Point Multipliers with Variable Accuracy for Error-Tolerant Applications', *J. Signal Process. Syst.*, vol. 90, no. 4, pp. 641–654, Apr. 2018, doi: 10.1007/s11265-017-1280-4.

[12] V. Leon, T. Paparouni, E. Petrongonas, D. Soudris, and K. Pekmestzi, 'Improving Power of DSP and CNN Hardware Accelerators Using Approximate Floating-point Multipliers', *ACM Trans. Embed. Comput. Syst.*, vol. 20, no. 5, pp. 1–21, Sep. 2021, doi: 10.1145/3448980.

[13] W. Ahmad and I. Hamzaoglu, 'An Efficient Approximate Sum of Absolute Differences Hardware for FPGAs', in *2021 IEEE International Conference on Consumer Electronics (ICCE)*, Las Vegas, NV, USA: IEEE, Jan. 2021, pp. 1–5. doi: 10.1109/ICCE50685.2021.9427756.

[14] G. A. Gillani, M. A. Hanif, B. Verstoep, S. H. Gerez, M. Shafique, and A. B. J. Kokkeler, 'MACISH: Designing Approximate MAC Accelerators With Internal-Self-Healing', *IEEE Access*, vol. 7, pp. 77142–77160, 2019, doi: 10.1109/ACCESS.2019.2920335.

[15] Y. Dou, C. Wang, R. Woods, and W. Liu, 'ENAP: An Efficient Number-Aware Pruning Framework for Design Space Exploration of Approximate Configurations', *IEEE Trans. Circuits Syst. Regul. Pap.*, pp. 1–12, 2023, doi: 10.1109/TCSI.2023.3252483.

[16] Z. Liu, K. Jia, W. Liu, Q. Wei, F. Qiao, and H. Yang, 'INA: Incremental Network Ap- proximation Algorithm for Limited Precision Deep Neural Networks', (2019), IEEE/ACM International Conference on Computer-Aided Design (ICCAD), (pp. 1-7)

[17] M. Courbariaux, I. Hubara, D. Soudry, R. El-Yaniv, and Y. Bengio, 'Binarized Neural Networks: Training Deep Neural Networks with Weights and Activations Constrained to +1 or -1'. arXiv, Mar. 17, 2016. Accessed: May 09, 2023. [Online]. Available: http://arxiv.org/abs/1602.02830

[18] S. Xu and B. C. Schafer, 'Approximate Reconfigurable Hardware Accelerator: Adapting the Micro-Architecture to Dynamic Workloads', in *2017 IEEE International Conference on Computer Design (ICCD)*, Boston, MA: IEEE, Nov. 2017, pp. 113–120. doi: 10.1109/ICCD.2017.25.

[19] J. Castro-Godinez, S. Esser, M. Shafique, S. Pagani, and J. Henkel, 'Compiler-Driven Error Analysis for Designing Approximate Accelerators', in *2018 Design, Automation & Test in Europe Conference & Exhibition (DATE)*, Dresden, Germany: IEEE, Mar. 2018, pp. 1027–1032. doi: 10.23919/DATE.2018.8342163.

[20] A. Gebregirogis and M. Tahoori, 'Approximate Learning and Fault-Tolerant Mapping for Energy-Efficient Neuromorphic Systems', *ACM Trans. Des. Autom. Electron. Syst.*, vol. 26, no. 3, pp. 1–23, May 2021, doi: 10.1145/3436491.

[21] A. Siddique and K. A. Hoque, 'Exposing Reliability Degradation and Mitigation in Ap- proximate DNNs under Permanent Faults', *IEEE Trans. Very Large Scale Integr. VLSI Syst.*, vol. 31, no. 4, pp. 555–566, Apr. 2023, doi: 10.1109/TVLSI.2023.3238907.

[22] T. Nomani, M. Mohsin, Z. Pervaiz, and M. Shafique, 'xUAVs: Towards Efficient Ap- proximate Computing for UAVs—Low Power Approximate Adders With Single LUT Delay for FPGA-Based Aerial Imaging Optimization', *IEEE Access*, vol. 8, pp. 102982–102996, 2020, doi: 10.1109/ACCESS.2020.2998957.

12 Virtual Memory Management Techniques

Hajira Noor, Hamayun Khan, Irfan Ud Din, Muhammad Imran Tarq, Muhammad Nabeel Amin, and Mehak Fatima

12.1 INTRODUCTION

By transferring information from random access to disc storage, the OS can employ the use of hardware and software to compensate for a lack of physical resources. Virtual memory is a technique that enables a computer to access memory beyond what is actually installed on the system. It does this by using a section of the hard disk to simulate RAM, allowing programs to run even if they require more memory than what is available in physical memory. [1] This research paper provides an overview of the various virtual memory management techniques used by modern operating systems. The paper begins by explaining the concept of virtual memory and its importance in modern computing systems. It then discusses the basic methods of mapping virtual memory to physical memory, namely paging and segmentation, and their advantages and disadvantages. The paper then examines the various virtual memory management techniques, including demand paging, page replacement, working set, and their implementation in modern operating systems. It also discusses the challenges associated with virtual memory management, such as the need to balance the competing demands of memory usage and memory allocation. The goal of this research study is to present an in-depth analysis of virtual memory management techniques and their importance in modern operating systems. It is a valuable resource for computer science students and researchers who seek to understand the intricacies of memory management in operating systems.

12.2 LITERATURE REVIEW

Virtual memory is a technique that enables a computer to access memory beyond what is actually install on the system. It does this by using a section of the hard disk to simulate RAM, allowing programs to run even if they require more memory than what is available in physical memory. [7]

There are two main benefits of virtual memory. The first one is to extend the use of physical memory by allowing data to be swapped in and out of disk storage as needed, which can help prevent programs from running out of memory. [8] The second is that it provides memory protection by mapping virtual addresses to physical addresses, which helps prevent programs from accessing or modifying memory that they should not have access to. [9]

Overall, virtual memory is an important technique that helps improve system performance and stability, allowing computers to start bigger programs and several programmes to run simultaneously without running out of memory or experiencing crashes due to memory-related issues. [11]

There are several situations where it may not be necessary to load an entire program into main memory. Here are some examples:

DOI: 10.1201/9781003497851-12

TABLE 12.1
Comparison B/W New and Previous Research in Virtual Memory Management Techniques

Previous Research Techniques	Newly Proposed Techniques		References
Demand Paging	Machine learning-based page prediction and proactive prefetching	fault page	[2]
Page Replacement Algorithms (e.g. LRU, LFU)	Segmented LRU, Multi-Level Replacement	Page	[3]
Page Fault Handling	Hardware-assisted page management, improved page management.	error error	[4]
Working Set Models	Hybrid Memory Systems (combining traditional memory and emerging memory technologies like PCM and mersisters)		[5]
Virtual Memory Management in Cloud Environments	Virtual Machine Migration for Balancing and Latency Reduction	Load	[6]

TABLE 12.2
Comparison of Virtual Memory Management Techniques

Techniques	Pros	Cons	References
Paging	Non-contiguous allocation is supported memory is used effectively, and pages may be swapped out easily.	Fragmentation, maintenance costs associated with page tables, and an increase in memory use for the page table.	[2]
Segmentation	Allows sharing of code and data segments, secures address space against unauthorized access, and permits non-contiguous allocation.	Finding free segments is necessary due to external fragmentation, complex implementation, and dynamic allocation.	[10]
Virtual Memory	Supports efficient memory use, a broader address space than physical memory, and supports demand paging.	High overhead from page swapping and address translation, longer access times from disc I/O, and the possibility of thrashing.	[10]

When a program is too large to fit into physical memory: Programs that require more memory than what is available in physical memory may not be able to load entirely into main memory. In this case, virtual memory can use to swap portions of the program in and out of memory as needed, allowing the program to run without running out of memory. When a program uses dynamic memory allocation, Programs that use dynamic memory allocation (e.g., with pointers) may not require all of their memory to be loaded into main memory at once. Instead, the program may allocate memory as needed during runtime, and virtual memory can be used to swap portions of the program in and out of memory as needed. [12]

When a program uses shared libraries: Programs that use shared libraries (i.e. libraries that are loaded into memory and shared among multiple programs) may not require all of their code to be loaded into memory at once. Instead, the program may only need to load the portions of code that it actually uses, and virtual memory can use to swap unused portions of the shared library out of memory.

When a program uses paging or demand loading: Some operating systems use techniques like paging or demand loading to load portions of a program into memory only when needed. This can help reduce the amount of memory needed for a program to run and improve system performance [13].

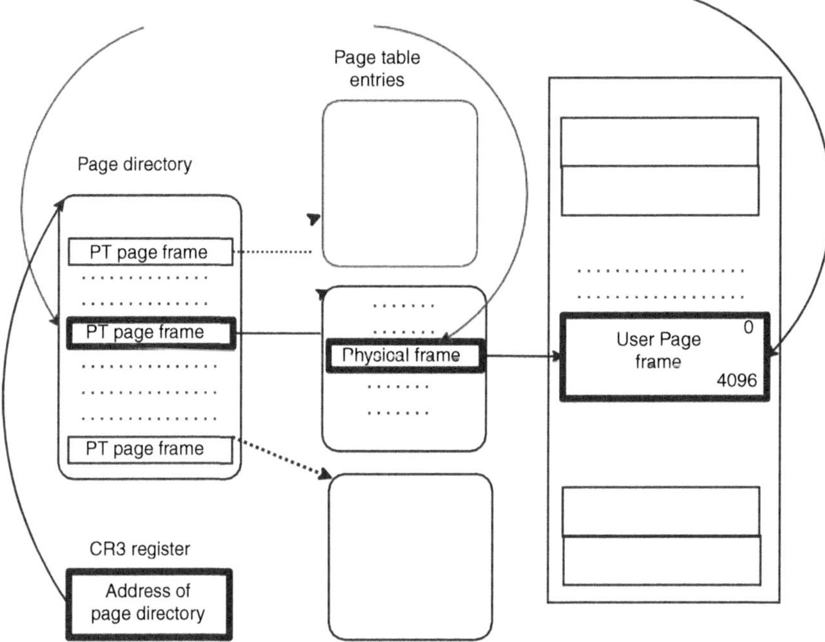

FIGURE 12.1 An illustration of MMU-based address translation using 4 KB pages in 32-bit x86 paradigm [11].

In all of these situations, virtual memory can be used to swap portions of a program in and out of memory as needed, allowing the program to run efficiently even if it cannot be loaded entirely into physical memory at once. [14] MMU plays a crucial role in implementing virtual memory by managing the equivalent of virtual addresses to physical addresses, facilitating efficient CPU usage by methods such as demand paging and segmentation, and allowing applications to use more memory than what is actually stored on the system [15] (Figure 12.1).

12.2.1 Paging

A procedure virtual address space is split up into fixed-size pages using the virtual memory management strategy known as paging that is mapped into physical memory through a process called page allocation. [16] Every page is a fixed-size block of memory that may be moved in and out of physical memory as necessary. Paging allows efficient use of physical memory by swapping out pages that a process is not currently using from physical memory and swapping in the pages that a process needs. [17]

12.2.2 Demand Paging

Pages for a process are not put into memory in a demand paging system until they are required or referenced. [18] This is in contrast to a pre-paging system where all the pages of a process are loaded into memory in advance. When a process is loaded into memory for the first time, only the first few pages of the process are loaded. As the process executes and references new pages, these pages are loaded from the disk into the physical memory as needed Calling this "demand paging". [19] When a system uses demand paging, the "OS" maintains track of the pages that are currently in memory and those that are not. When the CPU tries, a page fault occurs when you attempt to access a page that is not present in memory. The operating system then retrieves puts the necessary page into memory from the disc. Once the page is in memory, the processor can access it as usual. [20]

When a context switch occurs, the operating system is not required to transfer any pages from the new application or any pages from the old program into the main memory to disc. Instead, it only initiates the execution of the new program by putting the first page into the memory and retrieving additional pages as needed. [5] Demand paging has several advantages, including the ability to run larger programs and more programs simultaneously. [11] However, it also has some disadvantages, such as increased overhead due to page faults and slower performance when accessing pages that are not in memory. [17] Here is a pseudocode for the demand paging technique:

```
page_table page_fault page_num page_offset physical_address
frame_number page_table_entry memory
  while (true) {
   virtual_address = get_virtual_address();
   page_num = extract_page_number(virtual_address); page_
   offset = extract_page_offset(virtual_address); page_table_
   entry = page_table[page_num];
   if (page_table_entry.valid_bit == true) {frame_
   number = page_table_entry.frame_number;
   physical_address = (frame_number * PAGE_SIZE) + page_offset;
  } else {
    page_fault = true;

    frame_number = find_free_frame(memory); if (frame_number =
    = -1) {
    frame_number = page_replacement_algorithm(memory, page_table);
    }

   load_page_from_disk(page_num, memory[frame_number]);

   page_table_entry.valid_bit = true;
   page_table_entry.frame_number = frame_number;

   physical_address = (frame_number * PAGE_SIZE) + page_offset;
 }
 access_memory(physical_address);
}
```

12.2.3 Segmentation

An application's process' virtual address space is divided via segmentation as represented in Figure 12.2, a virtual memory management approach into variable-sized segments that are mapped into physical memory through a process called segment allocation. Each segment represents a logical division of the program, [22] such as: B. Code, data, and stack segments. Segmentation uses memory more efficiently than paging because it allows the operating system to allocate only the amount of memory it needs to each segment [20] (Figure 12.3).

12.2.4 Page Replacement Algorithm

If the desired number of pages cannot be stored in available memory, a page replacement algorithm is used to choose which pages to erase from memory. Page replacement algorithms are essential for improving the efficiency of virtual memory management. [3]

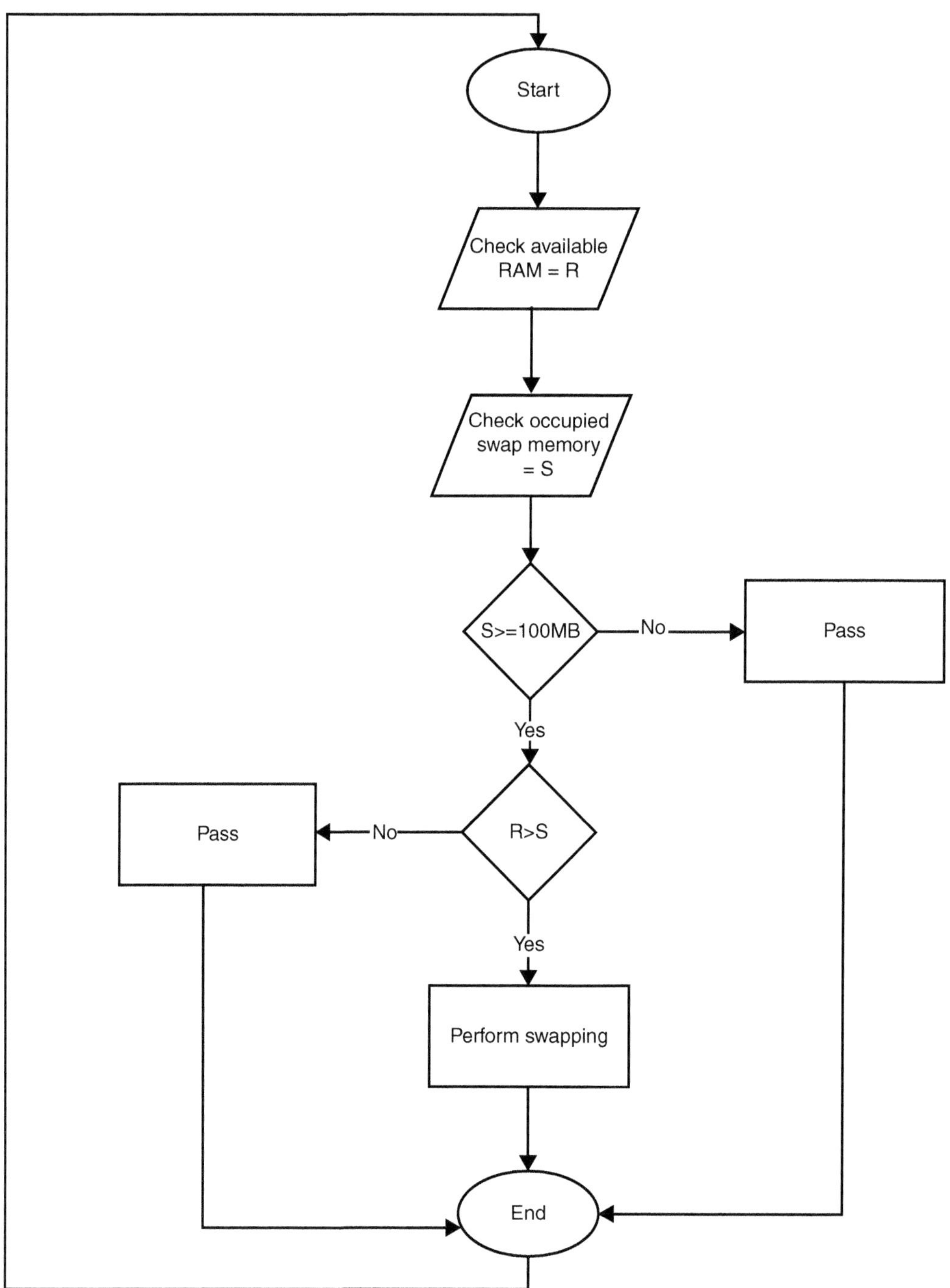

FIGURE 12.2 Flow-chart diagram [21].

12.2.4.1 Maximum Useful Life (LRU)

The LRU algorithm is a page replacement method built on the principle that the pages that have been used the longest are the ones that are least likely to be used again soon. Therefore, the operating system selects the pages that haven't been utilized in a while and deletes them from memory when a page fault occurs [8] (Figure 12.4).

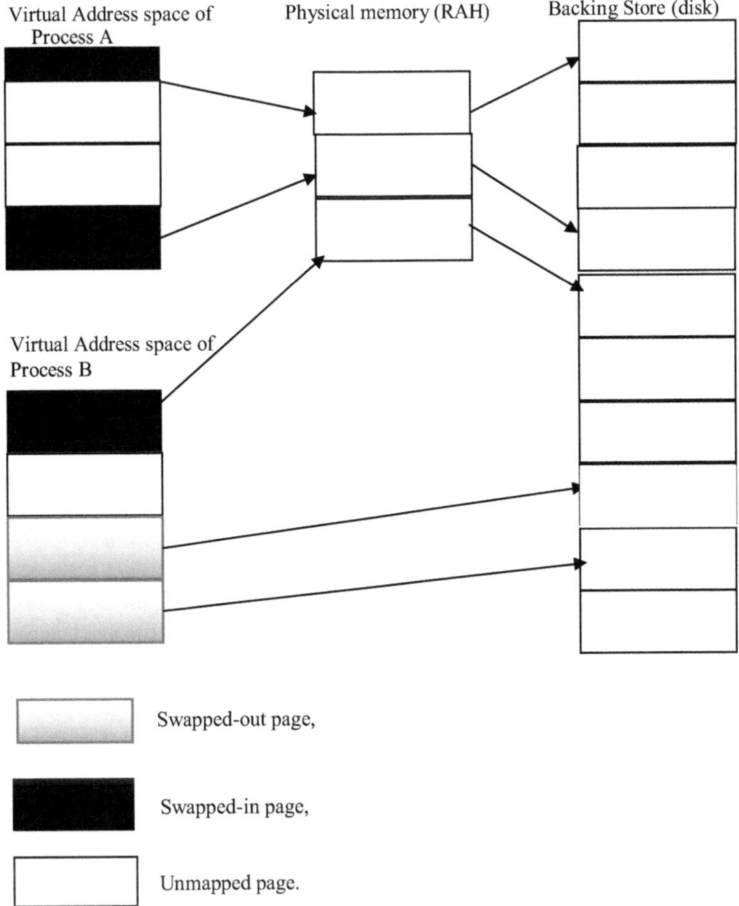

FIGURE 12.3 Simulation of storage space.

12.2.4.2 First In First Out (FIFO)

The FIFO algorithm is a PR algorithm based it on the principle that the pages loaded into memory first are removed first. Therefore, the operating system selects the oldest page in memory and deletes it when a page fault occurs [12] (Figure 12.5).

12.2.4.3 Optimal Page Replacement Algorithm

The most effective page-replacement method is a theoretical one based on the principle of removing the least-used pages. This algorithm requires knowledge of future memory references, making it impractical in real systems. [9] Segmentation is a virtual memory management technique that divides a program into logical segments such as code, data, and stack segments. Each segment is assigned a separate memory address space, and page tables map the logical address space to the physical address space. [23]

Segmentation allows programs to use more memory than is physically available by dividing it into smaller logical segments. This technique also provides better memory protection by ensuring that each segment is assign a separate memory address space. **Working set:**

The working set is a virtual memory management technique used to determine which pages are kept in physical memory. The collection of pages that a process is now using is known as a working

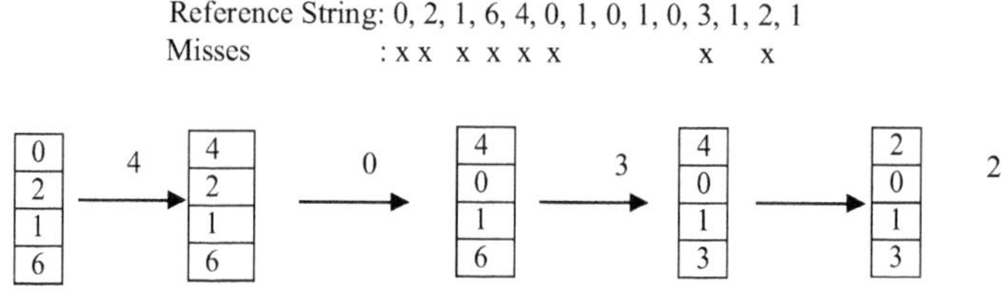

Fault Rate = 8/12 = 0.67

FIGURE 12.4 LRU process.

Reference String: 0, 2, 1, 6, 4, 0, 1, 0, 1, 0, 3, 1, 2, 1
Misses :X X X X X X X X

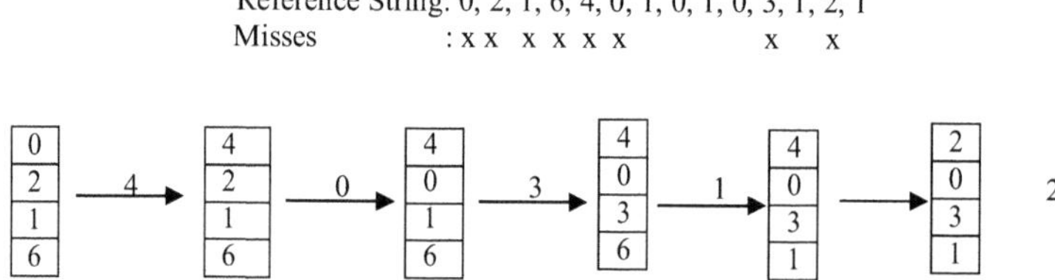

Fault Rate = 9/12 = 0.75

FIGURE 12.5 FIFO process.

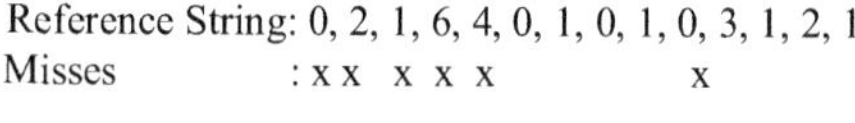

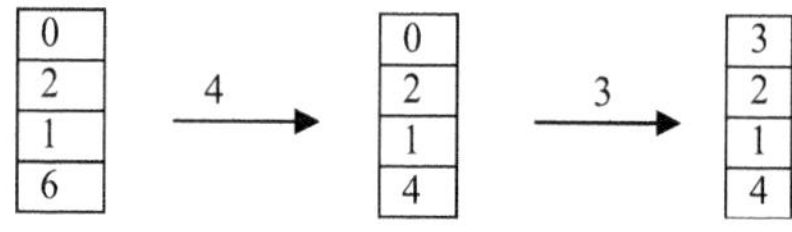

Fault Rate = 6/12 = 0.50

FIGURE 12.6 OPR algorithm.

set. The operating system makes decisions about which pages to stay in physical memory and which to switch to disc using the working set [12] (Figure 12.6).

12.2.5 Sub Paging

Sub paging is the process through which virtual memory is use with flash memory. Due to the restricted size of the flash memory, the real victim page, which will swapped in or out first, is divide into a number of subpages. Additionally, each page is give a dirty bit so that in the event of a page fault, only filthy subpages are stored into flash memory rather than the whole victim page. The idea behind a "dirty bit" is that a recently changed subpage has its dirty bit set in order that the cache or running gadget can use it later. [9]

TABLE 12.3
Techniques for Managing Memory in Virtualized Systems

Optimization	Aspects	Challenges	Solutions
Utilizing different memory demands	Darkness and dynamism (indirect)	Estimation of WSS and control interval	Demand paging via hypervisor, ballooning, and memory hotplug are examples of dynamic allocation enablers.
Delete duplicate content	Shadows, Diversity	Transparency and sharing costs	Page level deduplication using out-of-band scanning (VMWare)
Guest and VMM work together to manage memory.	Darkness, movement, and duality	changes to the guest operating system	Transcendent memory (tmem), which is enabled for distributed applications, allows for effective memory management for disc block caching.

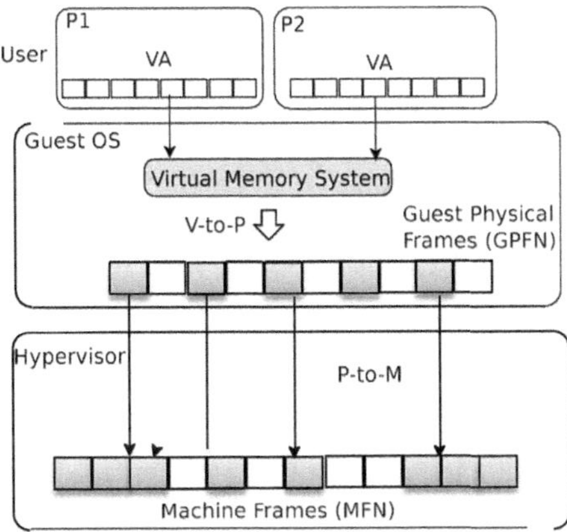

FIGURE 12.7 Dynamic pseudo (guest) physical pages to machine pages mapping with variable page- level memory allocation in virtualized systems.

12.3 METHODOLOGY

Although the DVM approach is used to boost system throughput, there are certain disadvantages, including high traffic density and other issues that will be covered later. By segmenting the system into clusters, each of which has its own cache (memory), our technology adds a second level of memory known as the cluster cache. We shall outline the criteria step-by-step based on improvements to earlier methods mentioned in the references. [15] Figure 12.7 displays factors that affect memory management in systems in virtualization and demonstrates our methodology while contrasting it with CVM and DVM methods. When a CVM process requires a page during execution, it does it right away, pulling it from the cache memory. When a cache miss occurs, it will then ask for the pertinent page from physical memory. [23] Since memory and the second storage device are involved, pages in CVM are exchanges. [14] In our scenario, the procedure will request the page through the 1, 2, 3, and 5 routes.

12.3.1 Page Replacement Algorithm Problem 1

Suppose we have a page replacement algorithm that uses LRU to determine which pages to evict from memory. The memory may be visualized as a graph, with each vertex denoting a page and each edge denoting the order in which the pages were accessed. We can then use the Floyd Warshall

TABLE 12.4
Floyd Warshall Algorithm

	W	X	Y	Z
W	0	1	2	3
X	1	0	1	2
Y	2	1	0	1
Z	3	2	1	0

algorithm the goal is to determine the shortest route between each pair of vertices, which is the bare minimum number of page faults that would occur if those pages were accessed in that order.

By analyzing, the results of the Floyd Warshall algorithm; we can determine how well the page replacement algorithm is performing. If there in number pairs of vertices with a high shortest path length, this indicates that the page replacement algorithm is not performing well and is causing many page faults. If the shortest path lengths are consistently low, this indicates that the page replacement algorithm is performing well and is minimizing the number of page faults.

Let's walk through an example of using the Floyd War shall algorithm to analyze how well a page replacement method performs using a graph.

Suppose we have a memory with four pages, labeled W, X, Y, and Z. The graph representing the memory might look like this:

W -> X -> Y -> Z

According to this graph, page "W" accessed first, then X, Y, and finally Z. Assume we are managing this memory with the LRU page replacement mechanism. To use the Floyd War shall algorithm to analyze the LRU page replacement algorithm's performance, we first need to assign weights to the edges in the graph. In this case, we can assign a weight of one to each edge, since each page access counts as one unit of work. The weighted graph now looks like this:

W -1-> X -1-> Y -1-> Z

Using the Floyd Warshall algorithm; we can compute for each pair of vertices in the graph, take the shortest route. The output matrix for the shortest path lengths is below:

This matrix tells us the minimum number of page faults that would occur if each pair of pages were access in that order. For example, the shortest path between pages W and Y is "2", which means that if we accessed W followed by Y, we would experience two page faults. Similarly, the shortest path between pages X and Z is 2, indicating that accessing X followed by Z would also result in 2 page faults.

By analyzing the matrix of the shortest path lengths, we can see that the LRU page replacement algorithm is performing reasonably well in this scenario. There are no pairs of pages for which the shortest path length is very high, indicating that the algorithm is not causing many page faults. However, if the graph were more complex or if the page access the pattern was distinct, we might see different results and be able to find locations where the page replacement algorithm needs work, use the Floyd Warshall algorithm.

12.3.2 Page Replacement Algorithm Problem 2

Here using the Floyd-Warshall method, determine the minimal number of page faults required to load all pages for a given sequence of page requests.

Here is how it works:

1. Create a 2D array Dist. of the number (m) and size (m + 1) x (n + 1) of pages in memory and the total number of pages is n. in the sequence of page requests. Each element of Dist. will represent the smallest number of page errors required to load the first "i" pages of the sequence into the first "j" memory slots.
2. Initialize the first row and first column of Dist. to be zero, since loading zero pages into zero memory slots requires zero page faults.
3. For every page p in the list of page requests, consider all memory slots m and update Dist. as follows: dist[m][p] = min(dist[m][p-1], dist[m-1][p-1] + (m!= page[p])).
4. After all pages have been handled, min (dist. [m][n] for m in range (1, m+1)) provides the minimal number of page faults needed to load all pages into memory.

Here's the implementation of the algorithm in Python:

```
def page_faults(pages, m):
  n = len(pages)
  dist = [[0] * (n + 1) for _ in range(m + 1)] for i in range(1, n + 1):
    dist[0][i] = i
  for j in range(1, m + 1):
    dist[j][0] = 0
  for i in range(1, n + 1):
    for j in range(1, m + 1):
      dist[j][i]  =  min(dist[j][i-1],  dist[j-1][i-1]  +  (j!=
      pages[i-1]))
return min(dist[m][n] for m in range(1, m+1))
```

Here, "pages" is a list of page requests and "m" is the count of memory slots. The function outputs how few page faults are necessary to load all pages into memory. For illustration, let's think about the following sequence of page requests:

Pages = [1, 2, 3, 4, 1, 2, 5, 1, 2, 3, 4, 5]

If we have 3 memory slots, running "page faults (pages, 3)" will return 9, which means that we need at least 9 page faults to load all pages into memory using a page replacement algorithm with 3 slots for memory.

12.4 CONCLUSION

In modern computer systems, virtual memory management techniques play a critical role in enhancing system performance and providing efficient memory utilization. Virtual memory allows a computer system to use more memory than is available by giving the appearance that there is more memory accessible. To achieve this, a virtual address space that corresponds to actual memory locations is created. Virtual memory management techniques have evolved over the years to meet the requirements of the changing computing landscape. The purpose of this review paper is to explore the various virtual memory management techniques that developed over the years. We have discussed the fundamentals of virtual memory management, including the paging and segmentation techniques and their variations. We have also discussed memory-mapping techniques, including Algorithms for demand paging, copy-on-write, and page replacement. One of the earliest virtual memory management techniques was paging, in which the Virtual address space is split into pages, which are fixed-size blocks that are mapped to specific physical memory locations. Paging enables a system to use more memory than is physically available; the size of the page is critical as it determines the granularity of memory allocation. Paging also enables a system to share memory between processes, leading to efficient use of resources. Another virtual memory management technique is segmentation, which divides the virtual address space into variable-sized segments; each segment represents a logical unit of the program, such as code data or stacks. Segmentation allows for dynamic allocation of memory, enabling a program to grow or shrink as needed; it also facilitates sharing of memory between processes. Demand paging is a memory mapping technique that loads pages into memory only when they are in need; this reduces the amount of memory needed to run a program, enabling a system to run more programs simultaneously. Copy and paste is another memory mapping technique that creates a copy of a page when it altered, this lessens the amount of memory needed to store duplicate pages. Page replacement algorithms used to determine which pages to swap out of memory when the physical memory is full. There are several instances of page replacement algorithms, including FIFO, LRU, and clock. Each algorithm has its own pros and cons, and its choice is totally up to the specific use case. Virtual memory management techniques have their advantages and disadvantages. Paging and grouping into effective at enabling a system to use more memory than is physically available, yet they require additional hardware resources to implement, leading to increased cost and demand. Paging and copy-on-write are effective at reducing memory usage, but they can also lead to increased page faults, and declining efficiency. Page replacement algorithms can influence system performance, with some algorithms performing better than others depending on the workload.

REFERENCES

[1] "Main and Virtual Memory Management," *CodeAhoy*, May 22, 2022. https://codeahoy.com/learn/computersos/ch4/ (accessed Apr. 18, 2023).

[2] "Operating Systems: Three Easy Pieces." https://pages.cs.wisc.edu/~remzi/OSTEP/ (accessed Apr. 19, 2023).

[3] "What are the Page Replacement Algorithms?" https://afteracademy.com/blog/what-are-the-page-replacement-algorithms (accessed Apr. 18, 2023).

[4] A. Ullah, N. M. Nawi, and S. Ouhame, "Recent Advancement in VM Task Allocation System for Cloud Computing: Review from 2015 to2021," *Artif Intell Rev*, vol. 55, no. 3, pp. 2529–2573, Mar. 2022, doi: 10.1007/s10462-021- 10071-7.

[5] R. Gilmiyarov, L. Galimzyanova, and S. Porshnev, "The Concept of Allocating Access Rights to Virtual Memory Area in Linux-based Operating Systems," in *2021 Ural Symposium on Biomedical Engineering, Radioelectronics and Information Technology (USBEREIT)*, May 2021, pp. 0423–0426. doi: 10.1109/USBEREIT51232.2021.9454999.

[6] P. Kumari and A. S. Saxena, "Advanced Fusion ACO Approach for Memory Optimization in Cloud Computing Environment," in *2021 Third International Conference on Intelligent Communication Technologies and Virtual Mobile Networks (ICICV)*, Feb. 2021, pp. 168–172. doi: 10.1109/ICICV50876.2021.9388492.

[7] G. Lee *et al.*, "A Case for Hardware-Based Demand Paging," in *2020 ACM/IEEE 47th Annual International Symposium on Computer Architecture (ISCA)*, May 2020, pp. 1103–1116. doi: 10.1109/ISCA45697.2020.00093.

[8] X. Long, X. Gong, B. Zhang, and H. Zhou, "Deep Learning Based Data Prefetching in CPU-GPU Unified Virtual Memory," *Journal of Parallel and Distributed Computing*, vol. 174, pp. 19–31, Apr. 2023, doi: 10.1016/j.jpdc.2022.12.004.

[9] O. Weisse *et al.*, "Foreshadow-NG: Breaking the Virtual Memory Abstraction with Transient Out-of-Order Execution," Aug. 2018, Accessed: Apr. 18, 2023. [Online]. Available: https://lirias.kuleuven.be/2089352

[10] K. P. Kumar, T. Ragunathan, and D. Vasumathi, "Virtual Machine Consolidation Using Enhanced Crow Search Optimization Algorithm in Cloud Computing Environment," in *Distributed Computing*

and Optimization Techniques, S. Majhi, R. P. de Prado, and C. Dasanapura Nanjundaiah, Eds., in Lecture Notes in Electrical Engineering. Singapore: Springer Nature, 2022, pp. 841–851. doi: 10.1007/978-981-19-2281-7_77.

[11] "A Survey of Memory Management Techniques in Virtualized Systems - ScienceDirect." www.sciencedirect.com/science/article/abs/pii/S1574013716301186 (accessed Apr. 18, 2023).

[12] P. Vogel, A. Marongiu, and L. Benini, "Exploring Shared Virtual Memory for FPGA Accelerators with a Configurable IOMMU," *IEEE Transactions on Computers*, vol. 68, no. 4, pp. 510–525, Apr. 2019, doi: 10.1109/TC.2018.2879080.

[13] "An Energy-Efficient Virtual Memory System with Flash Memory as the Secondary Storage | Request PDF," in *ResearchGate*, Feb. 2023. doi: 10.1109/LPE.2006.4271879.

[14] C. Shao, J. Guo, P. Wang, J. Wang, C. Li, and M. Guo, "Oversubscribing GPU Unified Virtual Memory: Implications and Suggestions," in *Proceedings of the 2022 ACM/SPEC on International Conference on Performance Engineering*, in ICPE '22. New York, NY, USA: Association for Computing Machinery, Apr. 2022, pp. 67–75. doi: 10.1145/3489525.3511691.

[15] A. Oliveri and D. Balzarotti, "In the Land of MMUs: Multiarchitecture OS- Agnostic Virtual Memory Forensics," *ACM Trans. Priv. Secur.*, vol. 25, no. 4, pp. 27:1–27:32, Jul. 2022, doi: 10.1145/3528102.

[16] A. Kundu *et al.*, "Virtual Multi-view Fusion for 3D Semantic Segmentation," in *Computer Vision – ECCV 2020*, A. Vedaldi, H. Bischof, T. Brox, and J.-M. Frahm, Eds., in *Lecture Notes in Computer Science*. Cham: Springer International Publishing, 2020, pp. 518–535. doi: 10.1007/978-3-030-58586- 0_31.

[17] "What is Paging? A Definition from WhatIs.com," *WhatIs.com.* www.techtarget.com/whatis/definition/paging (accessed Apr. 14, 2023).

[18] A. Kurth, P. Vogel, A. Marongiu, and L. Benini, "Scalable and Efficient Virtual Memory Sharing in Heterogeneous SoCs with TLB Prefetching and MMU- Aware DMA Engine," in *2018 IEEE 36th International Conference on Computer Design (ICCD)*, Oct. 2018, pp. 292–300. doi: 10.1109/ICCD.2018.00052.

[19] N. Hajinazar *et al.*, "The Virtual Block Interface: A Flexible Alternative to the Conventional Virtual Memory Framework," in *2020 ACM/IEEE 47th Annual International Symposium on Computer Architecture (ISCA)*, May 2020, pp. 1050–1063. doi: 10.1109/ISCA45697.2020.00089.

[20] J. Van Bulck *et al.*, "Breaking Virtual Memory Protection and the SGX Ecosystem with Foreshadow," *IEEE Micro*, vol. 39, no. 3, pp. 66–74, May 2019, doi: 10.1109/MM.2019.2910104.

[21] S. Sharma and S. A. Sajidha, "Implementing Smart Swapping Algorithm to Boost the Performance of Linux PC's.," *Procedia Computer Science*, vol. 165, pp. 406–414, Jan. 2019, doi: 10.1016/j.procs.2020.01.031.

[22] P. P. Deshmukh and S. Y. Amdani, "Virtual Memory Management using Memory Ballooning in OpenStack Cloud Platform," in *2020 11th International Conference on Computing, Communication and Networking Technologies (ICCCNT)*, Jul. 2020, pp. 1–5. doi: 10.1109/ICCCNT49239.2020.9225435.

[23] "Virtual Memory Palaces: Immersion Aids Recall | SpringerLink." https://link.springer.com/article/10.1007/s10055-018-0346-3 (accessed Apr. 18, 2023).

13 A Comprehensive Survey on Software Defined Networking (SDN) Security

Nouman Mabood, Noshina Tariq, Farrukh Aslam Khan, and Muhammad Ashraf

13.1 INTRODUCTION

The advent of the Internet and computer networks was a defining moment for businesses, schools, and households worldwide. Traditional computer networks, however, have limitations such as difficult-to-reprogram hardware, complex networks, and reliance on unaltered protocols, among others. In addition, the proliferation of networking-capable devices and the Internet of Things (IoT) introduces new challenges to network management. A new network paradigm, Software-Defined Networking (SDN), has emerged in response to these challenges. It makes it possible to surmount the limitations of conventional networks by enabling the software-based centralization of network control. This centralization enables the programmability and abstraction of network control, thereby disintegrating the vertical integration of network planes. Network management responsibilities are divided into three planes: management, control, and data, with management responsibilities falling under the management plane, control responsibilities falling under the control plane, and routing data responsibilities falling under the data plane.

Adopting SDN has generated numerous difficulties, including compatibility with traditional network paradigms, scalability, dependability, and security. The Open Network Foundation, an industry-driven organization, and the Open-Flow Network Research Centre, an academic initiative, promote SDN and standardize its application. Using SDN, network intelligence is administered by an external device called a controller, whose functions are programmed using the Northbound API. The Southbound API (e.g., OpenFlow) facilitates secure communication between the controller and network devices. Notably, industry stalwarts such as Google and Microsoft have adopted SDN, demonstrating the potential of the technology.

While SDN has opened networks to applications and provides numerous benefits, it also confronts significant security risks. The innovative design of its architecture also introduces new security issues and vulnerabilities that must be addressed to assure network reliability, availability, and privacy. In recent years, SDN security has emerged as an active study area, with researchers proposing various security mechanisms to mitigate the threats and attacks that SDN networks encounter. The separation of the SDN architecture's control plane and data plane generates new challenges, such as the potential for malicious actors to exploit control plane vulnerabilities. In addition, attackers can utilize the programmability features of SDN to launch various attacks, such as Distributed Denial of Service (DDoS) and network reconnaissance attacks. As a result of these obstacles, SDN networks are vulnerable to various assaults, making it imperative to develop effective security mechanisms to ensure network dependability and privacy.

Despite the increasing interest in SDN security, this paper seeks to address the main deficiencies by presenting a comprehensive overview, emphasizing its key characteristics, security issues, and potential solutions. This survey aims to provide researchers and practitioners with a road map

DOI: 10.1201/9781003497851-13

for understanding the current state of SDN security research and identifying the research gaps. This survey makes several substantial contributions to the field of SDN security. It begins with an overview of the fundamental characteristics of SDN architecture and its primary components, including centralized flow management and network programmability. Secondly, it identifies the security issues and threats SDN networks face due to the architecture's innovations and exploitable security architecture defects. Thirdly, the paper discusses the countermeasures taken against these threats and attacks, highlighting the importance of a secure SDN implementation to ensure network reliability and privacy. It identifies the unexplored and challenging areas of SDN that require additional research, emphasizing the potential for future work in this area. The significant contributions of this survey are as follows:

1. An overview of the main characteristics of SDN architecture and its primary components is presented.
2. The major security issues and threats encountered by SDN are identified.
3. The countermeasures taken against these threats and attacks are discussed.
4. The unexplored and challenging areas of SDN for future research are highlighted.

The paper's organization is as follows: Section 13.2 briefs SDN and its main components. Section 13.3 covers the vulnerabilities faced by SDN and their categorization. Section 13.4 is about SDN security, and Section 13.5 covers SDN security solutions. Section 13.6 is about open issues and challenges, while Section 13.7 concludes this paper.

13.2 SOFTWARE DEFINED NETWORKING (SDN)

The SDN defines networks differently from legacy networks because SDN splits the network intelligence from the control of the network. The control plane exercises network control. Here, the management of the entire network is done, and centralized control of the network is employed.

Switches are the network elements bound to follow the rules dictated by the network controller and implement the data-forwarding instructions made by the controller. The network is centrally controlled, so the network controller is on top of all the traffic going through it. To determine the definition of the data forwarding behavior, the traffic flows go through the network controller at least once. The best advantage that can be achieved through such centralized control nature of the SDN is the programmability of the whole network. Hence, software applications are made for the different functions intended for the web and can be easily embedded and deployed on the network controller and as separate data consumer functions. The software program defines network programmability, so the integration and data structure algorithms belong to the software development environment. Such features can benefit the network's security, a more complex challenge in conventional and legacy networks and network elements. The software programs embedded in the controller can enforce security features and policies.

On the other side, the SDNs also carry some security problems causing risks and security threats, some of which are known and others are unknown, some addressed and some yet to be addressed, as the interfaces of SDN architecture are still evolving. The main categories of these threats can be identified. We can make the network secure using SDN architecture, but at the same time, we must know all the security requirements. Therefore, these approaches are gaining the research community's attention by implying that until the time security of the SDN is secure and complete, it will not fulfill the requirements of the ISPs and enterprises. It can only be ensured with strict security measures. The very first work was published in 2008 for the security of SDNs. After that, numerous research articles, reports and proposals were published. All this work revolves around devising mechanisms for enhancing the security of SDN; some propose improvements in the existing features while others advocate a new framework for the deployment of the SDN. A lot of research has been conducted in the already addressed domains. This work mainly focuses on the newly published proposals and research papers, which have proposed and suggested innovative approaches to deal with the security issues of the SDN, especially those already used in older or conventional networks.

13.3 THREATS AND VULNERABILITIES IN SDN

With the advent of the SDN, new security-related vulnerabilities have been found, which have never been seen in conventional networks. These vulnerabilities are associated with the unique architecture of the SDN. As mentioned earlier, the network and data forwarding planes are centrally controlled, and if that central control is compromised, the entire network becomes at risk and can be controlled by hostile elements. The logical buildup of this work would be first to present a general rundown of the threat vectors explained earlier and the attacks identified so far in the interfaces and the planes defined in the introduction of this paper [1]. Then in the middle of the article, we move down to the issues identified so far, the effects of these issues, and threat vectors. At the end of this work, we discuss the most frequent attacks and the behavior of SDN architecture.

13.3.1 Threats of Attacks in SDN

The SDN and conventional networks can be subject to threats and misuse, intentionally or unintentionally, even after it has matured over decades of evolution. The protocols used, the network instances, the devices used, or the layers participating in the SDN can be at risk for the reasons mentioned above. It means that every element and part that form the SDN architecture is a threat vector and layer, especially when configuration issues and deployment inconsistencies can emerge as a threat to the SDN.

13.3.2 Security Issues in SDN Architecture

The main objective of the attack defines its category. For example, unauthorized access, disclosure of information, unauthorized modification of network information, destruction of network information, service disruption, misconfigurations, and the poor configuration of authentication, trust, and verification mechanisms can be some of the attacks. Table 13.1 presents the threats and vulnerabilities in an SDN architecture.

TABLE 13.1
Threat and Vulnerabilities of the SDN Architecture

SDN Plane	Reasons for Attack
Management Plane	Misconfiguration of Policies, Vulnerability in Administrative Stations, and Poor Authorization/ Authentication procedures
Application Plane	Malware in Apps, Poor Authorization/ Authentication procedures, Bugs in Apps, and Poor Authentication in Northbound Interface
Northbound Interface	Vulnerability in APIs, Poor protocols, and un-encrypted exchange of data
Control Plane	Poor Authentication in Northbound Interface, Vulnerability in Network Controllers, Poor policy implementation, Insufficient hardware, and Poor Authorization/ Authentication procedures in network devices.
Control Channel	Poor reliability and misconfiguration in protocols and un-encrypted exchange of data
Data Plane	Poor Authentication in the Controller, Forging packets, Poor network implementation, Insufficient hardware, and Vulnerability in network devices

13.3.3 Attacks on the SDN Architecture

Layers and interfaces of the SDN are sensitive to certain kinds of attacks. These attacks can either target or affect the components resident to that specific layer of the element of the other layer depending on the feature. The most common attacks are enlisted and briefly discussed below.

- **Application Layer.** Each application has fixed privileges, authority, and permissions. These can be abused with force by attacks and even terminated by third-party applications that exercise public authority. Particular execution of commands may lead to the disconnections of certain network APIs and critical applications [2].
- **Control layer.** The attackers may find themselves in a position where they have bypassed the firewall. From there on, they will be able to instruct overlap and conflict in the flow rules because the controller may not be able to identify the implicit conflict caused due to the guidelines issued afresh and differentiate them from the current regulations as per specified policies. It is called *Dynamic flow rule tunneling* [3].
- **Control Channel.** The control channel can be attacked by eavesdropping or a Man-in-the-Middle attack [2].
- **Infrastructure layer.** This layer can be compromised through numerous means. *Through ARP poisoning*, an attacker can pose itself as a controller. *Flow-rule modification/ Flushing* in which the information installed on the switches is prone to changes by the attackers. By flooding *Flow-rule and* utilizing side-channel attacks, the controller may be approached to register new flow rules. In *malformed control packet injection,* exceptionally and cleverly crafted packets with malformed or misused headers can change the behavior of the network. *Side-channel attacks* can be used to gain information about the network's behavior in a particular situation and how it responds to different network scenarios [2].

13.4 SDN SECURITY

SDNs have introduced modern and intelligent features like the programmability of the network, mechanism of forwarding the packets, centralized network control, and essential intelligence of the network flow. While, on the one hand, these features have addressed some security issues, the architecture has brought more and new security issues as well. These new issues were never present in the previously used conventional networks. Keeping all the new features, the SDN offers two different implications. Firstly, we can use all these advanced features to detect, analyze, react to them, and then mitigate the security issues that have been observed. It can be conveniently done by introducing and adding new applications or improving the existing ones. Secondly, the new threats perceived may be incorporated into defined networks so that they can proactively alleviate the risks of hazards [4].

13.4.1 Improving Security through SDN

This study suggests the implementation of three SDN features that can bring in many features related to security in the SDN, keeping different attacks in view; the significant attacks and their description can be found in Table 13.2. The following elements may be used to implement security solutions in SDN [5]:

Dynamic Flow Control. The rules for the flow control of data instructions through the network can be enforced through the middle-security boxes. Network applications may be installed directly on the controller, or the application is bound to the controller through northbound API. In this case, there is no need to deploy new and separate hardware, and the security rules can be implemented

TABLE 13.2
Attacks against SDN Architecture

A: Application NB: Northbound C: Control SB: Southbound D: Data

Attack	Source	A	NB	C	SB	D
Abuse of privileges & authority [2]	Vulnerable applications	x	x	x		
Service disruption [7]	Malware	x		x		
Application shutdown [7]	Vulnerability in Northbound APIs	x	x	x		
Dynamic flow rule tunneling [2]	Malware and vulnerability in switches	x		x		x
Poisoned network view [5]	Malware, the vulnerability in network services and protocols	x	x	x	x	x
NOS misuse [2]	Vulnerability in controller	x		x	x	x
Packet in flooding [3]	Faulty controller, compromised switches			x	x	x
Switch table flooding [3]	Faulty controller, compromised switches			x	x	x
Eavesdropping [3]	Un-ciphered control channel			x	x	x
Man in the Middle [1]	Un-ciphered control channel, compromised southbound interface, vulnerable data links			x	x	x
Flow table flooding [1]	Vulnerability in switches			x	x	x
Switch shutdown/ exploitations & forced disconnections	Packet injection, fuzzing techniques, the vulnerability in switches and protocols					x

for packet forwarding. Hence, the dynamic flow control in SDNs can deploy perimeter and internal firewalls, comprehensive access control lists, and traffic re-direction mechanisms. Here, we can also distinguish between normal (authorized) and suspicious (unauthorized) traffic [6].

Network-wide Visibility with Centralized Flow Control. It is the main feature of SDN that makes them different from conventional networks, and it defines the wholesome orientation of the framework from which the SDN is made [8]. It translates the meaning of visibility of the entire network at all times [9].

Network Programmability. The administrators/users do not have the authority to change the devices or applications on the conventional networks but can only be re-configured in some defined and preset manner [10]. On the contrary, the set of rules programmed by the customer to implement specific functionalities, for example, traffic filtering schemes and ACLs, to enforce drop or deny against certain packets. Through Southbound, the data plane can be configured for parameters in the backdrop of a security scenario.

13.4.2 Improving SDN Security

SDN is prone to new attack surfaces due to the changes introduced in the network components, which demands correct sanitization against the threats while leveraging its novel features [11]. Information flows through the channels and interfaces, which are to be cyphered to deny factual information to attackers if they can access the network. Message authentication codes, asymmetric keys, and signed certificates are to be used for authentication besides encryptions [12]. This way, the untrusted and rogue elements would be kept out of the network, and only trusted networks will be allowed. However, special consideration may be given while choosing the protocols and services for encryption mechanism as it may leave the channel open to vulnerabilities [13].

A significant shortcoming of SDN is its inability to detect the issuing of flow rules issued by some other applications. It can only detect conflicts of flow rules issued by itself and other applications against the policy. To avoid this, mediators are used for conflict rules and policy checking [12].

Network state monitoring mechanisms are implemented to record the network state at one point and time for correlation with the subsequent state at another time to find inconsistency in the network state. It mitigates the possibility of attacks directed towards the network state, such as topology poisoning, DoS, malicious flows rule, and infiltration of evil devices [14]. Virtualization at the functional level of the network and adaptability to a cloud-based environment enables the network to capitalize on the security solutions offered by different developers and implement them simultaneously to varying layers of the network [15]. It also gives the added advantage of engaging and releasing any security-based application from any security-based cloud, relieving the network elements from the processing load of the security function.

13.5 SDN SECURITY SOLUTIONS

After covering the introduction of SDN, its security threats and vulnerabilities, security provisions in SDN, and the security aspects it offers, we now classify the security solutions for a better understanding of the current set of SDNs. Only a brief overview is provided for each classification.

13.5.1 Threats Detection

In SDN, the control plane is in charge of the network architecture. It can demand traffic flow patterns and network status for correlation with previous records for analysis by security apps to detect threats. Programmability offers the flexibility to deploy middlebox solutions like Learn2Defend, which capitalize on machine learning techniques for threat detection [15,16]. In the same manner, [17] and [18] offer Network Intrusion Detection System (NIDS), which incorporates signature-based intrusion detection system well-paced in network topology and another machine learning-based system installed in the network controller to safeguard against the threats that are rarely detected. Another solution named Athena provides a framework for developers to build anomaly detection applications.

13.5.2 Network Function Virtualization

Instead of having a single security application, Network Function Virtualization (NFV) and cloud-based security provide an option to have multiple applications and utilize them for the security situation. The NFV hosts the application in network resources (like controller), and the cloud-based security solutions, where the apps are stored in external hardware and made available on demand. It enables the network designers to focus on the network architecture and interfaces while the security experts and vendors can focus on developing security solutions for SDNs.

13.5.3 Remediation of Attack

Since the controller holds the network information, solutions can be built around this information to issue flow rules and other countermeasures to mitigate the attack.

13.5.4 Management of Identity and Access

The sensitive data in an SDN is sent through the control channel. While the data is cyphered, the intruders may find a way into the network through a compromised entity. To avoid such intrusion, Authentication schemes are developed on a trust basis. Fingerprinting devices may be added to the network so that the data is blocked to any network entity that has been added recently until the authentication process is completed. The mechanism of secure sessions is also devised to check the strength of encryption algorithms used and set up expiration time. IEEE 802.1x protocols,

Extensible Authentication Protocols, RADIUS Server Implementation, KISS Framework for device registration, and SM-ONOS for administrative Permission System [15] are used for authentication purposes.

13.5.5 Network State Monitoring

A data-centric approach like the Path Class approach can be used to map the packets. The mapping will provide a flow path with defined security requirements like confidentiality, integrity, and availability. Global Flow Tables have been defined to inform the network resources and the third-party users about the flow mechanisms.

13.5.6 Forensics

Forensics plays a critical role in reconstructing a crime or a security violation and trying to identify the origin. SDN uses a multilayered approach to assist forensics by treating the evidence data, detecting an anomaly, and using AI to identify the source and raise of alarm.

13.5.7 Utilization of Blockchain Technology

Blockchain technologies can be implemented to enhance the network's capacity with an acceptable performance level [19]. It would assist in scaling up the network [20]. Guarding against threats and protection against attacks would become more accessible, especially against DDoS [21] and DoS attacks [22, 23].

13.6 OPEN CHALLENGES

Numerous open challenges are still there to be addressed in the SDN security domain. A few of the challenges are mentioned below.

13.6.1 Non-Standardized Interfaces for NFV Integration

Network Function Virtualization is a feature that enhances the security of SDN at a distributed level and reduces the workload of the controller and other network entities. However, the interfaces for implementing NFV are not standardized. SDN security enhancement can benefit from the standardization of NFV interfaces.

13.6.2 Security Assessment

The programmability and scalability of SDN provide an initiative to the users and administrators to implement new features and applications. However, the effects such enhancements have on the other aspects of the SDN, especially security, should be considered as it is hard to assess.

13.6.3 Forensics

SDN does not encompass robust evidence-collection mechanisms for determining the root cause of the attacks. SDN capitalizes on the attack detection engines that, when compromised, will provide insufficient evidence for forensic analysis. Foolproof mechanisms are required for forensic data collection and identifying the root cause of the attack.

13.6.4 Resilience

SDN depends on a control plane for network management, and controllers can be prone to system crashes, power outages, and data disruption, especially under specific types of attacks. There is room for research in detecting and mitigating the attacks so that the controller has complete visibility of the SDN and controls the network effectively.

13.6.5 Trusted Applications

As discussed, the controller cannot detect applications that are maliciously interfering with the flow rules. Moreover, the controller is also unable to detect the vulnerabilities posed by "innocent" Apps due to misconfigurations or poor design and programming of the application. To counter this, authentication mechanism may be implemented in the control plane for handling such issues [24, 25].

13.7 CONCLUSION

The advent of SDN is a breakthrough in network management. It has brought to the world of networking new features that can be utilized to improve the network performance. The programmability of SDN has offered features of conventional networks, which can be implemented through software in the SDN. Apart from the elements, it has brought new dimensions of security along with numerous challenges. The security paradigm of SDN is evolving at a faster pace. Despite the new tools implemented for security and the new mechanism devised, information security cannot be guaranteed on SDN. New vulnerabilities and threats are emerging as the new layers are introduced. In general, two focus groups are researching the SDN. One group is focused on the features of the SDN with added security. In contrast, the other group focuses on safety while the new layers and interfaces evolve. Since the threats and attacks against the SDN are complex, numerous approaches are being developed for capitalizing on the AI and machine learning methodologies to detect and counterfeit the attacks. The SDN is still evolving, and multiple security aspects must be addressed while some remain dormant. The scope of research in this field is enormous. More than a decade has passed since the first security-related research work was published; still, there exist numerous aspects of SDN security that require more investigation, like forensics, policy mediators, debuggers, and detection systems for identifying vulnerabilities.

REFERENCES

1 C. Yoon *et al.*: Flow Wars: Systemizing the Attack Surface and Defenses in Software-Defined Networks. *IEEEACM Trans. Netw.*, vol. 25, no. 6, pp. 3514–3530 (Dec. 2017).
2 C. Ropke.: SDN Malware: Problems of Current Protection Systems and Potential Countermeasures. p. 12. 2016.
3 K. Benton, L. J. Camp, and C. Small.: OpenFlow Vulnerability Assessment. *Proceedings of the Second ACM SIGCOMM Workshop on Hot Topics in Software Defined Networking - HotSDN'13*, Hong Kong, China: ACM Press, (2013).
4 S. Shin, L. Xu, S. Hong, and G. Gu.: Enhancing Network Security through Software Defined Networking (SDN). *2016 25th International Conference on Computer Communication and Networks (ICCCN)*, Waikoloa, HI, USA: IEEE, (Aug. 2016).
5 T. H. Nguyen, and M. Yoo.: Analysis of Link Discovery Service Attacks in SDN Controller. *2017 International Conference on Information Networking (ICOIN)*, Da Nang, Vietnam: IEEE (2017).
6 C. F. T. Pontes, M. M. C. de Souza, J. J. C. Gondim, M. Bishop, and M. A. Marotta.: A New Method for Flow-Based Network Intrusion Detection Using the Inverse Potts Model. *IEEE Trans. Netw. Serv. Manag.*, vol. 18, no. 2, pp. 1125–1136 (Jun. 2021).
7 D. Butts.: SDN Security Attack Vectors and SDN Hardening | Network World. p. 5. *Artikkeli Network*, 2014

8 M. C. Dacier, H. Konig, R. Cwalinski, F. Kargl, and S. Dietrich.: Security Challenges and Opportunities of Software-Defined Networking. *IEEE Secur. Priv.*, vol. 15, no. 2, pp. 96–100 (Mar. 2017).
9 O. Yurekten, and M. Demirci.: "SDN-Based Cyber Defense: A Survey.: *Future Gener. Comput. Syst.*, vol. 115, pp. 126–149 (Feb. 2021).
10 L. Schehlmann, S. Abt, and H. Baier.: Blessing or Curse? Revisiting Security Aspects of Software-Defined Networking. *10th International Conference on Network and Service Management (CNSM) and Workshop*, Rio de Janeiro, Brazil: IEEE (Nov. 2014).
11 S. Scott-Hayward, S. Natarajan, and S. Sezer.: A Survey of Security in Software Defined Networks. *IEEE Commun. Surv. Tutor.*, vol. 18, no. 1, pp. 623–654 (2016).
12 M. Liyanage, M. Ylianttila, and A. Gurtov.: Securing the Control Channel of Software-Defined Mobile Networks. *Proceeding of IEEE International Symposium on a World of Wireless, Mobile and Multimedia Networks 2014*, Sydney, Australia: IEEE. (Jun. 2014)
13 D. V. Bernardo, and B. B. Chua.: Introduction and Analysis of SDN and NFV Security Architecture (SN-SECA). *2015 IEEE 29th International Conference on Advanced Information Networking and Applications*, Gwangiu, South Korea. (2015).
14 S. Ali, N. Tariq, F. A. Khan, M. Ashraf, W. Abdul, and K. Saleem.: BFT-IoMT: A Blockchain-Based Trust Mechanism to Mitigate Sybil Attack Using Fuzzy Logic in the Internet of Medical Things. *Sensors*, vol. 23, no. 9, p. 4265. (Apr. 2023).
15 E. Tantar, A.-A. Tantar, M. Kantor, and T. Engel.: On Using Cognition for Anomaly Detection in SDN. in *EVOLVE - A Bridge between Probability, Set Oriented Numerics, and Evolutionary Computation VI*, A.-A. Tantar, E. Tantar, M. Emmerich, P. Legrand, L. Alboaie, and H. Luchian, Eds., in Advances in Intelligent Systems and Computing, vol. 674. Cham: Springer International Publishing. (2018).
16 F. A. Khan, and A. Gumaei.: A Comparative Study of Machine Learning Classifiers for Network Intrusion Detection in *Artificial Intelligence and Security*. X. Sun, Z. Pan, and E. Bertino, Eds., in Lecture Notes in Computer Science, vol. 11633. Cham: Springer International Publishing. (2019).
17 A. Abubakar, and B. Pranggono.: Machine Learning Based Intrusion Detection System for Software Defined Networks, *seventh international conference on emerging security technologies (EST)* pp. 138–146. (2017).
18 G. A. Ajaeiya, N. Adalian, I. H. Elhajj, A. Kayssi, and A. Chehab.: Flow-based Intrusion Detection System for SDN. *2017 IEEE Symposium on Computers and Communications (ISCC)*, Heraklion, Greece: IEEE. (Jul. 2017).
19 N. Tariq, M. Asim, F. A. Khan, T. Baker, U. Khalid, and A. Derhab.: A Blockchain-Based Multi-Mobile Code-Driven Trust Mechanism for Detecting Internal Attacks in Internet of Things. *Sensors*, vol. 21, no. 1, p. 23. (Dec. 2020).
20 M. P. Novaes, L. F. Carvalho, J. Lloret, and M. L. Proença.: Adversarial Deep Learning Approach Detection and Defense Against DDoS Attacks in SDN Environments. *Future Gener. Comput. Syst.*, vol. 125, pp. 156–167 (Dec. 2021).
21 M. Imran, M. H. Durad, F. A. Khan, and H. Abbas.: DAISY: A Detection and Mitigation System against Denial of Service Attacks in Software Defined Networks. *IEEE Syst. J.*, vol. 14, issue 2, pp. 1933–1944. (June 2020).
22 M. Imran, M. H. Durad, F. A. Khan, and A. Derhab.: Toward an Optimal Solution against Denial of Service Attacks in Software Defined Networks. *Future Gener. Comput. Syst.*, vol. 92, pp. 444–453. (March 2019).
23 N. Tariq *et al.*: The Security of Big Data in Fog-Enabled IoT Applications Including Blockchain: A Survey. *Sensors*, vol. 19, no. 8, p. 1788 (Apr. 2019).
24 M. Imran, M. H. Durad, F. A. Khan, and A. Derhab.: Reducing the Effects of DoS Attacks in Software Defined Networks using Parallel Flow Installation. *Human-centric Comput. Inf. Sci.*, vol. 9, no. 1, pp. 1–19. (May 2019).
25 N. Tariq, M. Asim, Z. Maamar, M. Z. Farooqi, N. Faci, and T. Baker.: A Mobile Code-Driven Trust Mechanism for Detecting Internal Attacks in Sensor Node-Powered IoT. *J. Parallel Distrib. Comput.*, vol. 134, pp. 198–206 (Dec. 2019).

14 Access Control Techniques for Cloud Computing

Review and Recommendations

Mannan Javed, Noshina Tariq, Farrukh Aslam Khan, and Muhammad Ashraf

14.1 INTRODUCTION

Access control methods are used in cloud computing to regulate who or what has access to an organization's cloud-based resources. It is an important security measure to guard against unauthorized access to systems and information. Access control enables businesses to manage who has access to and how to use the cloud resources. Access control in cloud computing can be implemented through authentication, authorization, and encryption [1]. Authentication guarantees only legitimate users access to the cloud resources, while authorization determines the type of access each user is allowed. Cloud computing has enabled businesses and users to reduce costs by renting Internet-based cloud services provided by different cloud service providers [2]. Thus, the users or businesses are not required to own expensive hardware and develop their human resources comprising professionals for maintaining the same. On the one hand, cloud computing has enhanced the growth of businesses by offering low-cost services through CSPs, whereas on the other hand, and it has introduced new risks to the security of user data [3]. Therefore, the CSPs must ensure adequate security to protect users' data. Any breach or lapse in the security mechanism would result in information leakage, compromising the data security and resulting in system failure.

In the access-control model, a system bases the decision to grant access based on the access-control model implemented. In light of consumers' ever-increasing reliance on cloud services, access control solutions based on proper rules are essential for information security. Therefore, to guarantee the security of users' private/sensitive data, a firm security policy that explicitly covers all aspects must be defined [4]. The resources are shared through the Internet or isolated network in the cloud platforms. The services in cloud computing include access to storage, programs, and applications [5]. The users may utilize personal computers, laptops, mobile devices, and servers for accessing the cloud service through CSP.

Similarly, cloud computing can give IoT the privilege of cost-effective on-demand services for intensive processing and big data storage [6]. IoTs produce and share enormous amount of sensitive data, yet both the devices and the data are subject to numerous privacy and security risks [7]. This survey on access control methods in cloud computing summarizes the access control strategies presently implemented on cloud platforms. It describes the pros and cons of each strategy and provides recommendations for implementing secure access control measures for cloud computing services. In addition, the article provides a comprehensive review of cloud models based on NIST evaluation criteria. It is crucial in academia and industry because it explains the various approaches to access control on cloud systems and provides guidance on the types of access control mechanisms that can facilitate secure access to cloud resources.

DOI: 10.1201/9781003497851-14

Additionally, the article suggests potential research areas that could contribute to developing a more secure method of accessing cloud resources. In addition, the paper identifies prospective future research areas that can contribute to creating more secure access control mechanisms in cloud computing. However, other state-of-the-art surveys (e.g., [4], [5], and [6]) analyzed cloud computing access control techniques but did not present recommendations for access control strategies. This survey paper has the following contributions:

1. A comprehensive survey of access control techniques and mechanisms for cloud computing systems is presented.
2. Recommendations and future research directions are proposed for implementing a secure access control mechanism on cloud computing resources.
3. Comprehensive comparison of cloud models based on NIST evaluation requirements is conducted.

The layout of this paper is as follows: Section 14.2 provides a comprehensive insight into cloud service, deployment models, access control languages, and access control techniques. Section 14.3 comprehensively describes the access control models for the cloud. Recommendations for implementation in the cloud are given in Section 14.4, followed by the future research directions and conclusion in Sections 14.5 and 14.6, respectively.

14.2 CLOUD SERVICE TYPES

As per the study carried out during this survey paper, three types of cloud services have been identified, as tabulated in Table 14.1.

14.2.1 Cloud Deployment Models

Based on the literature reviewed, the cloud deployment models have been categorized into four types. These types vary based on the user/organization requirements and their specific application scenarios; however, the utilization of each type of cloud deployment model is not uniform. The summary is given in Table 14.2.

14.2.2 Access Control Language

Four primary access control languages have been identified in the literature review [11]. The first is Security Assertion Markup Language (SAML), which allows users to access multiple services with a single verification of the credentials [12]. This language uses Extensible Markup Language (XML)

TABLE 14.1
Types of Cloud Services

Cloud Service Type	Description
Software as a Service (SaaS)	SaaS facilitates users with online access to cloud-based applications—for example, Amazon Web Service, Dropbox, and G Suite [8].
Platform as a Service (PaaS)	PaaS facilitates users to develop applications through online cloud platforms—for example, Google App Engine, Red Hat, Openshift, and Oracle cloud platform [9].
Infrastructure as a service (IaaS)	IaaS allows users to run their applications through online cloud resources, including computation, storage, and network—for example, Microsoft Azure and Web Services by Amazon [10].

TABLE 14.2
Types of Cloud Deployment Models

Cloud Deployment Model	Description
Public cloud	In this type of cloud, the service provider owns the cloud. The users can access cloud services online—for example, Google App Engine [15].
Private cloud	This type of cloud is similar to an intranet owned by an organization. The authorized users of the organization can access the online cloud services—for example, Oracle cloud services [16].
Hybrid cloud	This type of cloud is composed of several clouds having independent infrastructure that is a mix of public and private clouds—for example, Amazon and Google [17].
Community cloud	This type of cloud is composed of collaboration amongst several cloud computing solutions working on a shared project—for example, US-Based dedicated IBM soft layer cloud for federal agencies [18].

for communication between IdP and Service Provider (SP). However, it cannot exercise any control over the access to the data. The second is the Service Provisioning Markup Language (SPML), an open-source language based on XML. SPML automates sharing users' identities amongst different cloud organizations and exchanging data between users [13]. The third is Extensible Access Control Markup Language (XACML), which facilitates cloud service providers with access controls required for implementation on federated models of cloud computing [14]. It comprises four essential parts, including Policy Decision Point (PDP), Policy Information Point (PIP), Policy Enforcement Point (PEP), and Policy Access Point (PAP). The access control policy language defines access control policy, and request/response language defines answers to queries for permissions, and reference architecture ensures the implementation of security policies through appropriate software [20, 21].

14.2.3 Access Control Techniques in Cloud

Cloud data is vulnerable to a number of security risks and assaults, including privacy and confidentiality breaches [15, 29]. Therefore, access control measures are necessary for cloud systems to guarantee that only authorized users can access and edit information and resources [26]. Techniques for access control shield cloud computing systems from harmful intrusions, unauthorized access, and unlawful disclosure of data. A brief description of access control techniques for cloud computing is given in Table 14.3.

14.3 ACCESS CONTROL MODELS

The access control models were initially focused on two types of requirements: military-specific, focused on confidentiality of data, and commercial-specific, focused on data integrity [8,30]. The following subsections briefly cover the fundamental types of access control models and their comparison based on the NIST requirement/properties.

14.3.1 Discretionary Access Control (DAC)

In the DAC model, the owner of an object can assign permissions related to the object to any other subject (for example, read/write permission). A subject with the right to some resources can grant read/write permissions to other subjects who have access to the same resources [31]. DAC is sometimes called Identity-Based Access Control (IBAC). In DAC, an owner can create a group and assign permissions to users of that group. Thus, the owner would control the group permissions. However,

TABLE 14.3
Access Control Techniques in Cloud Computing

Technique	Description
Identity and Access Management (IAM)	IAM solutions enable organizations to manage the identities and access of their users. These solutions enable organizations to execute control over who can access the cloud and what they can do with it [19].
	Cryptographic keys may be extracted illegally, enabling attackers to steal any individual's sensitive information [20]. Cloud providers can guarantee that data is secure and reliable throughout its lifecycle, from storage to retrieval, by utilizing blockchain-based encryption and smart contract capabilities [21].
Data Encryption	Data encryption allows organizations to secure their data stored on the cloud by encryption. Since the user would need the key to decrypt the data, it is harder for unauthorized users to access it [22].
Multi-Factor Based Authentication	By binding the users to give more than one form of verification, such as a password and a one-time code delivered to their phone, multi-factor authentication adds an extra layer of security. As a result, it is more challenging for attackers to access the cloud system [23].
Virtual Private Networks (VPNs)	VPNs provide a dedicated secure path between end users and services. It makes it difficult for attackers to access the system as the connection is encrypted [24].
Application Firewalls	As multiple attack vectors and stealthy techniques are often employed to avoid detection, advanced persistent threats (APTs) in the cloud can be especially challenging to detect and mitigate [25]. Therefore, Firewalls can be used to filter traffic to the cloud system, blocking malicious requests and preventing attackers from gaining access [26], [27].
Access Control Lists (ACLs)	ACLs allow administrators to specify which users can access the services / resources [28].

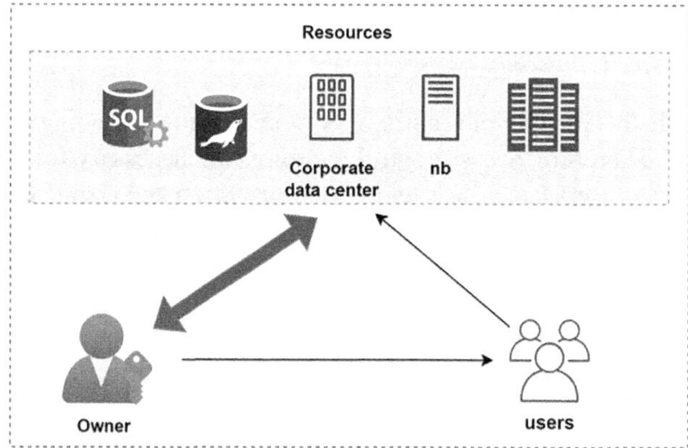

FIGURE 14.1 DAC model.

if the owner is not trustworthy, then this would be a security risk. The DAC model lacks control over the flow of information. Thus, an unauthorized user can read a copy of the file without permission authorized by the owner. Figure 14.1 illustrates a DAC model.

14.3.2 Mandatory Access Control (MAC)

In MAC, the administrator can assign permission to subjects over any objects. In this model, the users cannot modify their permissions. In MAC, only the system administrator manages the security

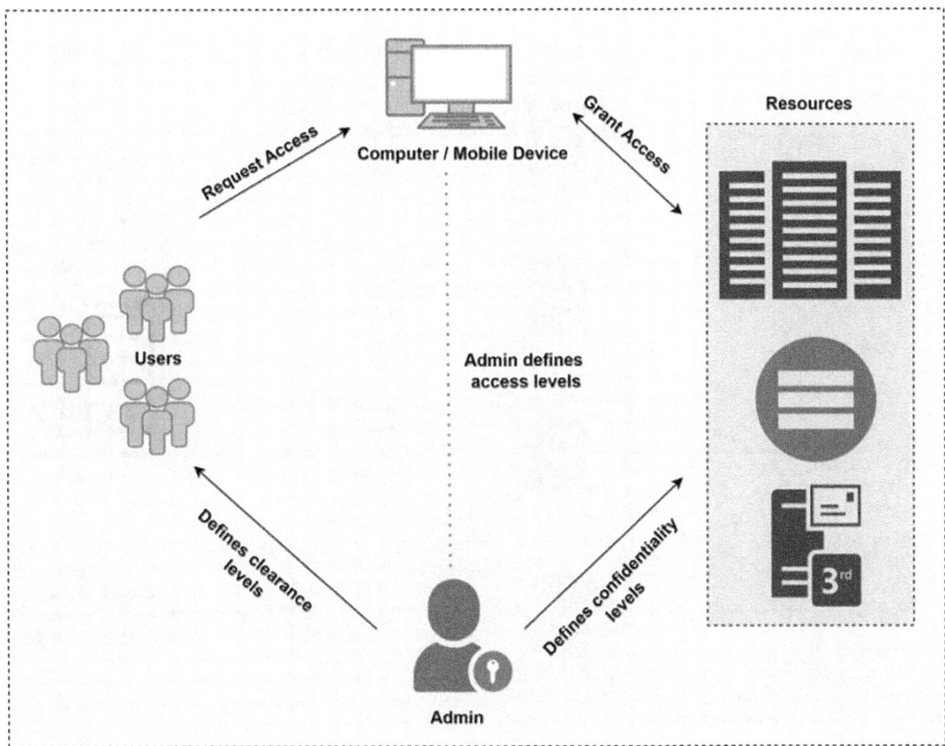

FIGURE 14.2 MAC model.

policy, and the operating system implements the defined policy for subjects/users [32]. Thus, the administrator trusts the parts of the OS implementing the security policy. The MAC model does not allow the users to modify their permissions and thus ensures the integrity of the information. Therefore, MAC-based systems are considered secure but costly and difficult to use due to constraints applied by the OS. In MAC, subjects and objects are categorized based on their security levels, for example, Top Secret, Secret, Confidential, and Unclassified. Figure 14.2 illustrates a MAC model.

14.3.3 Role Based Access Control (RBAC)

The RBAC approach strongly emphasizes limiting users' access to resources inside an organization based on their role/function/job description. The role of the users is predefined by the system administrator based on duties assigned to an individual [33]. For example, in a University, the users can be divided into different roles like faculty members, administration, and students. Therefore, a user assigned with a student role will have different access rights than a faculty member or administration. The main elements in the RBAC model are users/subjects, objects/resources, roles, operations, and permissions. Figure 14.3 illustrates an RBAC model.

14.3.4 Attribute Based Access Control (ABAC)

The ABAC model bases its choice of granting access on the attributes/characteristics of users and resources. The system administrator makes the security policy on a predefined set of attributes [34]. These attributes may include but are not limited to the user's location, time, date, authentication level, and role. The ABAC model has a policy decision point (PDP), which has a set of values for

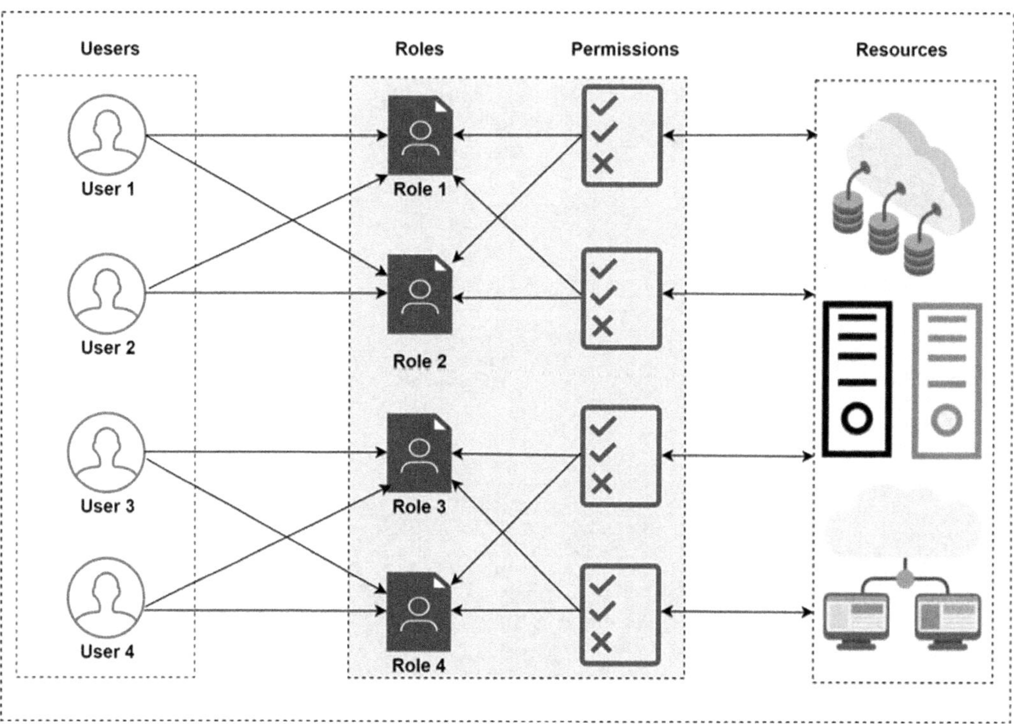

FIGURE 14.3 RBAC model.

each attribute and compares it to the attributes of requesting user before making an access decision. In the ABAC model, the attributes are not required to be related. The main elements of the ABAC model are users, subjects, objects, subject/object attributes, user attributes, permissions, and authorization policy. Figure 14.4 illustrates an ABAC model.

14.3.5 Attribute-Based Encryption (ABE)

When a user saves his data on the Cloud, the data is simultaneously saved at different locations as per the architecture of the CSP. Even if a user deletes his data from the cloud, the same data can be recovered through other servers of the same platform by the CSP [35]. Thus, a large amount of users' personal information resides on the cloud, and a compromise in the security of the cloud may result in the leakage of confidential information. The said problem can be addressed if the user's information is stored in the encrypted form. Thus, if an unauthorized user gets access to the data, he cannot view the data. However, the encrypted data would be accessible by sharing the key with the requesting entity. Thus, after getting the security key, the user would get access to all information, including that information which is not desired to be shared [36]. Attribute-Based Encryption was designed to solve this issue. ABE is further divided into five types [37], including Key Policy ABE (KP-ABE), simple ABE, ABE with Non-monotonic Access, Cypher-Text Policy ABE (CP-ABE), and Hierarchical ABE (HABE). Figure 14.5 illustrates an ABE model.

14.3.6 Federated ID Management (FIM)

Cloud platforms have increased users` access to various websites pertaining to their areas of interest. When a user requests access to a website, his user ID/password is initially verified, and the decision

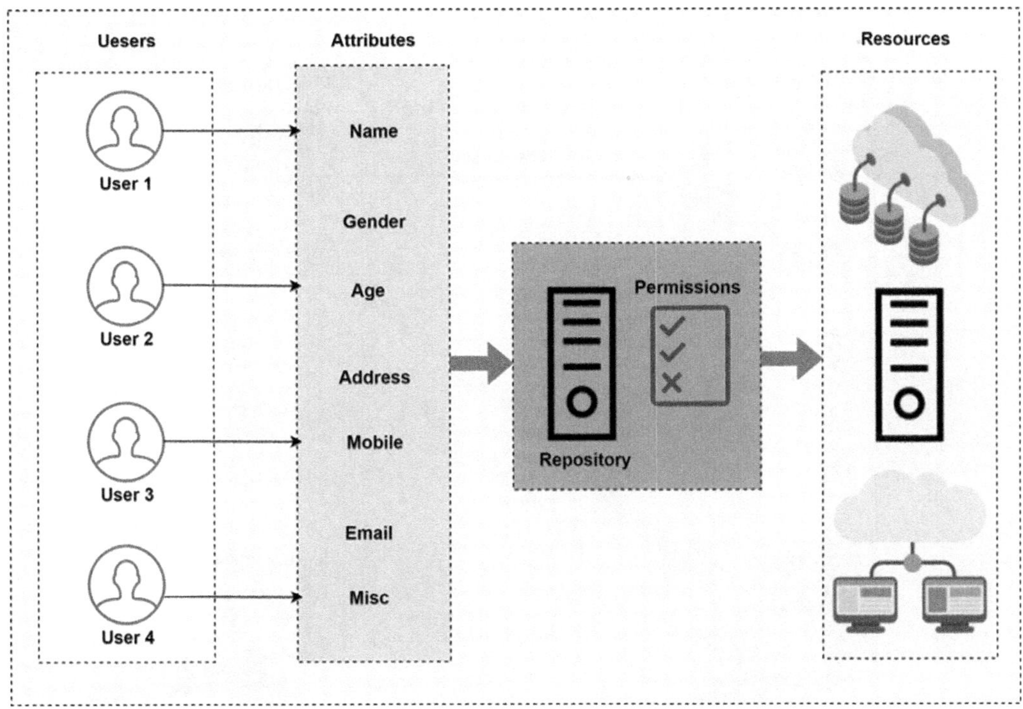

FIGURE 14.4 ABAC model.

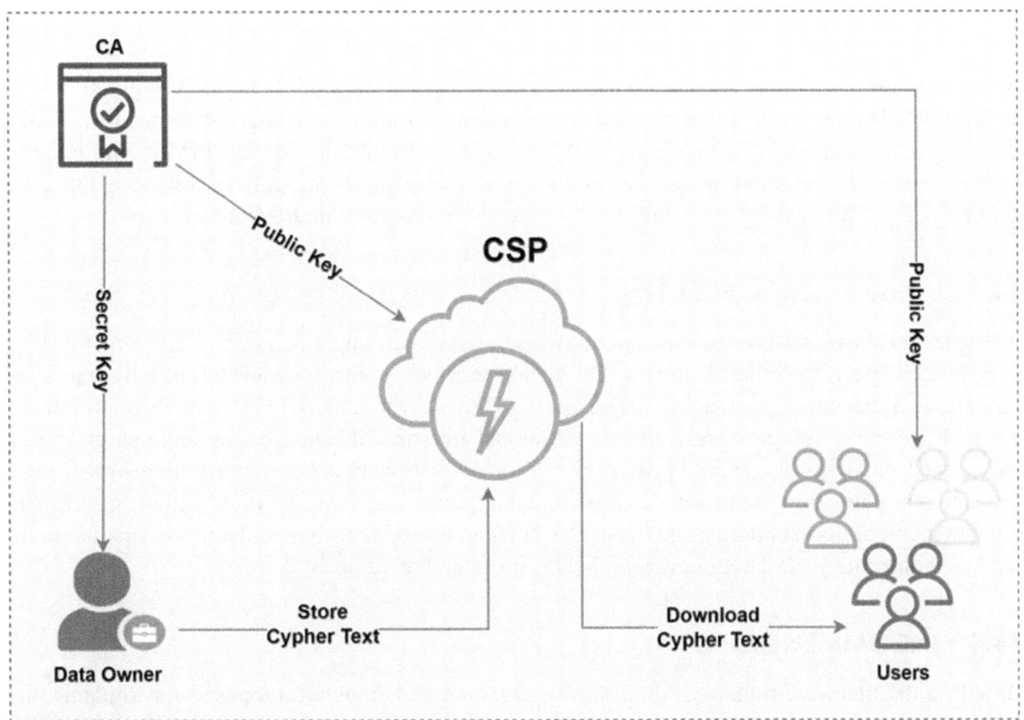

FIGURE 14.5 ABE model.

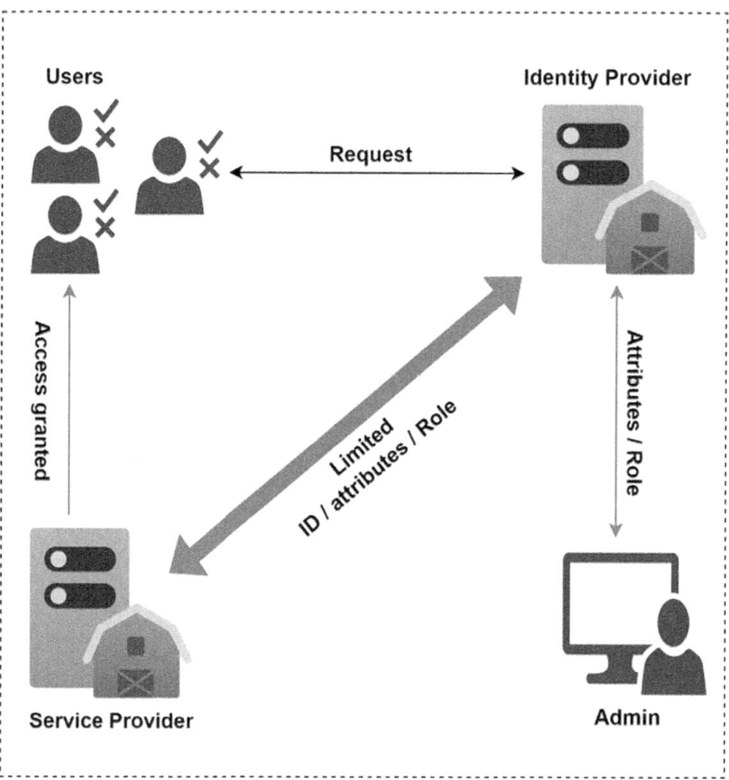

FIGURE 14.6 FIM model.

is made to allow/deny access. When a user requests access to multiple sites, he is required to enter his user details repeatedly. To resolve this repetitive entry of user credentials on domains hosted by the same SP or separate SPs, Federated ID Management was introduced [38]. FIM is a middleware between the users and applications that stores the user credentials for shifting from one domain to another without re-entering the details [39]. Figure 14.6 illustrates an FIM model.

14.3.7 Comparison of ACM Types

As explained above, different access control models have been implemented by CSPs/organizations to meet their specific requirements. Each of the above access control models has its advantages and limitations. It becomes challenging for an entity to assess which model best suits its requirements. To facilitate the developers and CSPs, the National Institute of Science and Technology (NIST) has defined standards that should be considered for evaluating an access control model [40]. These evaluation requirements are broadly classified into four categories, i.e., administration, enforcement, implementation, and support properties. The comparison of the above-mentioned access control models (concerning NIST requirements) [41] is presented in Table 14.4.

14.4 RECOMMENDATIONS

Based on the literature reviewed, the following recommendations are proposed for implementing robust access control on the cloud.

1. A combination of Role Based Access Control (RBAC) and Attribute Based Access Control (ABAC) may be implemented.
2. Multi-factor Authentication (MFA) with at least two or more authentication factors could be implemented.
3. To ensure that only users with legitimate authorization can use the system, comprehensive Identity and Access Management (IAM) guidelines may be instituted and implemented [23].
4. Data encryption (Public Key Encryption) may be implemented to ensure that data is protected even if it is intercepted or stolen.
5. If possible, Token-Based Authentication may be implemented for Mission-Critical/Sensitive applications. Each user would be required to provide a unique token for each session. This token may be a randomly generated code or a one-time password.
6. Security Policies and Procedures are documents that specify the rules and protocols that govern access to a system. These security policies should be regularly updated and reviewed.
7. Data loss prevention techniques must be implemented by the cloud service provider.
8. The Service-Level Agreement (SLA) [42] should be elaborated to clearly define the rights of the customer and the responsibilities of the cloud service provider. Disaster recovery procedures, incident response, data security, and performance should be included in the SLA.

14.5 FUTURE RESEARCH DIRECTIONS

Based on the survey, the following research directions are proposed for improving the cloud's ease of implementation and security.

1. A unified security policy in a cloud environment would facilitate users to judge their cloud service provider better. The unified technical solution should comprise reference access control mechanism architecture and security policy.
2. RBAC and ABAC models have their specific advantages over each other. However, merging these two models to make a new model could combine their unique advantages and reduce their limitations.
3. Adding device signature in the access control to enhance cloud security, especially for IoT environments [43] (Model, IMEI/MAC, IP Address, OS, Firmware, Location, Network Element, and access history).
4. Automatic user revocation after a defined idle time.

14.6 CONCLUSION

Data integrity and confidentiality are critical elements in cloud environments. The choice of an appropriate access control mechanism is crucial for CSPs to ensure the security of the data. For private users, the confidentiality of the data is of utmost importance where data is stored on the cloud, and it directly impacts the reliability of the cloud. In this survey, a comprehensive review of cloud deployment models, cloud service types, access control languages, access control models, and their comparison concerning NIST evaluation requirements and recommendations has been carried out. This study found that several access control models are used for various functions and that cloud environments have no unified access control solution. Due to its encryption capabilities and wide range of applications in various industries, attribute-based encryption has emerged as today's most exciting access control paradigm.

TABLE 14.4
Comparison of Access Control Models

		Properties (NIST Guidelines 7874)												
		Administration								Support				
		Auditing	Privileges Capabilities Discovery	Ease of Privilege Assignments	Supporting Syntax and Semantics for Specifying AC Rules	Policy Manage-ment	Delegation of Administrative Capabilities	Flexibility of Configuration into Existing System	Horizontal Scope of Control (across platforms and applications)	Policy Import and Export	OS compatibility	Policy Source Management	User Interfaces and API	Verification and Compliance Function Support
Type of Access Control Model	**DAC** [31]	N/A	G	G	W	S	G	G	G	N/A	S	N/A	S	W
	MAC [32]	N/A	W	S	W	S	W	W	W	N/A	G	N/A	S	S
	RBAC [33]	N/A	S	G	S	W	G	G	G	W	N/A	N/A	S	W
	ABAC [34]	N/A	N/A	G	G	G	W	W	G	N/A	S	W	W	G
	ABE [1], [35]	N/A	W	S	G	G	S	W	W	S	S	N/A	N/A	S
	FIM [38], [39]	N/A	N/A	S	G	S	S	G	S	S	S	N/A	S	W

Abbreviations: Good = G, Satisfactory = S, Weak = W, Not Available = N/A

REFERENCES

[1] P. Kumar, P. J. A. Alphonse.: Attribute Based Encryption in Cloud Computing - A Survey, Gap Analysis and Future Directions. *Journal of Network and Computer Applications*, vol. 108, pp. 37–52 (2018).

[2] V. Maheshwari, S. Sahana, S. Das, I. Das, A. Ghosh.: Factors Influencing Security Issues in Cloud Computing. In: Advanced Communication and Intelligent Systems, Springer, (2023).

[3] Z. Tari.: Security and Privacy in Cloud Computing. *IEEE Cloud Computing*, vol. 1, pp. 54–57 (2014).

[4] Y. S. Abdulsalam, M. Hedabou.: Security and Privacy in Cloud Computing - Technical Review. *Future Internet*, vol. 14, no. 1, p. 11 (2021).

[5] S. El Kafhali, I. El Mir, M. Hanini.: Security Threats, Defense Mechanisms, Challenges, and Future Directions in Cloud Computing. *Archives of Computational Methods in Engineering*, vol. 29, pp. 223–246 (2022).

[6] N. Abbas, M. Asim, N. Tariq, T. Baker, S. Abbas.: A Mechanism for Securing IoT-enabled Applications at the Fog Layer. *JSAN*, vol. 8, no. 1, p. 16 (2019).

Enforcement										
The Vertical Scope of Control (between application, DBMS & OS)	Policy Combination, Composition and Constraint	Bypass	Least Privilege Support Principal	Separation of Duty	Safety (Confinements & Constraints)	Conflict Resolution or Prevention	Operational / Situational Awareness	Granularity of Control	Expression (policy/ model) properties	Adaptable to the enactment and development of AC Policies
S	W	S	L	S	G	G	S	S	N/A	S
S	W	N/A	S	G	G	S	S	S	N/A	S
G	W	N/A	S	S	S	G	G	S	N/A	G
S	S	N/A	S	G	G	G	S	S	N/A	S
W	W	N/A	G	G	G	G	N/A	S	N/A	S
S	S	N/A	W	N/A	S	S	N/A	S	N/A	S

[7] U. Farooq, N. Tariq, M. Asim, T. Baker, A. Al-Shamma'a.: Machine Learning and the Internet of Things Security: Solutions and Open Challenges. *Journal of Parallel and Distributed Computing*, vol. 162, pp. 89–104, (2022).

[8] R. El Sibai, N. Gemayel, J. Bou Abdo, J. Demerjian.: A Survey on Access Control Mechanisms for Cloud Computing. *Transactions on Emerging Telecommunications Technologies*, vol. 31, no. 2, pp. e3720 (2020).

[9] P. Srivastava, R. Khan.: A Review Paper on Cloud Computing. IJARCSSE, (2018).

[10] G. Karataş and A. Akbulut.: Survey on Access Control Mechanisms in Cloud Computing. *Journal of Cyber Security and Mobility*, vol. 7, no. 3, pp. 1–36 (2018).

[11] S. Murugesan, I. Bojanova.: *Encyclopedia of Cloud Computing*. John Wiley & Sons, (2016).

[12] N. Naik, P. Jenkins.: An Analysis of Open Standard Identity Protocols. In: Cloud Computing Security Paradigm. *2016 IEEE 14th Intl Conf on Dependable, Autonomic and Secure Computing*, Auckland, (2016).

[13] K. Munir, S. Palaniappan.: Framework for Secure Cloud Computing. *IJCCSA*, (2013).

[14] R. Laborde, F. Barrère, A. Benzekri.: Toward Authorization as a Service - a Study of the XACML Standard. In: *Proceedings of the 16th Communications & Networking Symposium, CNS '13*. San Diego, CA, USA, (2013).
[15] M. Srilakshmi, C. L. Veenadhari, I. K. Pradeep.: Deployment models of Cloud Computing: Challenges. *International Journal of Advanced Research in Computer Science*, (2010).
[16] V. Davidovic, D. Ilijevic, V. Luk, I. Pogarcic.: Private Cloud Computing and Delegation of Control. *Procedia Engineering*, vol. 100, pp. 196–205 (2015).
[17] A. Srinivasan, M. A. Quadir, V. Vijayakumar.: Era of Cloud Computing - A New Insight to Hybrid Cloud. *Procedia Computer Science*, vol. 50, pp. 42–51 (2015).
[18] A. Marinos, G. Briscoe.: Community Cloud Computing. In: Cloud Computing, Berlin, Heidelberg: Springer Berlin Heidelberg, (2009).
[19] E. Sturrus, O. Kulikova.: Identity and Access Management. In: Encyclopedia of Cloud Computing, S. Murugesan, I. Bojanova, Eds., Chichester, UK: John Wiley & Sons, (2016).
[20] S. Ali, N. Tariq, F. A. Khan, M. Ashraf, W. Abdul, K. Saleem.: BFT-IoMT - A Blockchain-Based Trust Mechanism to Mitigate Sybil Attack Using Fuzzy Logic in the Internet of Medical Things. *Sensors*, vol. 23, no. 9. (2023).
[21] N. Tariq, M. Asim, F. A. Khan, T. Baker, U. Khalid, A. Derhab.: A Blockchain-Based Multi-Mobile Code-Driven Trust Mechanism for Detecting Internal Attacks in Internet of Things. *Sensors*, vol. 21, no. 1, Art. no. 1. (2021).
[22] R. Arora, A. Parashar.: Secure User Data in Cloud Computing Using Encryption Algorithms. *International Journal of Engineering Research*, vol. 3, no. 4, pp. 1922–1926 (2013).
[23] R. K. Banyal, P. Jain, V. K. Jain.: Multi-factor Authentication Framework for Cloud Computing. In: *Fifth International Conference on Computational Intelligence, Modeling and Simulation*, Seoul, Korea (South): IEEE, (2013).
[24] K. Jyothi, B. I. Reddy.: CSEIT1835225 | Study on Virtual Private Network (VPN), VPN's Protocols And Security. (2023).
[25] N. A. S. Mirza, H. Abbas, F. A. Khan, J. Al Muhtadi.: Anticipating Advanced Persistent Threat (APT) Countermeasures Using Collaborative Security Mechanisms. (ISBAST), (2014).
[26] E. B. Fernandez, N. Yoshioka, H. Washizaki.: Patterns for cloud firewalls. (2014).
[27] U. Farooq, M. Asim, N. Tariq, T. Baker, A. I. Awad.: Multi-Mobile Agent Trust Framework for Mitigating Internal Attacks and Augmenting RPL Security. *Sensors*, vol. 22, pp. 4539 (2022).
[28] M. Mulimani, R. Rachh.: Analysis of Access Control Methods in Cloud Computing. Preprints, (2016).
[29] S. A. Moqurrab et al.: A Deep Learning-Based Privacy-Preserving Model for Smart Healthcare in IoMT Using Fog Computing. *Wireless Personal Communications*, vol. 126, pp. 2379–104 (2022).
[30] F. Cai, N. Zhu, J. He, P. Mu, W. Li, Y. Yu.: Survey of Access Control Models and Technologies for Cloud Computing. *Cluster Computing*, vol. 22, no. S3, pp. 6111–6122 (2019).
[31] D. D. Downs, J. R. Rub, K. C. Kung, C. S. Jordan.: Issues in Discretionary Access Control. In: *IEEE Symposium on Security and Privacy*, Oakland, CA, USA, (1985).
[32] Lindqvist, Hakan.: Mandatory Access Control. Master's thesis in computing science, Umea University, Department of Computing Science, SE-901 87, (2006).
[33] M. J. Moyer, M. Abamad.: Generalized Role-Based Access Control. In: *Proceedings of 21st International Conference on Distributed Computing Systems*, Mesa, AZ, USA (2001).
[34] V. C. Hu, D. R. Kuhn, D. F. Ferraiolo.: Attribute-Based Access Control. *Computer*, vol. 48, no. 2, pp. 85–88 (2015).
[35] V. Goyal, O. Pandey, A. Sahai, B. Waters.: Attribute-based encryption for fine-grained access control of encrypted data. In: *Proceedings of the 13th ACM conference on Computer and communications security*, Alexandria Virginia USA, (2006).
[36] D. Arshad, M. Asim, N. Tariq, T. Baker, H. Tawfik, D. Al-Jumeily, OBE.: THC-RPL - A Lightweight Trust-Enabled Routing in RPL-Based IoT Networks Against Sybil Attack. *PloS one*, vol. 17, no. 7. (2022).
[37] J. Bethencourt, A. Sahai, B. Waters.: Ciphertext-Policy Attribute-Based Encryption. In: *IEEE Symposium on Security and Privacy*, Berkeley, CA, (2007).
[38] J. Jensen.: Federated Identity Management Challenges. In: *2012 Seventh International Conference on Availability, Reliability and Security*, Prague, TBD, Czech Republic, (2012).

[39] E. Ghazizadeh, M. Zamani, J. Ab Manan, A. Pashang.: A Survey on Security Issues of Federated Identity in the Cloud Computing. In: *4th IEEE International Conference on Cloud Computing Technology and Science Proceedings*, Taipei, Taiwan, (2012).

[40] V. C. Hu, K. Scarfone.: Guidelines for Access Control System Evaluation Metrics. National Institute of Standards and Technology, Gaithersburg, MD, NIST IR 7874, (2012).

[41] V. C. Hu, D. F. Ferraiolo, D. R. Kuhn.: Assessment of Access Control Systems. National Institute of Standards and Technology, Gaithersburg, MD, NIST IR 7316, (2006).

[42] W. Halboob, H. Abbas, M. K. Khan, F. A. Khan, M. Pasha.: A Framework to Address Inconstant User Requirements in Cloud SLAs Management. *Cluster Computing (Springer)*, vol. 18, no. 1, pp 123–133 (2015)

[43] A. Derhab, M. Belaoued, M. Guerroumi, F. A. Khan.: Two-Factor Mutual Authentication Offloading for Mobile Cloud Computing. *IEEE Access*, vol. 8, pp. 28956–28969 (2020).

15 A Jurisprudence for Pakistan's Digital Forensic Investigation Framework Regarding Cybersecurity

Ehtisham Ul Haque and Waseem Abbasi

15.1 INTRODUCTION

Technology-driven services, especially cloud-based storage solutions have gained immense popularity over time, majorly because they offer a secure and user-friendly approach. This fact has allowed more people nowadays to get their critical data stored online on open-source platforms like Dropbox, One Drive, Google Drive etc. Some concerns arise with this growth in use since there is always an associated potential risk for a cyber-attack. The result of such a scenario is that sensitive important private data, i.e. bank details or credit/debit card information, may end up being stolen or interfered with publicly.

One way through which Law Enforcement Agencies (LEAs) are attempting to control the damage of such breaches is by taking steps to minimize these vulnerabilities and educating the public about probable risks associated with cybercrime. What's more, the increasing prevalence of anti-forensic methods is proving challenging for investigators engaging in digital forensic investigations. While considering all the different elements which are integral to a comprehensive investigative approach that encompasses preparation, collection, examination, analysis, identification and digital evidence interpretation [1], it's important that the current anti-forensic techniques do not eventually decay it and critical evidence should be collected ensuring strict adherence to legal guidelines to make it capable of holding up under scrutiny before a judge.

Globally, the increased use of digital platforms for the storage of data in many organizations has led to a rapid rise in cybercrime [2,3]. To combat cybercrimes, law enforcement and security organizations worldwide have devised various mechanisms. In recent times, there has been a notable increase in the number of studies aimed at organizing and defining the various stages of the digital forensic investigation process. The field of Digital Forensic Investigations (DFIs) encompasses a wide range of intricate tasks, activities, processes, and subdomains [4]. In 2022, there was a significant rise in cybercrime and cyber-attacks due to the advancement of anti-forensic techniques [5]. Perpetrators now only require an internet connection and computer; thus, the widespread availability of free computer software and tools facilitates their exploitation of vulnerabilities. Researchers in this field are actively exploring diverse strategies and tools to enhance the resilience of digital content against unauthorized access and deceptive practices, thereby mitigating the risks posed by cyber-attacks.

Organizations have at their disposal various measures such as procedures, guidelines, policies, baselines, and global practices to impose restrictions on cyberspace misuse. The National Institute of Standards and Technology (NIST) has developed a framework and set of standards that assist organizations in proactively countering potential cyberattacks and evaluating their security structure

DOI: 10.1201/9781003497851-15

[6]. These standards and frameworks are frequently revised every three years, with mandatory upgrades for corporations. An effective cyberattack often necessitates the cooperation of several threat actors, and the escalating threat landscape makes data theft a persistent risk that is challenging to prevent. Attackers employ basic attacks as the most convenient means of gaining access to system assets, subsequently leveraging their impact to execute more sophisticated hybrid attacks. In Pakistan, an example of such policies is the Prevention of Electronic Crimes Act (PECA) 2016 [7]. The Government of Pakistan mandates compliance with this act by all residents of Pakistan, including foreign nationals within the country's borders.

The purpose of this article is to perform an in-depth analysis of Pakistan's current cybercrime investigative infrastructure. The research aims to investigate the country's cyber threat landscape and assess the capacity of the existing infrastructure to deal with such attacks. The research will look at many areas of cybercrime investigative infrastructure, such as the legislative framework, available resources, and the obstacles that relevant agencies confront. Nevertheless, it is important to note that the research conducted focuses solely on analyzing the cybercrime investigation infrastructure in Pakistan and does not encompass broader dimensions landscape of the cybersecurity. Factors such as national- and domestic-level cyber security strategies, policies, and socioeconomic influences that shape the cyber threat landscape are not explored in this study. Furthermore, it should be recognized that the findings and conclusions are specific to the examined country and may not be directly applicable to other regions. It is essential to consider these limitations as the research heavily relies on publicly available data and literature, thus being constrained by the sources utilized. It is crucial to emphasize that the present body of research on DFI infrastructure is limited.

By emphasizing the following important factors, this research aims to significantly advance our knowledge:

- In-depth analysis of the infrastructure's challenges, scope, processes, and policies, with a focus on policy, jurisdictional, technical, legal, academic organizational and cooperative ventures.
- The primary objective of this study is to establish an empirical foundation to assist policymakers in formulating a comprehensive framework for cybercrime investigation in Pakistan. Additionally, it provides valuable insights and recommendations to enhance the existing infrastructure and effectively combat cybercrime within the country.
- The study endeavors to develop robust legal actions that can bolster the DFI framework and align it with internationally recognized global best practices and standards. The study strives to optimize the global effectiveness of digital investigation and publicize responsible utilization of digital technologies.
- This study is designed to address the prevailing deficiencies in the legal structure governing DFI and propose recommendations for its enhancement. By doing so, it aims to contribute to the ongoing initiatives aimed at combating cybercrime in Pakistan.

The study examines different aspects of DFIs in Pakistan, including regulations, institutional scope, scrutiny and development, procedures and mechanisms, and applied aptitude to deal with e-crime investigations. In order to gather data for this research, primary sources such as administrative papers, previous studies from local and international organizations, reports focusing on cybercrime and digital forensics in Pakistan, as well as demonstrations by governing authorities and publications from the International Telecommunication Union (ITU) have been taken into account.

The paper is organized as follows: Section 15.2 discusses related work; Section 15.3 highlights the current cyber landscape of Pakistan and identifies some key issues in Pakistan for which DFI would be helpful; Section 15.4 introduces the contextual framework and detailed dimensions for an operational DFI in Pakistan; and Section 15.5 concludes the paper.

15.2 RELATED WORK

In recent years, significant advancements have been made in the development, formalization and standardization of digital forensic procedures and processes [8]. The NIST plays a decisive role by providing valuable standards for conducting computer and network forensic investigations within the IT domain [9]. Similarly, the National Criminal Justice Reference Service (NCJRS) [10] provides rules and data on the digital forensic procedures used by law enforcement, such as identifying, gathering, obtaining, and storing digital evidence [11,12]. Nevertheless, further research in this field is necessary [13]. Digital forensics finds its primary applications in law enforcement and incident response (IR). Incident response entails managing and addressing incidents within cyber (IT) environments, while digital forensic analysis in law enforcement involves examining digital evidence to determine the occurrence of criminal activity and, if established, attributing responsibility and prosecuting offenders. While both disciplines employ digital forensics in similar ways, this study specifically concentrates on the utilization of digital forensics within law administration.

In developed nations such as the United States and the United Kingdom, the increasing development of technology has led to the establishment of DFI as a highly respected professional and academic discipline [14]. These countries have established a well-defined infrastructure for conducting digital investigations, encompassing the remote sector, police, and covert services. They have comprehensive legislation addressing computer misuse and widely accepted recommendations for the contribution and handling of digital testimony. In this study, we refer to the collective legislative bodies, educational programs, research documents, procedures, standards, guidelines and baselines as the framework for DFIs. Developed nations have implemented strategies such as fostering cooperation between criminal LEAs and research laboratories and institutes at universities. This collaboration aims to strengthen the Criminal Justice System (CJS) by leveraging expertise in forensic sciences and research. The researcher society in Pakistan holds the potential to serve as a valuable resource for updating outdated practices and advancing forensic science. Such advancements would contribute to an overall improvement in crime detection and the prosecution of criminal cases throughout the country [15].

In developing Asian countries, digital forensics has gotten insufficient attention. According to the study of literature, DFI research has mostly been undertaken in developed countries. There is a scarcity of accurate information about the state of digital forensics in Pakistan. Although the internet is widely used in the country, it is also commonly exploited for illicit purposes. The government has not dedicated enough resources to tackle cybercrime, resulting in a backlog of cases and an increase in complaints. There was no formal framework for investigating cybercrime in Pakistan until a few years ago, and there was a dearth of experience in the sector [16]. The National Response Center for Cybercrime (NR3C) was established by the government to address these issues and is overseen by the Federal Investigative Agency (FIA). The administration is tasked with addressing various aspects of digital forensics, cyber fraud, and technical investigations [17, 18]. However, despite efforts, digital crime remains a persistent issue in Asia, with notable instances of credit card fraud, false identification, SIM-box fraud, money laundering, and financing terrorism showing significant increases. Considering the substantial rise in cybercrime cases within the country from 2020 to 2022, along with the overall state of digital governance, it is anticipated that the figure of cybercrime suitcases will remain to escalate as Pakistan's digital landscape progresses.

The field of digital forensics is relatively new compared to traditional forensics [11]. However, in the past 35 years, law enforcement investigators have made significant efforts to develop innovative strategies for analyzing computers and other electronic devices in criminal cases [19]. As a result of the increasing demand for digital evidence to be presented in court, law enforcement agencies were forced to disseminate their findings to a larger audience. Nonetheless, the effectiveness and efficiency of traditional forensic investigation techniques remain a concern. Some investigators have been criticized for their cynicism and resistance to modern methods and tactics and have failed to keep up with contemporary practices. Moreover, adequate training to acquire the necessary skills is often lacking [21].

15.3 CURRENT CYBER LANDSCAPE OF PAKISTAN

In 2020, the International Telecommunication Union (ITU) [22] released the Global Cybersecurity Index (GCI), which rated Pakistan seventy-ninth in terms of cybersecurity. The ITU unit nations' devotion to cybersecurity is assessed using five criteria: legal mechanism, organizational measures, capacity measures, technical measures, and cooperation. Figure 15.1 depicts the GCI's rating. It is worth noting that Pakistan has declined from sixty-sixth to seventy-ninth in 2020. This decline can be attributed to the lack of progress in the five pillars of commitment, whereas other countries have shown commitment to these areas. Unfortunately, Pakistan is not adequately prepared for cyberattacks as it lacks both a National Computer Emergency Response Team (CERT) and a sectorial CERT [20,21]. The absence of adequate preparedness is not unexpected, considering the general vulnerability of the public to digital threats resulting from an absence of extensive cybersecurity information and a free-for-all online circumstances.

Institutions that heavily rely on computer technology face inherent vulnerabilities to malicious cyberattacks. Even armed forces have embraced network-enabled technology, exposing them to potential disruptions by adversaries. The growth of the communications industry has fostered an unprecedented level of global interconnectedness through modern communication devices, amplifying the potential impact of any system compromise. While a few organizations are separately striving to establish lay-off measures in this domain, there is a deficiency of coordination between

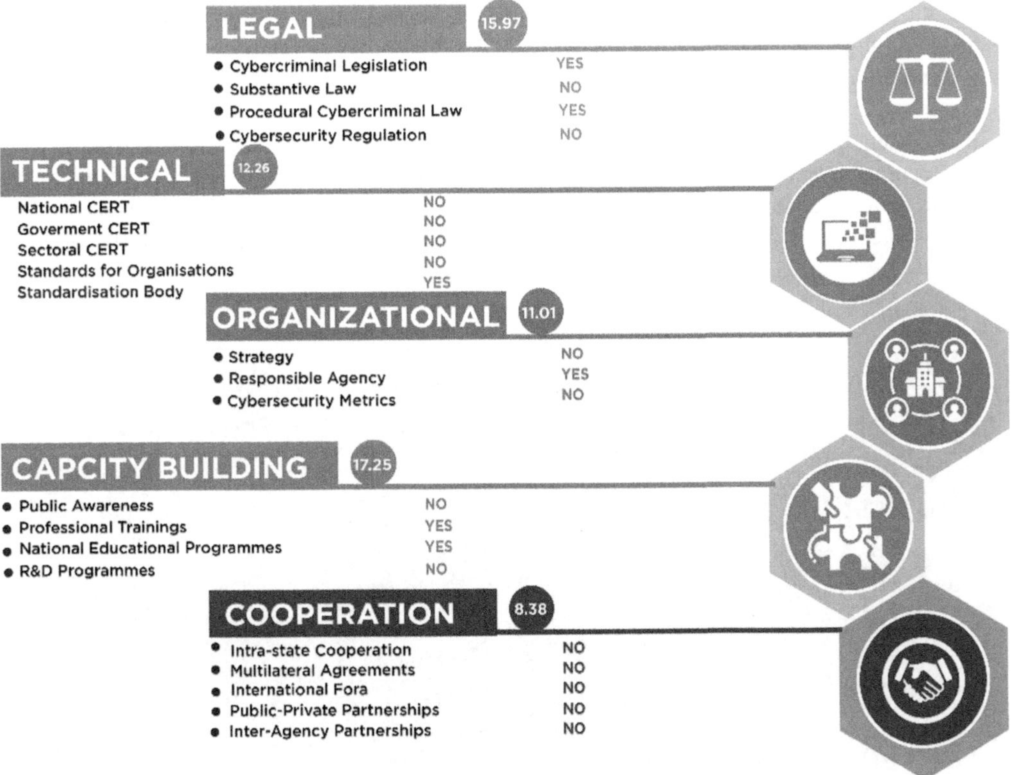

FIGURE 15.1 Pakistan current cyber landscape.

national efforts. Additionally, security corporations' function within their respective sectors and necessitate enhanced collaboration to ensure a comprehensive response. Despite the prevalence of cybercrimes in contemporary society, a substantial number of incidents go unreported. It is also progressively frequent for perpetrators to employ innovative techniques to hack into systems and unlawfully extract funds from collection accounts.

Until recently, Pakistan had an underdeveloped framework for managing cyberspace, resulting in the absence of technological corporations like the national CERT and sectoral CERTs. As a result, organizations such as telecommunication, defense, banking and gas relied on cooperative efforts to mitigate cyberattacks. However, this lack of coordination poses a threat to the country's infrastructure, as successful cyberattacks targeting one industry could have a cascading effect on others that depend on their services. To effectively counter complex cyberattacks and hybrid warfare, the Pakistani military lacks an integrated tri-services command, commonly referred to as an "Inter-Services Cyber Command" (ISC2). Additionally, the limited availability of efficient cybersecurity assets is a significant challenge in Pakistan. This can be attributed to various aspects, comprising the absence of a firm career footpath in the cybersecurity field, limited opportunities for professional development, accreditation, and the retention of highly skilled cybersecurity professionals.

15.4 CONTEXTUAL MODEL

The examination of DFI will center around six primary components, which are derived from the 2020 Global Cybersecurity Index [22]. These elements encompass the framework for multi-stakeholder collaboration in cybersecurity as outlined by the International Telecommunication Union (ITU), aiming to foster synergy between existing and future initiatives. The six foundational pillars consist of policy, legal framework, technical framework, organizational aspects, capacity building, and cooperation. These dimensions form the basis of the investigation, as they delineate the critical elements of digital investigation on a global scale. Digital forensics can be applied in almost every sector and field. The details of the six dimensions of the digital forensic investigation framework can be found in Figure 15.2.

In Pakistan, the current state of DFI position remains weak, lacking a comprehensive and proactive security program. The existing counter-cybercrime measures seem reactive in nature and focus on damage control rather than prevention [20]. These measures are inadequately detailed, under-resourced, and mostly superficial in nature. The traditional "security box centric" approach

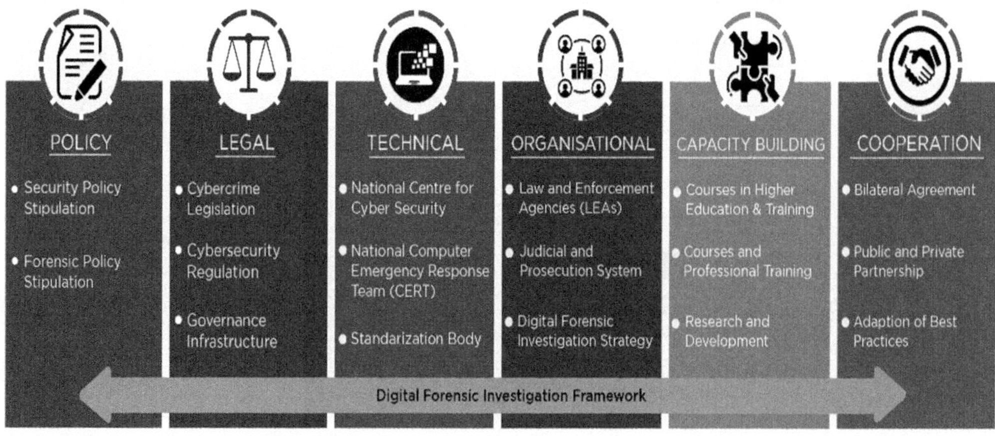

FIGURE 15.2 Digital forensic investigation framework.

of addressing DFI concerns is too narrow-minded and fails to consider innovative solutions. Furthermore, there is a lack of consensus within the same organization, resulting in a wastage of time and resources. Governance and documentation are also problematic, with excessive academic policies and procedures but poor execution strategies. In most cases, only a small percentage of the approved policies are implemented.

It is important for various stakeholders within Pakistan's administrative landscape, including the CJS, LEAs, educational institutes, departments, and ministries to collaborate effectively in order to foster the advancement of digital forensic skills at the national level. This collaboration can be accomplished through initiatives such as forming public–private partnerships, encouraging private-sector cooperation, facilitating inter-state and intra-state partnership speeding up efforts to embrace and integrate digital forensics on a broader scale. The study will focus on key aspects related to the establishment of a robust digital framework. The first pillar entails the development of policies and procedures, which will establish a comprehensive framework for defining the properties and characteristics of digital forensics through a forensic policy. The second pillar addresses the legal aspects, studying the judicial structures and organizations in position to discourse cybercrime and facilitate digital forensic investigations. The third pillar centers around technical aspects, assessing the responsiveness of technical organizations, particularly standards bodies, in addressing the challenges of digital forensics. The fourth pillar encompasses the structure of nationwide digital forensics organizational frameworks and programs to facilitate policy expertise in this field. Capacity building is the focus of the pillar dimension, encompassing academic degree programs, R&D initiatives, and accreditations aimed at promoting authority development. Finally, the sixth pillar explores the mechanisms in place to foster global alliance and cooperation in the realm of digital forensics.

15.5 CONCLUSION

Pakistan's current cyber landscape is confronted with a range of challenges, including deficit of cybersecurity comprehension, trained personnel, and an underdeveloped investigative structure. The country lags in high-tech and administrative initiatives, rendering it susceptible to cybercrime threats. The adoption of DFI framework holds immense capability in tackling these issues and strengthening the overall cybersecurity of Pakistan. DFI offers a constituted and broad approach to recognizing and probing cybercrimes, thereby fostering a more robust and secure cyber environment in the country. Nonetheless, concerns regarding the effectiveness and feasibility of DFI persist. Adequate resource allocation and funding for DFI initiatives may pose a challenge, particularly considering the constrained resources available for cybersecurity endeavors. Moreover, the scarcity of skilled professionals may impede the capability to efficiently tackle complicated cyber risks. The implementation of DFI must navigate the legal and ethical implications surrounding data protection and privacy rights, ensuring that it supports the ethics and morality of the nation. Proper policy approach, legal measures, technical and organizational measures must guide the implementation of DFI to safeguard the rights of individuals, privacy and organizations. In conclusion, DFI signifies a fundamental stride towards fortifying Pakistan's cybersecurity posture and focusing the encounters confronting its current state of the cyber landscape.

REFERENCES

1. Graeme Horsman and Nina Sunde. Unboxing the digital forensic investigation process. *Science & Justice*, 62(2):171–180, 2022.
2. Vitalii Fedorovich Vasyukov and Zarina Ilduzovna Khisamova. Investigation and seizure of electronic media in the production of investigative actions. *Revista de Direito, Estado e Telecomunicacoes*, 13(2):79–88, 2021.

3. Ghulam Muhammad Kundi, Allah Nawaz, Robina Akhtar, and IER MPhil Student. Digital revolution, cyber-crimes and cyber legislation: A challenge to governments in developing countries. *Journal of Information Engineering and Applications*, 4(4):61–71, 2014.
4. Nickson M Karie, Victor R Kebande, HS Venter, and Kim-Kwang Raymond Choo. On the importance of standardising the process of generating digital forensic reports. *Forensic Science International: Reports*, 1:100008, 2019.
5. Rapuluchukwu Ernest Nduka and Vinesh Basdeo. The need for harmonised and specialised global legislation to address the growing spectre of cybercrime. *Southern African Public Law*, 36(2):22, 2021.
6. M Barrett. Framework for Improving Critical Infrastructure Cybersecurity Version 1.1, NIST Cybersecurity Framework, [online], 2018, https://doi.org/10.6028/NIST.CSWP.04162018, www.nist.gov/cyberframework Accessed on 14-October-2022
7. Eesha Arshad Khan. The Prevention of Electronic Crimes Act 2016: An analysis. *LUMS LJ*, 5:117, 2018.
8. Humaira Arshad, Aman Bin Jantan, and Oludare Isaac Abiodun. Digital forensics: review of issues in scientific validation of digital evidence. *Journal of Information Processing Systems*, 14(2):346–376, 2018.
9. Paul Owen and Paula Thomas. An analysis of digital forensic examinations: Mobile devices versus hard disk drives utilising ACPO & NIST guidelines. *Digital Investigation*, 8(2):135–140, 2011.
10. National Institute of Standards, Technology (NIST), and United States of America. Forensic examination of digital evidence: A guide for law enforcement, 2004. https://nij.ojp.gov/library/publications/forensic-examination-digital-evidence-guide-law-enforcement
11. Karen Kent, Suzanne Chevalier, and Tim Grance. Guide to integrating forensic techniques into incident. Technical Report, 800–86, 2006.
12. ISO-IEC. 14882: 2011 information technology—programming languages—C++. International Organization for Standardization, Geneva, Switzerland, 27:59, 2012.
13. International Telecommunication Union. Next-generation digital forensics: Challenges and future paradigms. In *2019 IEEE 12th International Conference on Global Security, Safety and Sustainability (ICGS3)*, pages 205–212. IEEE, 2019.
14. Francis Kwabena Boachie. Ict infrastructure required for sustainable library services in the 21st century issues and challenges from a developing country's perspective. In *2018 5th International Symposium on Emerging Trends and Technologies in Libraries and Information Services (ETTLIS)*, pages 12–15. IEEE, 2018.
15. Ahmed Farooq and Usman Waheed. Forensics: Science of justice in pakistan. *Asian Journal of Biological and Life Science*, 2(2), 2013.
16. Jibran Jamshed, Waheed Rafique, Khurram Baig, and Waqas Ahmad. Critical analysis of cybercrimes in pakistan: Legislative measures and reforms. *International Journal of Business and Economic Affairs*, 7(1):10–22, 2022.
17. Jonathan Lusthaus. Front Matter. In *Cybercrime in Southeast Asia: Combating a Global Threat Locally*, pages [i]-01. Australian Strategic Policy Institute, 2020, www.jstor.org/stable/resrep25127.1
18. Erum Irfan, Yousaf Ali, and Muhammad Sabir. Analysing role of businesses' investment in digital literacy: A case of pakistan. *Technological Forecasting and Social Change*, 176:121484, 2022.
19. Eva A Vincze. Challenges in digital forensics. *Police Practice and Research*, 17(2):183–194, 2016.
20. Ehtisham Ul Haque, Waseem Abbasi, Sathishkumar Murugesan, Muhammad Shahid Anwar, Faheem Khan, and Youngmoon Lee. Cyber forensic investigation infrastructure of Pakistan: An analysis of the cyber threat landscape and readiness. *IEEE Access*, 11: 40049–40063, 2023, doi: 10.1109/ACCESS.2023.3268529.
21. Muhammad Abrar Khan. Nadra database hack in context of cyber kill chain and overview of pakistan's cyber security. Technical report, EasyChair, 2021.
22. Global Cybersecurity Index. International Telecommunication Union (ITU) Publication, 2020, www.itu.int/en/ITU-D/Cybersecurity/Documents/GCIv5/513560_2E.pdf Accessed on 15-December-2022

16 LockPath
Logic Locking Technique Based on Longest Path Constraint for Hardware Security

Muhammad Younas, Haroon Waris, Ahsan Iqbal, Waqar Ahmad, Muhammad Yasir Qadri, and Arif Zafar

16.1 INTRODUCTION

Outsourcing of the IC design to external industries leads to the injection of hardware trojan insertion and theft of intellectual property (IP) because of the cost factor most of the IC are outsourced to untrusted foundries. This state-of-the-art technique is a gateway to the security vulnerabilities like IP piracy, overbuilding, reverse engineering, and malicious modification [1]. Integrated circuits can be recycled, noted, and unlawfully sold. In this design flow, the functionality of an IC/IP may be cloned. The IP can then be stolen and claimed as one's own. An untrustworthy IC foundry may overproduce ICs and unlawfully sell them. IC are vulnerable to different attacks [2] that extract information about the IC design and some useful information from the device. In the modern era of technology, everyone is concerned about its security but some organizational level security is much more important for commercial companies, government, and military agencies that cannot be compromised [3]. The approximate loss of IP piracy is an estimated $4 billion per year to the semiconductor industry [4].

Different strategies are being proposed by researchers to obfuscate these sorts of assaults. The hardware obfuscation approach is one of these strategies. Hardware obfuscation approaches mostly address the change of the IC's description or circuit architecture to make reverse engineering more difficult [5]. Watermarking is one type of obfuscation method [6], split manufacturing [7], gate camouflaging [8], and logic locking [9, 10] proposed in the literature.

16.1.1 Watermarking

Watermarking is a technique for adding hidden information that functions as a tag widely used in hardware and software. These tags help protect the data or design against tampering and detect it if it occurs. Watermarking is typically used in software, but it has moved to hardware because of the advantages of low power consumption, limited space utilization, and the dependability of creating hardware for specific tasks [11].

Design Marker is an IP protection platform based on watermarking technology. This platform has both a netlist-level and a layout-level watermarking module. Using this method, watermarking messages may be inserted into the IP at the netlist or layout level with little overhead costs [12]. The downside of using watermarking is that it can only detect viruses in data and cannot effectively fight attacks.

DOI: 10.1201/9781003497851-16

16.1.2 Split Manufacturing

To acquire access to innovative technology, many firms outsource their design to untrustworthy foundries. It is vital to protect the design during the manufacturing process. The split manufacturing method can protect the design. Split manufacturing is performed by separating the design into two layers: FEOL (Front End of Layer), which includes transistor specifications, and BEOL (Back End of Layer) (Back end of Layer). The FEOL component, which consists of a transistor level and a thin metal layer, must be manufactured in sophisticated or untrustworthy foundries, but the BEOL component may be manufactured in low-trust foundries, safeguarding the overall design [13].

16.1.3 Gate Camouflaging

Gate camouflaging is another approach for safeguarding designs against theft, intellectual property infringement, or overproduction. In this strategy, more adjustable gates are inserted into the design to make it tough for opponents to understand the layout [14, 15]. Because it is a layout method, IC camouflaging cannot provide security protection for the 3PIP/IC gate-level netlist. The experimental results show that IC camouflage also has significant overheads [16].

16.1.4 Logic Locking

Logic locking is seen to be the most effective approach against attacks such as malicious modification, IP infringement, and reverse engineering. Extra gates are added to the initial design in logic locking to protect it from intruders. These additional gates serve as key gates, ensuring that the netlist functions properly by using the right pattern as input; otherwise, the incorrect output is produced. Logic Locking enhances the design to be secured by modifying it with extra key gates, and the changed design is known as a locked netlist. It is more difficult for adversaries to reverse engineer the design or add hardware trojans to it when logic locking is used. [17, 18].

Logic locking is a method that involves inserting supplementary gates into the circuit that provide worthless output until the proper key is given. Inserting more gates acts as a key, and if the right key is not provided, the circuit is believed to be locked. Circuit parts such as XOR/XNOR gates and Mux are utilized to lock the circuit and serve as buffers by supplying the right input to the netlist. The circuit will output incorrect values by giving incorrect patterns, rendering the circuit inoperable. As an illustration, Figure 16.1 will help you comprehend the notion of logic locking.

Addition of logic gates to the netlist, such as one with a key bit value input. On the chip, key values are kept in tamper-proof memory. By increasing the circuit's security, these extra gates increase power consumption, area, and power overhead. Logic locking and gate camouflage are two methods of hardware protection.

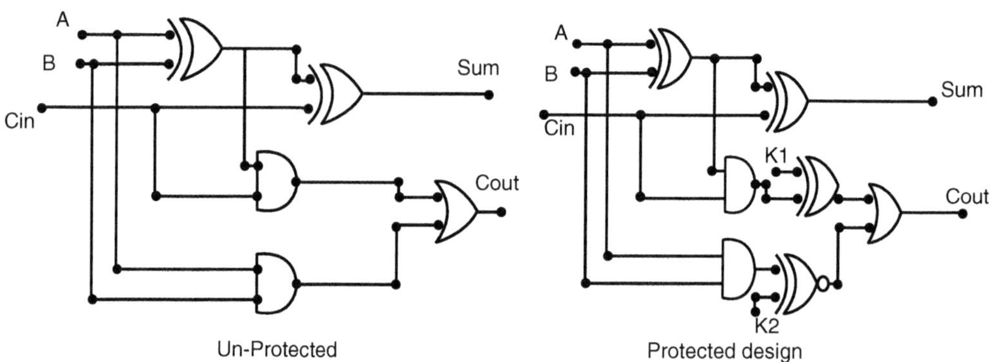

FIGURE 16.1 Logic locking.

TABLE 16.1
Truth Table of Full Adder along with Protected Design Output

A	B	C in	K1	K2	Sum	C out	Sum	C out
Input			**Key inputs**		**Original Output**		**Protected Output**	
0	0	0	0	0	0	0	0	1
0	0	1	1	0	1	0	1	1
0	1	0	1	1	1	0	1	1
0	1	1	0	1	0	1	0	1

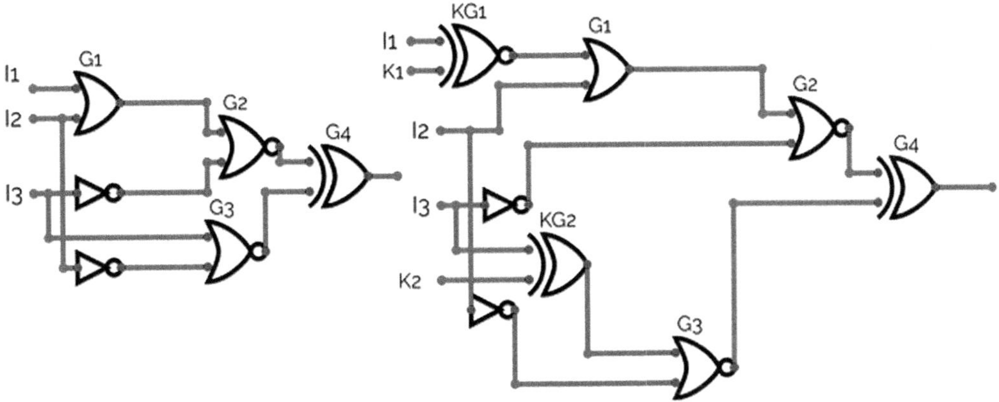

FIGURE 16.2 Diagram of logic locking including both the original and locked circuits. The right answer is 100.

Let's look at the full adder example to better understand logic locking and SAT attacks. As can be seen in Table 16.1, the full adder's original output is adjacent to the protected design. In an SAT attack, the intruder tries various distinguishing patterns (DIPs) to obtain the proper key pattern able to obtain the correct output from the design, which operates incorrectly until and unless the exact key pattern is delivered.

The key for a specific design is provided in a reliable foundry or distantly through a method for safe key exchange [19, 20] proposed logic locking, in which extra gates are put at random locations to safeguard the design. Later on, the researcher proposes other logic-locking strategies to improve its efficiency. Sequential and combination logic locking is also possible. Additional states are established in sequential logic locking to secure the design, and the design will only operate by going through the right states; if the states are not followed correctly, the design will provide erroneous output [21, 22]. In combinational logic locking additional gates like XOR/XNOR and Mux are used for the protection of design against different attacks [18, 23–26].

16.2 ATTACKS UPON LOGIC LOCKING

In this part, we will discuss the attacks against logic-locking approaches. The SAT attack, which changes the logic-locking field, is the most effective.

16.2.1 Sensitization Attack

Individual key bits are discovered via an oracle-guided attack by designing and applying arrangements that make them sensitive to the principal main output of a functional IC [27]. The key-bit K3 in the locked netlist in Figure 16.2 can be sensitized to the output O if the higher input to the gate G4 is set to 0 (its noncontrolling value). This can be done by making I3 and I2 independent of K1 and K2. As a consequence, anytime the inputs are set to 000, the result O is the same as K3. K3 is set to the value established in the preceding step to propagate K1 to the output. Similarly, K2 is found by adjusting K1 & K3 to the provided values. This approach avoids RLL (Random Logic Locking), but it cannot break SLL (Strong Logic Locking) [4].

16.2.2 Logic Cone Analysis Attack

Each circuit contains gates for the main outputs and may be divided into sections of circuits known as logic cones [28]. Because the number of gates utilized for defective logic locking and randomized logic locking is quite modest, brute-force assaults are possible. A similar strategy is used for DPA assaults in [29, 30].

16.2.3 Boolean Satisfiability Attack

One of the most successful attacks against logic-locking techniques is the Boolean Satisfiability Attack, often known as the SAT Attack. The primary idea behind an SAT assault is to undertake a brute force attack in which Distinguishing input patterns (DIPs) are provided at the input and the result is monitored. As a result, part of the input is processed, and attackers can learn about the proper pattern for the secure design to unlock it.

To better understand it, the procedure of SAT attack on the protected circuit and the result has been discussed in detail in [31].

Similarly, in oracle-less assaults, the assailants are not required to have the golden chip from the foundry and can disrupt the design, like in the Test Data Analyst approach, if the test data is not correctly obtained. Different data is delivered using the Hill climb [25] assault, and the hamming distance is measured to come closer to the key bits pattern. The Oracle-less De-Synthesis Attack uses just a locked netlist to recover the secret key [32]. In the literature, you may also find machine learning-based attacks [33].

16.3 PROPOSED METHODOLOGY

In numerous ways, researchers have presented their theories and demonstrated the findings that realistically safeguard hardware against various forms of assaults. Keeping the same principle in mind, we also want to secure the design against dangers such as reverse engineering and malicious modification, for which we suggested the Lock Path approach. We are first concerned in our suggested technique with information on the circuit parts in design under test (DUT). There are other techniques to obtain it, including parsing through the entire code and locating the relevant bits. Following that, the most difficult task was to determine the longest path constraint in our DUT, for which we utilized a recursive function that looped over all the pathways and returned the specifics of each one.

Because of the enormous number of circuit elements in the netlist, we choose the longest path as an ideal site for the installation of extra gates. We then compared the effectiveness of our techniques by performing an SAT attack on our protected design and discovered that it is more than twice as effective as a random logic-locking technique with the same number of additional gates while keeping the other constraints such as area, power, and delay unchanged.

16.3.1 Parsing

First, gather the essential information about the netlist that has been used for testing and insertion of extra key gates. To obtain these facts, we employed parser code that takes a Verilog file as input, parses it, and outputs information on the items indicated in the test netlist.

16.3.2 Longest Path

Once we get all of the information from the parser code, we want to locate the best spot for adding more key gates. For the placement of extra gates, we chose the longest path constraints. The recursive function is used to do this, and the longest path is presented at the output. In each iteration, the recursive function begins with input and proceeds to the output while keeping track of the gates in the lists. All of the lists from input to output are then compared, and only the ones with the most gates are chosen to include the key gates.

16.3.3 Key Gates Insertion

To insert extra gates, an algorithm was created to determine the number of gates to be placed in each tested design. Additional gates are added based on the results of the preceding algorithm. The number of gates can be changed or lowered to test the efficiency of the lock circuit against SAT attacks. The key size is determined by the number of additional gates inserted.

16.3.4 Locked Netlist Achieved

Finally, we fulfilled our aim of safeguarding our design from outsiders by using the LockPath logic-locking mechanism. The new file with the additional key gates has been stored in the Verilog files directory. This updated file has all of the original circuit parts, but with additional gates added to protect the circuit against assaults.

16.3.5 SAT Attack

One of the most successful attacks against logic-locking techniques has been undertaken to test the efficacy of our suggested approach. The SAT attack is a brute force assault that will verify every conceivable combination of input to key gates to breach the circuit and obtain the needed information. To breach the design information, a Verilog file is handed to it and various combinations of key gates are constructed. Each iteration of the SAT attack will locate the DIP (distinguishing input pattern) and will discard any wrong numbers (Figure 16.3).

16.4 IMPLEMENTATION

To protect the benchmark circuits, we developed Python code to implement a safeguarding technique. This script takes a Verilog benchmark netlist as input and generates a protected Verilog benchmark netlist. The protected circuit in the obfuscated design is unlocked by adding key gates with predefined keys. We created a script that includes a function to parse each netlist and extract the required information, such as input list, output list, and gate types. This information is then utilized to calculate the key size proportionally to the number of gates in the netlist. The second function inserts the key gates into the new netlist. For development purposes, we included the key components, gates, and key parameters in the header of the new netlist.

To generate and evaluate netlists before and after obfuscation, we utilized Synopsys Design Compiler. With the help of this tool, we were able to view the gate-level description of each standard during the analysis. It allows us to observe the position of logic elements and the placement of the implanted key gate after the complication process. Additionally, we used the Design Compiler to perform essential means evaluation based on our additional constraints. This tool facilitated timing analysis among various logic elements in the netlist, helping us identify critical paths and remove any points that were on them. The findings from Synopsys were incorporated by feeding them through a text file processed in the main script.

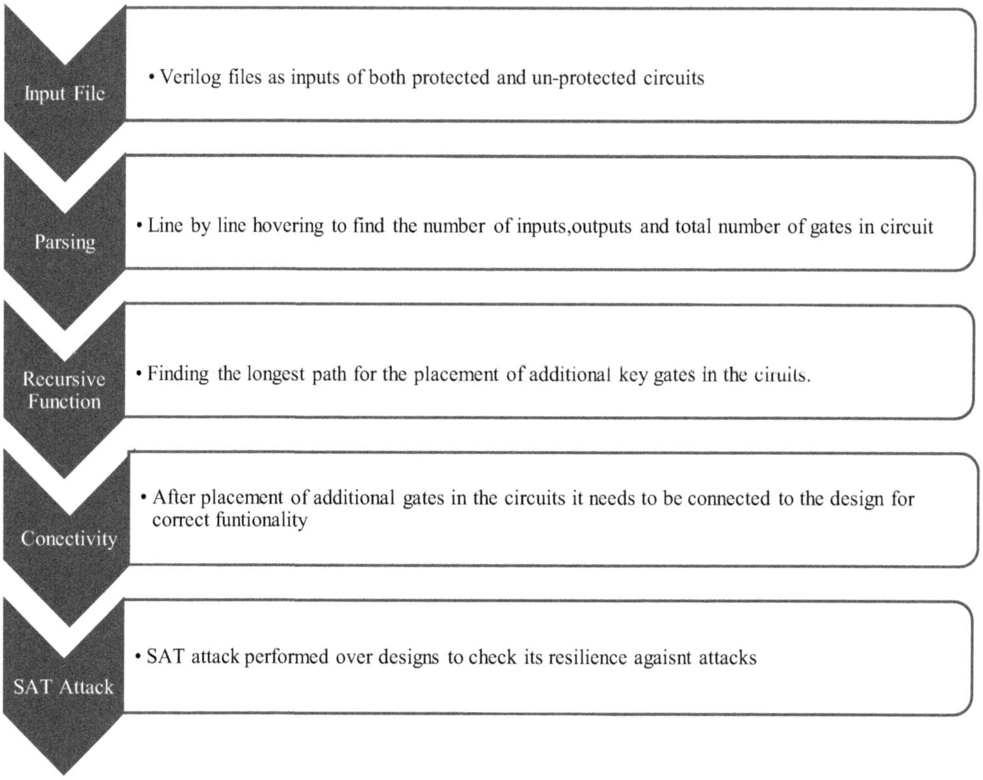

Input File
• Verilog files as inputs of both protected and un-protected circuits

Parsing
• Line by line hovering to find the number of inputs,outputs and total number of gates in circuit

Recursive Function
• Finding the longest path for the placement of additional key gates in the ciruits.

Conectivity
• After placement of additional gates in the circuits it needs to be connected to the design for correct funtionality

SAT Attack
• SAT attack performed over designs to check its resilience agaisnt attacks

FIGURE 16.3 Step-by-step understanding of the proposed method.

Algorithm for longest path constraints

Inputs: Gate List & Inputs List
Output: List of nodes for each possible path

Part I: Localizing the nodes in the Gate List array (e.g., not NOT1_1(N118, N1) → N118, N1)

```
    GATE_LIST ← []
    for I in GATE LIST:
  LIST SEP = Select the string elements only inside Parenthesis '(.)'
      for j in LIST SEP:
          LIST SEP = Remove the spaces (if present) before Gate names
      end
      GATE_LIST = GATE_LIST + LIST SEP[j]
  end
```

Part II: Search input nodes & their corresponding successors in GATE_LIST until a path is completed.

```
  global PATH← []
  for an in-INPUT LIST: % Total Inputs to be matched
      for b in GATE_LIST: % Total number of Gates where input is to be searched
          for c in GATE_LIST[b]: % Total nodes involved in one Gate
              If the input is matched at any instance of Gates
      NODE_1 = Next node to be searched, Gate output where the match occurred
                  end
                  def
RECURSIVITY (NODE_1): %Recursive function
      Match NODE_1, excluding the instances where it serves as an output
      Call 'RECURSIVITY' until the loop breaks
      PATH = Save every incremental path
                  return
PATH
              end
          end
      end
```

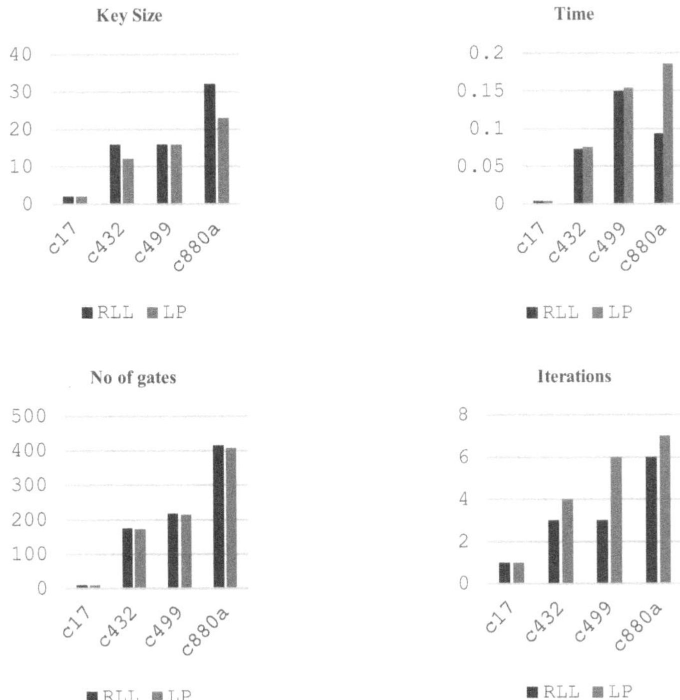

FIGURE 16.4 Comparison w.r.t to different parameters.

16.5 EXPERIMENTAL RESULTS

To evaluate our suggested technique for design protection the protected design has been subjected to a Satisfiability based attack. The SAT attack is a brute force assault that examines every possible way to penetrate the protected design. During the experiment, we first add the extra key gates i-e key size is controlled by the number of key gates, in addition to the random nodes, to the longest path list generated by our suggested longest path approach. We discovered this by comparing the random logic and just the longest path technique-protected designs while running the SAT attack technique.

The efficacy of longest path key nodes with fewer gates is about equivalent to the protection provided by random logic locking with more extra gates, increasing the area overhead and power overhead. We ran our suggested approach on multiple ISCAS'85 benchmark circuits to better understand the comparison, and the results are displayed in Tables 16.2 and 16.3 and Figure 16.4.

16.6 CONCLUSION AND FUTURE DIRECTIONS

We presented Lockpath as an obfuscation strategy since it uses fewer resources while boosting resistance to SAT attacks. This technique can be further modified to improve the results by combining the longest path and random logic-locking technique. Although it will increase the overhead to some instant the protection of the design can be improved.

TABLE 16.2
Comparison of RLL and LP Increase in Key Gates

Test bench	Type	Key Size	Time	No of gates	No. of iterations
c17	RLL	2	0.0041	8	1
c432	RLL	16	0.0728	175	3
c499	RLL	16	0.1498	216	3
c880a	RLL	32	0.0936	414	6
c17	LP	2	0.004	8	1
c432	LP	12	0.0752	172	4
c499	LP	16	0.1541	214	6
c880a	LP	23	0.1858	406	7

TABLE 16.3
Comparison of RLL and LP Increase in Key Gates

Test bench	Type	Key Size	Time	No of gates	No. of iterations
c17	RLL	4	0.0032	10	1
c432	RLL	64	0.1121	222	14
c499	RLL	64	0.1867	256	7
c880a	RLL	128	0.5185	505	15
c17	LP	4	0.0043	9	2
c432	LP	64	0.1883	219	16
c499	LP	64	2.5942	241	15
c880a	LP	128	1.9045	503	39

REFERENCES

1. Karri, R., et al., Trustworthy hardware: Identifying and classifying hardware trojans. *Computer*, 2010. 43(10): p. 39–46.
2. Mellor, J., et al. Attacks on logic locking obfuscation techniques. in *2021 IEEE International Conference on Consumer Electronics (ICCE)*. 2021. IEEE.
3. Adee, S., The hunt for the kill switch. *IEEE Spectrum*, 2008. 45(5): p. 34–39.
4. Yasin, M., et al., On improving the security of logic locking. *IEEE Transactions on Computer-Aided Design of Integrated Circuits and Systems*, 2015. 35(9): p. 1411–1424.
5. Forte, D., S. Bhunia, and M.M. Tehranipoor, Hardware protection through obfuscation. 2017: Springer.
6. Fan, Y.-C. and H.-W. Tsao, Watermarking for intellectual property protection. *Electronics Letters*, 2003. 39(18): p. 1316–1318.
7. Imeson, F., et al. Securing Computer Hardware Using 3D Integrated Circuit ({{{{IC}}}}) Technology and Split Manufacturing for Obfuscation. in *22nd USENIX Security Symposium (USENIX Security 13)*. 2013.
8. Yu, Q., et al., Hardware hardening approaches using camouflaging, encryption, and obfuscation. Hardware IP security and trust, 2017: p. 135–163.
9. Rostami, M., F. Koushanfar, and R. Karri, A primer on hardware security: Models, methods, and metrics. *Proceedings of the IEEE*, 2014. 102(8): p. 1283–1295.
10. Rajendran, J., O. Sinanoglu, and R. Karri, Regaining trust in VLSI design: Design-for-trust techniques. *Proceedings of the IEEE*, 2014. 102(8): p. 1266–1282.
11. SleetSleit, A. and N. Fetais. Watermarking: A Review of Software and Hardware Techniques. in 2018 *International Conference on Computational Science and Computational Intelligence (CSCI)*. 2018. IEEE.

12. Du, Y., et al. IP protection platform based on watermarking technique. in *2009 10th International Symposium on Quality Electronic Design*. 2009. IEEE.
13. Perez, T.D. and S. Pagliarini, A survey on split manufacturing: Attacks, defenses, and challenges. *IEEE Access*, 2020. 8: p. 184013–184035.
14. Magaña, J., et al. Are proximity attacks a threat to the security of split manufacturing of integrated circuits? *IEEE Transactions on Very Large Scale Integration (VLSI) Systems*, 2017. 25(12): p. 3406–3419.
15. Rajendran, J., O. Sinanoglu, and R. Karri. Is split manufacturing secure? in *2013 Design, Automation & Test in Europe Conference & Exhibition (DATE)*. 2013. IEEE.
16. Zhang, J., A practical logic obfuscation technique for hardware security. *IEEE Transactions on Very Large-Scale Integration (VLSI) Systems*, 2015. 24(3): p. 1193–1197.
17. Yasin, M., et al. Provably-secure logic locking: From theory to practice. in *Proceedings of the 2017 ACM SIGSAC Conference on Computer and Communications Security*. 2017.
18. Dupuis, S., et al. A novel hardware logic encryption technique for thwarting illegal overproduction and hardware trojans. in *2014 IEEE 20th International On-Line Testing Symposium (IOLTS)*. 2014. IEEE.
19. Yasin, M., et al. Activation of logic encrypted chips: Pre-test or post-test? in *2016 Design, Automation & Test in Europe Conference & Exhibition (DATE)*. 2016. IEEE.
20. Roy, J.A., F. Koushanfar, and I.L. Markov, Ending piracy of integrated circuits. *Computer*, 2010. 43(10): p. 30–38.
21. Chakraborty, R.S. and S. Bhunia, HARPOON: An obfuscation-based SoC design methodology for hardware protection. *IEEE Transactions on Computer-Aided Design of Integrated Circuits and Systems*, 2009. 28(10): p. 1493–1502.
22. Chakraborty, R.S. and S. Bhunia. Security against hardware Trojan through a novel application of design obfuscation. in *2009 IEEE/ACM International Conference on Computer-Aided Design-Digest of Technical Papers*. 2009. IEEE.
23. Rajendran, J., et al. Fault Analysis-Based Logic Encryption. *IEEE Transactions on Computers*, 2015. 64(2): p. 410–424.
24. Rajendran, J., et al. Logic encryption: A fault analysis perspective. in *2012 Design, Automation & Test in Europe Conference & Exhibition (DATE)*. 2012. IEEE.
25. Plaza, S.M. and I.L. Markov, Solving the third-shift problem in IC piracy with test-aware logic locking. *IEEE Transactions on Computer-Aided Design of Integrated Circuits and Systems*, 2015. 34(6): p. 961–971.
26. Kamali, H.M., et al. Advances in logic locking: Past, Present, and Prospects. Cryptology ePrint Archive, 2022.
27. Rajendran, J., et al. Security analysis of logic obfuscation. in *Proceedings of the 49th Annual Design Automation Conference*. 2012.
28. Lee, Y.-W. and N.A. Touba. Improving logic obfuscation via logic cone analysis. in *2015 16th Latin-American Test Symposium (LATS)*. 2015. IEEE.
29. Yasin, M., et al. Security analysis of logic encryption against the most effective side-channel attack: DPA. in *2015 IEEE International Symposium on Defect and Fault Tolerance in VLSI and Nanotechnology Systems (DFTS)*. 2015. IEEE.
30. Sengupta, A., et al. Logic locking with provable security against power analysis attacks. *IEEE Transactions on Computer-Aided Design of Integrated Circuits and Systems*, 2019. 39(4): p. 766–778.
31. Bhunia, Swarup and Tehranipoor, Mark M, Hardware security: a hands-on learning approach, Morgan Kaufmann, 2018.
32. Massad, M.E., et al. Logic locking for secure outsourced chip fabrication: A new attack and provably secure defense mechanism. arXiv preprint arXiv:1703.10187, 2017.
33. Chakraborty, P., J. Cruz, and S. Bhunia, SAIL: Machine learning guided structural analysis attack on hardware obfuscation. in *The 2018 Asian Hardware Oriented Security and Trust Symposium (AsianHOST)*. (pp. 56-61), 2018. IEEE.

17 A Review of the Various Methods of Communication Used in Ad Hoc Networks

Ahmed Hassan, Muhammad Arif, M. Arfan Jaffar, and Muhammad Ejazulghaffar

17.1 INTRODUCTION

WANETs and MANETs are two categories of autonomous wireless networks that can be established without any prior infrastructure. The primary distinction between the two entities lies in the mobility of their respective nodes. In a WANET, the constituent nodes are immobile and maintain a fixed position about each other. WANETs are commonly employed when the nodes remain stationary, such as in a sensor or residential network. A MANET is characterized by the mobility of its constituent nodes, which are capable of unrestricted movement within the network. MANET is utilized in diverse contexts, such as military missions, emergency management, and remote regions. The distinctive features and difficulties of WANETs and MANETs necessitate careful consideration of the design that best fits the particular demands and limitations of the circumstances [1] (Table 17.1).

The establishment of a MANET can present difficulties owing to the inherent limitations and constantly changing characteristics of the network. The dynamic nature of a MANET can make it challenging to provide each device with the necessary information to efficiently direct traffic. The use of ad hoc wireless networks is frequently observed in scenarios where the implementation of a permanent infrastructure is either unfeasible or unattainable, as exemplified by military operations, disaster response, and rural regions. The use of P2P overlay techniques is a viable approach for the decentralized arrangement of nodes within a VANET, thereby eliminating the requirement for a central server or infrastructure. In a Vehicular ad hoc network (VANET), nodes denote vehicles that are equipped with wireless communication devices, facilitating inter-vehicle communication and the establishment of a real-time network. Peer-to-peer (P2P) overlay techniques possess the capability to facilitate efficient communication and information sharing among nodes within the VANET [2]. The link layer within a Mobile Ad hoc Network (MANET) consists of interconnected links that facilitate communication between individual nodes. The link layer is responsible for the establishment of physical and data link connections between nodes, facilitating the transmission of data and information [3]. An ad-hoc network, commonly referred to as a peer-to-peer network, lacks a central access point or base station. In this network, every individual node function as a bidirectional transmitter and receiver, facilitating the direct exchange of packets with other nodes within the network. The implementation of a wireless base station or access point is a logical choice for the administration of a centrally managed wireless infrastructure network. This network is characterized by a central point of control and management, whereby devices are connected to the network through the aforementioned central point [4]. MANET or WANET are distributed networks where nodes or devices communicate with each other directly without the requirement of a central router or access point. The visual representation depicted in Figure 17.1 shows several fundamental attributes of wireless ad-hoc networks.

 DOI: 10.1201/9781003497851-17

TABLE 17.1
The List of Abbreviations Used in This Article

Abbreviation	Full Name	Abbreviation	Full Name
WSN	Wireless Sensor Network	**WMN**	Wireless Mash Network
WAP	Wireless Access Point	**VANET**	Vehicular Ad Hoc Network
P2P	Peer to Peer	**MANET**	Mobile Ad Hoc Network
AODV	Ad-Hoc On-Demand Distance Vector	**VOIP**	Voice over Internet Protocol
FANET	Flying Ad hoc Networks	**LAN**	Local Area Network
V2I	Vehicle to Vehicle	**IoT**	Internet of Things
AODV	Ad-hoc On-Demand Distance Vector protocol	**MAC**	Media access control

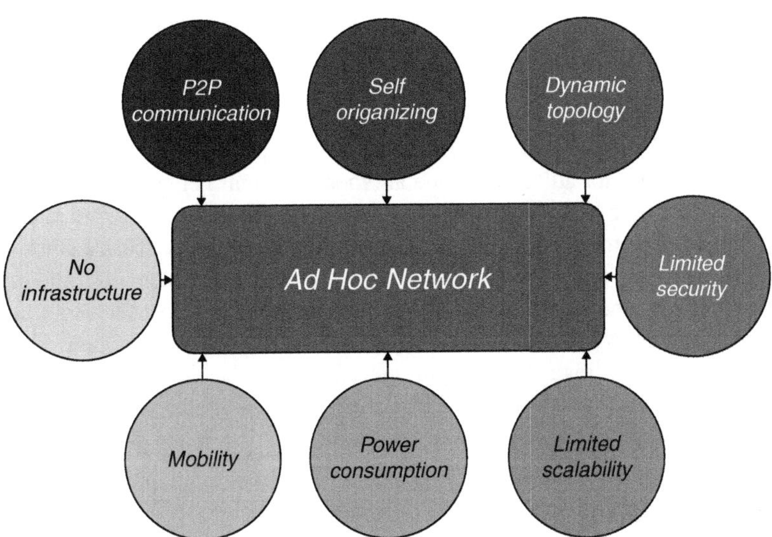

FIGURE 17.1 The characteristics of WANET.

A decentralized network enables direct communication without a central connection point, allowing for dynamic adaptation without centralized administration. Ad-hoc networks have a dynamic topology and limited security due to rapid changes and limited computational and power resources. Devices must expand to facilitate communication, and battery life is crucial in ad-hoc networks. An impromptu wireless network provides quick internet connectivity across numerous devices, but may be susceptible to security risks due to the lack of a centralized point of control. Users should employ robust encryption techniques and security protocols to protect the network [5].

17.1.1 Motivation

Technological advancements in mobile devices, such as laptops and smartphones, have facilitated the ease of accessing the internet and making phone calls while on the move [6]. The networks are recommended to users who engage in frequent file transfers between devices. The establishment of an impromptu network is feasible in virtually any setting. Hence, they possess potential benefits both for-profit and non-profit organizations, as well as for individual and household purposes [1].

The network mode is more convenient to configure than the infrastructure mode. They can communicate in what is known as ad-hoc mode if no bridge is present. The Wi-Fi standard is based on

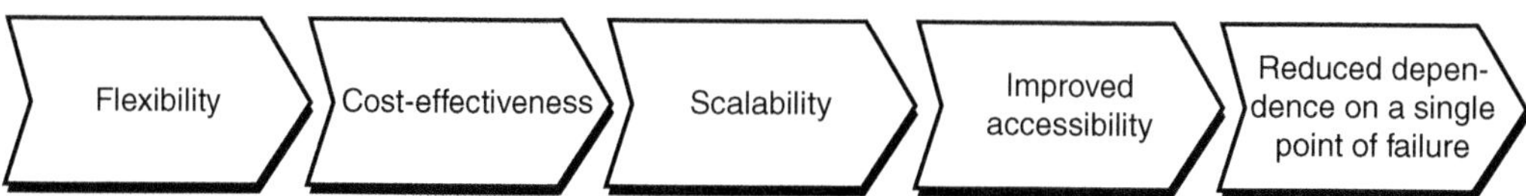

FIGURE 17.2 The benefit of WANET.

network mode, which lets users communicate with each other even without a network link [6]. In Figure 17.2, we can see the many additional benefits of cellular ad hoc networks.

17.1.2 Scope

Since there is no centralized administration or physical wiring in place, no one is controlled in the network. In a wireless network, neighbors can speak to each other if they are physically near enough. Even if they aren't in the same wireless area, they can still communicate with each other. That makes each one a Wi-Fi hub as well. As a result, the application may cease functioning if even a single component changes location or functionality. The most capable nodes are burdened with an increased share of the network's transportation and computation duties. They are used to link any two locations that are not adjacent to one another. A virtual network's backbone components are the network's building blocks. The process of creating and managing a simulated network path is easy to understand. Due to the lack of a graphical depiction of the network's components, ad hoc networks are blind to their existence. The network's components are linked via radio signals [7]. The network is decentralized, and communication occurs between sites. It connects to other nodes in the network automatically. Each node can create its own Wi-Fi network of other nodes that operate their online infrastructure. Low infrastructure requirements, and low barrier to entry [8]. Wireless ad hoc networks can take many forms and function in a variety of ways. It also talks about tough issues and gives a workable solution based on the newest ideas.

17.2 LITERATURE REVIEW

Ad-hoc networks allow devices to join and exchange data only when necessary. As the experts at the beginning of our conference discovered, its effects are temporary at best. Participants in her gatherings can now exchange data from their devices thanks to a network built specifically for this purpose [9]. Ad hoc networks are impossible to join because they lack the necessary networking gear. Without a central server or gateway, all wireless devices in the range can communicate with one another to form a MANET. It is not necessary to configure a Wi-Fi access point when two individuals are close to a Wi-Fi device allows users to establish internet connections [10].

17.3 TYPES OF AD HOC NETWORK

A wireless base station or gateway is not required for users to access the internet. Ad hoc networks are a type of LAN. Only a brief period is spent using it. It is possible to permanently establish an ad hoc network, at which point it becomes a cellular network [19]. Ad hoc networks have a variety of types, as shown in Figure 17.3.

17.3.1 Wireless Mesh Network

A wireless gateway serves as an entry point for wireless devices like laptops and smartphones to join a network. Individuals communicate with one another and share information via the backbone mesh network [20]. By connecting to a gateway, a wireless network can connect to a larger network,

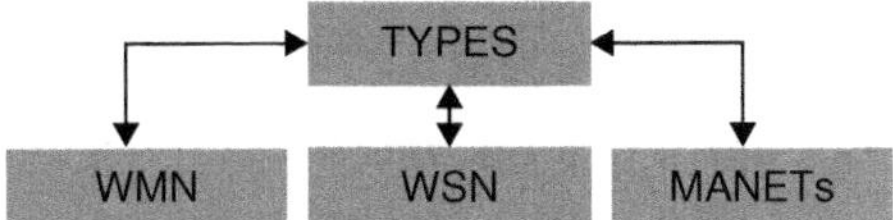

FIGURE 17.3 The ad hoc networks show different types.

like the Internet. WMN is used for the following purposes: For two or more wireless devices to talk to each other, they must first create a shared RF channel. All of the devices work together thanks to traffic algorithms. There is only one Internet Protocol address and single service set identifier for the entire system, regardless of the number of devices it employs. As a result, it can be readily distinguished from competing networks [21].

17.3.2 Wireless Sensor Network

The system's functionality, the user's surroundings, and additional portable devices are all displayed. WSNs are used for environmental management and monitoring in specific locations. There are many restrictions on how security can be implemented in WSNs. Because they have limited RAM, processing speed, battery life, and network throughput, it can be hard to keep them safe [22]. Individuals may become the target of an attack by those seeking to build upon their personal space, assert dominance, or impede their access to specific locations. Furthermore, it offers directions on techniques to guarantee the confidentiality of cryptographic keys. The authors participate in a scholarly discussion regarding a new routing methodology designed for Wireless Sensor Networks (WSNs), which considers their energy usage. To guarantee the security of the Internet of Things (IoT), a diverse range of devices will be employed, and the system will persist in an open state [23].

17.3.3 Mobile Ad Hoc Networking

The concept of multiple network communication (MANET) refers to the rapid metamorphosis of wireless communication systems, characterized by dynamism and adaptability to meet the needs of mobile employees and work teams. With the increasing number of autonomous mobile users using wireless communication systems, prompt integration is essential. MANETs have a decentralized structure, allowing nodes to interact without a central governing entity or router operation. They are easily configured and deployed, but practical implementation remains limited. Further research is being conducted on MANETs, but their practical implementation remains limited [24]. Efficient communication relies on appropriate routing protocols, which are crucial for individuals. To manage network topology changes, diverse routing methodologies are essential. Flat routing methods are suitable for small-scale networks, but they may cause noise and interference. In wired network environments, link bandwidth variability is a potential outcome. Each network serves as both a host and a routing facilitator, impacting the overall system's operation. Node inclusion, departure, and mobility are the causes of network topology variability, which results in temporal changes [25]. Table 17.2 presents a comparative analysis of the distinguishing features of VANET and MANET.

17.3.3.1 Vehicular Ad Hoc Network

Vehicular ad hoc networks are established in various places during their deployment. The transportation systems encompass a variety of modes of transportation and associated infrastructure, including structures such as bridges that enable connectivity. The individuals involved participate in wireless communication and exchange crucial information [26]. Intelligent transportation systems, including vehicle ad hoc networks, are being developed to enable automobiles

TABLE 17.2
Significant Ad Hoc Network Contribution between 2021 and 2022

Contributions	Ref	Year
A new strategy for MANETs improves service quality by using the AODV protocol and Ant Colony Optimization algorithm. The most efficient path for data transmission is determined by pheromone levels, with data packet transmission occurring via the route with the highest concentration of pheromones.	[11]	2021
Clustering in vehicular ad hoc networks improves stability and scalability by handling topological changes quickly. It addresses issues like inadequate responsibility, high growth rates, and insufficient data analysis. Multihop clustering is a technique used to address these challenges.	[12]	2022
Implementing a media access control protocol can resolve issues like overhearing, crashes, and passive listening, promoting network persistence. Quorum is a cost-effective solution to mitigate idle listening and overhearing, preventing unnecessary listening.	[13]	2022
Ad-hoc networks consist of devices like phones, laptops, iPads, and MacBooks, transmitting data and enabling remote contact. However, they face challenges in movement, power, pathfinding, and maintenance.	[14]	2022
The new way enhances the network to solve routing node movement.	[15]	2022
Vehicle ad hoc networks show efficacy in using Voice over Internet Protocol (VOIP) for high-quality audio calls. Optimized link-state routing schemes test the efficiency of these networks.	[16]	2022
The Internet of Things relies on in-flight Wi-Fi and aviation ad hoc networks for remote or marine areas. Efficient analysis is crucial for high service and precise communication models, while enhancing data transmission quality remains a challenging task.	[17]	2022
Clustering planar ad hoc networks minimizes overhead by integrating routing, setting up, and cluster head selection to evaluate network overhead efficiency in flat and grouped networks.	[18]	2022

TABLE 17.3
Difference between VANET and MANET

PARAMETERS	VANET	MANET
Node of Mobility, Network Density	High	Low
The node of the Mobility pattern	Regular	Irregular
Production Cost	Expensive	Cheap
Network Reliability	High	Medium
Multi-hop	Weak	Strong
Network address scheme	Location-based	Attribute base
Geographical	GPS Device	Ultrasonic device

to exchange information among themselves. The Vehicle Ad-Hoc Network, also known as the Intelligent Transportation Network, is expanding and may transition into an autonomous vehicle network in the near future. This requires faster topology changes due to the mobility of nodes, as more cars with computer systems and wireless communication devices are becoming prevalent [27] (Table 17.3).

The primary challenge faced by VANET pertains to the coordination of hubs, irrespective of the steering mechanism. The efficacy of guidebooks would be rendered futile in the absence of supportive seating arrangements that facilitate the dissemination of information to individuals. The concept of incorporating mobile social networking into conventional VANET collaboration tools has the potential to expand the pool of individuals able to contribute to a central hub [28]. Vehicular ad hoc networks will utilize dedicated short-range communications technology, such as Wi-Fi, to facilitate vehicle-to-vehicle connectivity. The term "vehicle-to-infrastructure" communication denotes a mode of communication. (V2I). Individuals from diverse backgrounds have exhibited a keen inclination towards information technology owing to the emergence of a network commonly referred to as VANET [29]. The network functions as the principal energy provider for a multitude of Intelligent Transportation System applications. VANET possesses the capability to augment road safety, optimize traffic flow, and enhance the commuter experience. VANET refers to a network comprising a multitude of vehicles that collaborate to address a specific requirement or circumstance [30].

17.4 ROUTING PROTOCOL

Protocols ensure data comprehension for connected machines using MANETs. There are three categories of routing protocols, designed to efficiently manage large nodes while minimizing space and resources. Intermittent node presence challenges routing protocols, so efforts should be made to expedite message transmission despite longer processing times [31].

17.5 AD HOC NETWORK ISSUES

This part will address the identified issues with the MANET. Additional information on these issues is provided below.

Wireless networks may send and receive less data than wired networks due to the narrower radio frequency. While minimizing delay, routers must utilize all available bandwidth. Routing protocols can't monitor position information due to bandwidth constraints. Control overhead is high due to topology changes.

The challenge lies in acquiring adequate competencies for wireless network devices, which require small, lightweight devices with power sources. As battery capacity and computing capabilities increase, nodes' relevance decreases, making them more difficult to move. To maximize the effectiveness of ad hoc wireless network routing techniques, users must consider limited resources.

Wireless network topology can change due to node movement, causing disruptions to ongoing sessions. Wired network protocols can find alternative routes in such cases, but determining the most efficient reconnection method can be time-consuming. Mobile wireless networks cannot use wired network routing protocols due to frequent topology changes and the need for high-performance protocols to handle these changes.

Ad hoc wireless networks face challenges due to the radio channel's broadcast nature, which can cause fluctuations in Wi-Fi connection response times. To optimize routing protocol and MAC layer performance, coordinated efforts are needed to discover and choose efficient routes. Ad hoc networks are sensitive to data and control packet intermingling, which arises from hidden terminals.

The term "hidden terminal problem" pertains to the situation where multiple nodes are not nearby but are within the range of transmission of a different node that is concurrently transmitting data. In scenarios where two nodes simultaneously transmit packets, the identities of the communicators are generally undisclosed. The parties involved lack familiarity with one another.

17.6 ANALYSIS AND DISCUSSION

This study compares the routing protocols used in mobile MANET, WMN, and WSN and identifies their distinctive features. The AODV is a more flexible solution since it works for both MANET and WSN. In comparison to WSN and WMN, the Mobile Ad-hoc Network exhibits fewer restrictions, making it more suited for customizing goods or services to meet the needs of particular users. Wireless sensor networks (WMNs) and MANETs lack specificity, whereas WMNs and MANETs

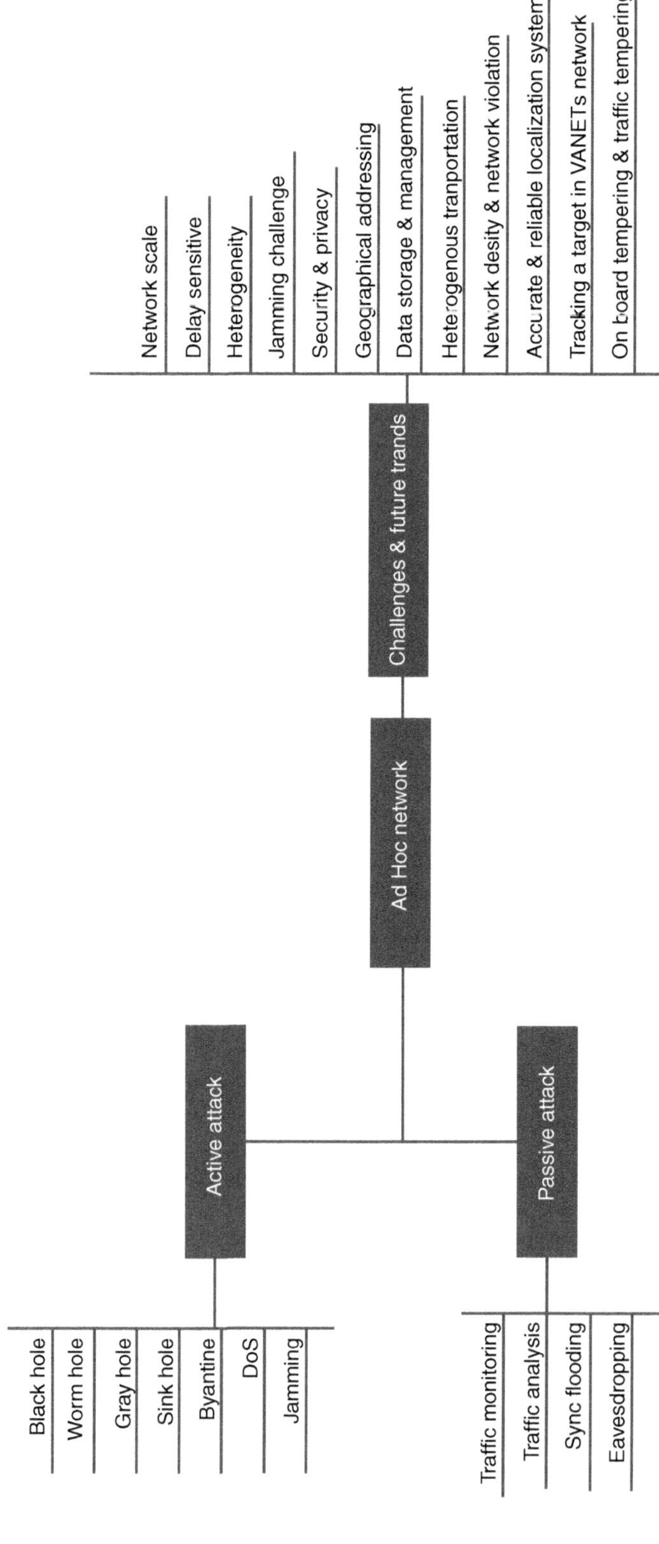

FIGURE 17.4 Challenges and future by VANETs.

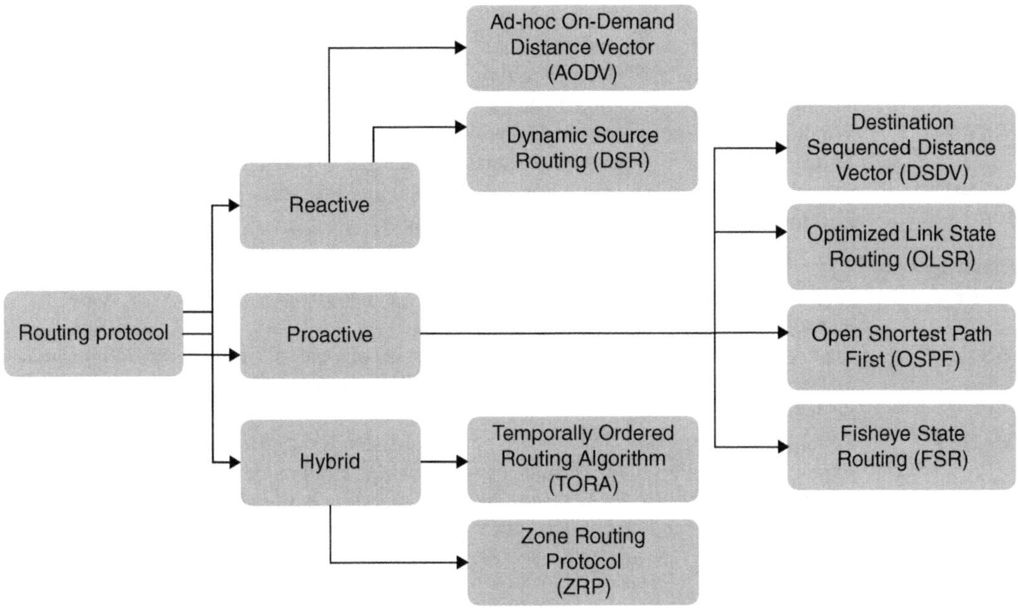

FIGURE 17.5 Different routing protocol in MANETs

exhibit variable capacity characteristics. The network and transport layers, as well as bandwidth restrictions and security flaws, are all limitations related to MANET, WMN, and WSN. In the health industry, WMN is constrained by GPRS and GPS, whereas WSN has limited architectural design, data aggregation, and localization capabilities. While WSN struggles with service quality, performance, and safety issues, MANET struggles with energy limitations and scalability. The potential of these networks for numerous applications is highlighted throughout the study.

Routing Protocol [32–33]	
WMN	Associativity, Predictive, Zone, Hazy Sighted, Dynamic Source, Hybrid wireless mash, On-demand distance vector
WSN	Featured, Energy Efficient, Secure, Delay Less, Reliable
MANET	Proactive, Reactive, Hybrid, Geographical, Hierarchical
Applications [34–36]	
WMN	High speed, voice over Internet protocol, Surveillance, Communication, Smart Home Devices, Electric Smart Meter
WSN	Monitoring, Noising, Agriculture detection, Internet of Things (IoT), Medical Applications, Environmental temperature
MANET	Industrial and Commercial Markets, For Military services, Education Gaming, and Entertainment Market
Characteristics [34–36]	
WMN	Enhanced Reliability, Random Deployment, Mobility, Non-line-of-sight (NLOS) Transmission, Fast Movement, Confidentiality
WSN	Application Specific Design (ASD), Dynamic Topologies, Node Failure, Communication Models, Scalability and Density, and Communication
MANET	Capacity Links, Bandwidth, Autonomous Behavior, Limited Security, Limited Bandwidth, No Centralized Control, Random Change

(*Continued*)

Limitation [37–38]	
WMN	General Packet Radio Service, Monitoring, Telemonitoring, Mobile Emergency Management, Emergency Telemedicine
WSN	Wireless Radio Communication Characteristics, Medium Access Scheme, Synchronization Middleware, Data Aggregation Architecture, Database, Programming Models, Security
MANET	Transmission Error, Processing Capability, Scalability Problems, Location, Limited Resources, Roaming Commercially Unavailable
Challenges [38–39]	
WMN	High Performance, Scalability, Security, Coverage, and External networks can cause interference, Power Transmission
WSN	Energy Efficiency, Performance, The issue, Service of Quality, Cross-Layer Optimization, Large scale to Scalability, Node failure, Cross-Layer Optimization, Large scale to Scalability, Node failure
MANET	Reliability, Survivability, limited wireless transmission Range, Broadcast, Hidden and Exposed Terminal Problems, Route changes resulting from mobility, Packet, Route changes resulting from mobility, Packet losses linked to mobility. Battery Constraints,

17.7 CONCLUSION

Ad-hoc networking faces challenges, and ongoing research is needed to find solutions. Current key management methods are costly, difficult, and ineffective. To meet modern network requirements, increased security and resilience are needed. This study reviews research on ad hoc networks from 2021 to 2022, focusing on features and advantages. It analyzes MANETs, focusing on network node movement, and provides a deeper understanding of MANET. Analysis of MANET, WMN, and WSN is performed in comparison across a range of factors, including constraints, difficulties, characteristics, benefits, and disadvantages as well. The primary objective is to enhance communication precision and efficiency across various networks.

REFERENCES

1. Das, S., Rao, R.S., Das, I., Jain, V., Singh, N.: *Cloud Computing Enabled Big-Data Analytics in Wireless Ad-Hoc Networks*. CRC Press (2022).
2. Trofimova, Y., Tvrdík, P.: Enhancing reactive ad hoc routing protocols with trust. *Future Internet*. 14, 28 (2022).
3. Qin, R., Li, X.: Parameter optimization for neighbor discovery probability of ad hoc network using directional antennas. In: *LISS 2021: Proceedings of the 11th International Conference on Logistics, Informatics and Service Sciences*. pp. 523–536. Springer (2022).
4. Malnar, M., Jevtic, N.: An improvement of AODV protocol for the overhead reduction in scalable dynamic wireless ad hoc networks. *Wirel. Netw*. 28, 1039–1051 (2022).
5. Li, Q., He, D., Yang, Z., Xie, Q., Choo, K.-K.R.: Lattice-based conditional privacy-preserving authentication protocol for the vehicular Ad Hoc network. *IEEE Trans. Veh. Technol*. 71, 4336–4347 (2022).
6. Liu, R., Li, X.: Research on reliability assurance mechanism of MAC layer control messages in wireless ad hoc networks. In: *Advances in Guidance, Navigation and Control: Proceedings of 2020 International Conference on Guidance, Navigation and Control, ICGNC 2020, Tianjin, China*, October 23–25, 2020. pp. 5301–5310. Springer (2022).
7. Kharchenko, V., Grekhov, A., Kondratiuk, V.: Traffic simulation in SAGIN air segment containing ad hoc Network of flying drones. *EEE*. 2, 33–42 (2022).
8. Hassani, M.M., Pashmforoush, S., Zarandi, A.A.E.: Effect of the spectrum sensing period on the TCP performance over cognitive radio ad-hoc networks. *Wirel. Pers. Commun*. 124, 2411–2426 (2022).
9. Li, H., Li, X. , Jing, T.: Design and analysis of low signaling overhead multiple access protocol for wireless ad hoc networks. In: *Advances in Guidance, Navigation and Control: Proceedings of 2020*

International Conference on Guidance, Navigation and Control, ICGNC 2020, Tianjin, China, October 23–25, 2020. pp. 5325–5336. Springer (2022).

10. Nagpal, S., Aggarwal, A., Gaba, S.: Privacy and security issues in vehicular ad hoc networks with preventive mechanisms. In: *Proceedings of International Conference on Intelligent Cyber-Physical Systems: ICPS 2021*. pp. 317–329. Springer (2022).
11. Sarkar, D., Choudhury, S., Majumder, A.: Enhanced-Ant-AODV for optimal route selection in mobile ad-hoc network. *J. King Saud Univ.-Comput. Inf. Sci.* 33, 1186–1201 (2021).
12. Katiyar, A., Singh, D., Yadav, R.S.: Advanced multi-hop clustering (AMC) in vehicular ad-hoc network. *Wirel. Netw.* 1–24 (2022).
13. Alzahrani, E., Bouabdallah, F., Almisbahi, H.: State of the art in quorum-based sleep/wakeup scheduling MAC protocols for ad hoc and wireless sensor networks. *Wirel. Commun. Mob. Comput.* 2022, 6625385 (2022).
14. Fatima, M., Khursheed, A.: Heterogeneous ad-hoc network management: An overview. *Cloud Comput. Enabled Big-Data Anal. Wirel. Ad-Hoc Netw.* 2, 103–123 (2022).
15. Pandey, M.R., Mishra, R.K., Shukla, A.K.: An improved node mobility patten in wireless ad hoc network. In: *Applied Information Processing Systems: Proceedings of ICCET 2021*. pp. 361–370. Springer (2022).
16. Dafalla, M.E.M., Mokhtar, R.A., Saeed, R.A., Alhumyani, H., Abdel-Khalek, S., Khayyat, M.: An optimized link state routing protocol for real-time application over Vehicular Ad-hoc Network. *Alex. Eng. J.* 61, 4541–4556 (2022).
17. Gurumekala, T., Indira Gandhi, S.: Toward in-flight Wi-Fi: a neuro-fuzzy based routing approach for Civil Aeronautical Ad hoc Network. *Soft Comput.* 26, 7401–7422 (2022).
18. Shang, J., Liu, Y., Tong, X.: Research on network overhead of two kinds of wireless ad hoc networks based on network fluctuations. In: *LISS 2021: Proceedings of the 11th International Conference on Logistics, Informatics and Service Sciences*. pp. 584–595. Springer (2022).
19. Benjbara, C., Habbani, A., Mouchfiq, N.: New multipath OLSR protocol version for heterogeneous ad hoc networks. *J. Sens. Actuator Netw.* 11, 3 (2021).
20. Chiejina, E., Xiao, H., Christianson, B., Mylonas, A., Chiejina, C.: A robust Dirichlet reputation and trust evaluation of nodes in mobile ad hoc networks. *Sensors*. 22, 571 (2022).
21. Seshasayee, G.: Fast fading model effects in vehicular ad-hoc network. Kauno technologijos universitetas (2022).
22. Kurumbanshi, S., Rathkanthiwar, S., Patil, S.: Packet scheduling algorithm to improvise the packet delivery ratio in mobile ad hoc networks. In: *Proceedings of Second Doctoral Symposium on Computational Intelligence: DoSCI 2021*. pp. 535–544. Springer (2022).
23. Gaurav, A., Gupta, B., Peñalvo, F.J.G., Nedjah, N., Psannis, K.: Ddos attack detection in vehicular ad-hoc network (vanet) for 5g networks. *Secur. Priv. Preserv. IoT 5G Netw. Tech. Chall. New Dir.* 5(1), 263–278 (2022).
24. Behura, A., Srinivas, M., Kabat, M.R.: Giraffe kicking optimization algorithm provides efficient routing mechanism in the field of vehicular ad hoc networks. *J. Ambient Intell. Humaniz. Comput.* 13, 3989–4008 (2022).
25. Theerthagiri, P., Gopala Krishnan, C.: Vehicular multihop intelligent transportation framework for effective communication in vehicular ad-hoc networks. *Concurr. Comput. Pract. Exp.* 34, e6833 (2022).
26. Deva Priya, M., Rajkumar, M., Karthik, S., Christy Jeba Malar, A., Kanmani, R., Sandhya, G., Anitha Rajakumari, P.: Emperor Penguin Optimization Algorithm and M-Tree-Based multi-constraint multicast ad hoc on-demand distance Vector Routing Protocol for MANETs. In: *3rd EAI International Conference on Big Data Innovation for Sustainable Cognitive Computing*. pp. 101–116. Springer (2022).
27. Chauhan, R., Rao, R.S., Das, S.: Cloud-based underwater ad-hoc communication: Advances, challenges, and future scopes. In: *Cloud Computing Enabled Big-Data Analytics in Wireless Ad-hoc Networks*. pp. 1–13. CRC Press.
28. Tabassum, K., Shaiba, H., Essa, N.A., Elbadie, H.A.: An efficient emergency patient monitoring based on mobile ad hoc networks. *J. Organ. End User Comput.* 34, 1–12 (2022).
29. Li, X., Yin, X.: Blockchain-based group key agreement protocol for vehicular ad hoc networks. *Comput. Commun.* 183, 107–120 (2022).

30. Janardana, A.P.: Implementasi Ad-Hoc On Demand Distance Vector (Aodv) Berdasarkan Faktor Kecepatan Dan Arah Di Lingkungan Vanets. (2022).
31. Verma, N., Soni, S.: A review of different routing protocols in MANET. *Int. J. Adv. Res. Comput. Sci.* 8, (2017).
32. Tsao, K.-Y., Girdler, T., Vassilakis, V.G.: A survey of cyber security threats and solutions for UAV communications and flying ad-hoc networks. *Ad Hoc Netw.* 133, 102894 (2022).
33. Srilakshmi, U., Veeraiah, N., Alotaibi, Y., Alghamdi, S.A., Khalaf, O.I., Subbayamma, B.V.: An improved hybrid secure multipath routing protocol for MANET. *IEEE Access.* 9, 163043–163053 (2021).
34. Grover, J.: Security of vehicular ad hoc networks using blockchain: A comprehensive review. *Veh. Commun.* 34, 100458 (2022).
35. Alladi, T., Chamola, V., Sahu, N., Venkatesh, V., Goyal, A., Guizani, M.: A comprehensive survey on the applications of blockchain for securing vehicular networks. *IEEE Commun. Surv. Tutor.* 24, 1212–1239 (2022).
36. Inedjaren, Y., Maachaoui, M., Zeddini, B., Barbot, J.-P.: Blockchain-based distributed management system for trust in VANET. *Veh. Commun.* 30, 100350 (2021).
37. Mathur, A., Newe, T.: Comparison and overview of Wireless sensor network systems for Medical Applications. *Int. J. Smart Sens. Intell. Syst.* 7, 1–6 (2020). https://doi.org/10.21307/ijssis-2019-014.
38. Yogarayan, S.: Wireless ad hoc network of MANET, VANET, FANET and SANET: A review. J. *Telecommun. Electron. Comput. Eng. JTEC.* 13, 13–18 (2021).
39. Goyal, A.K., Agarwal, G., Tripathi, A.K., Sharma, G.: Systematic study of VANET: Applications, challenges, threats, attacks, schemes and issues in research. *Green Comput. Netw. Secur.* 33–52 (2022).

18 Authentication of User Data for Enhancing Privacy in Cloud Computing Using Security Algorithms

Rashid Ali, Hamayun Khan, Muhammad Waleed Arif, Muhammad Imran Tariq, Irfan Ud Din, Amna Afzal, and Muhammad Azam Khan

18.1 INTRODUCTION

Cloud computing is the practice of accessing a set of computing resources that are owned and managed by a third party through the internet. While not a new concept, it is a delivery method for computing resources that builds on existing technologies such as server virtualization. The "cloud" encompasses various hardware, storage, network, interface, and service components that allow users to access on-demand infrastructure, computational power, applications, and services from anywhere. Cloud computing usually entails transferring, storing, and processing data on the provider's infrastructure, which is beyond the control of the customer's policy.

Cloud computing is linked to Information as a Service (IaaS), Platform as a Service (PaaS), and Software as a Service (SaaS), which are all based on a service-oriented architecture [1]. This technology provides a significant advantage by minimizing the need for expensive hardware at the user's end. By storing data remotely, there is no need for the user to maintain on-premises data storage, leading to cost savings. Users no longer need to purchase the entire infrastructure needed for executing operations and storing vast amounts of data. Instead, they can rent the necessary assets based on their specific needs.

All cloud networks share a similar concept which involves using remote services via a network utilizing various resources. The main objective is to achieve maximum output with minimal resources. In other words, users require minimal hardware resources but can leverage the maximum computing power. This is achievable only through this technology, which optimizes resource usage to its fullest potential [2].

Cloud computing is favored over traditional computing due to its benefits such as agility, lower cost of entry, device and independence of location, and scalability. To address concerns regarding checking of integrity of data, various approaches have been suggested across different systems and security models. These schemes aim to satisfy diverse demands such as efficient operation, stateless validation, limitless query usage, and ease of data retrieval. They are categorized into private auditability and public auditability based on the verifier's role in the model. While private auditability schemes are more efficient, public auditability schemes allow anyone to challenge the cloud server for data storage correctness without revealing any private information.

The focus of this paper is to analyze and evaluate the performance of various encryption algorithms and their impact on security in the context of cloud computing. The study was conducted through

DOI: 10.1201/9781003497851-18

a series of experiments, using different parameters such as correlation analysis, visual assessment, statistical analysis, time complexity, and execution time. Additionally, the paper provides detailed pseudocode and flowcharts to aid in understanding the algorithms [3]. To further elaborate on the topic, the paper also includes a comparison of the performance of commonly used encryption algorithms, namely DES, AES, RSA, Blowfish, Twofish, and Elliptic Curve Cryptography, as they relate to cloud service protection. Section 18.1 introduces the broader field while Sections 18.2 and 18.3 represent various security issues and challenges associated with cloud computing and the formulates the problem. Section 18.4 outlines the proposed work plan and Sections 18.5 and 18.6 provide an overview of security algorithms used in cloud computing, along with pseudocode, flowcharts, complexity analysis, implementation, and results and conclusion.

18.2 SECURITY ISSUES AND CHALLENGES OF CLOUD COMPUTING

Cloud computing, like everyday computing, relies heavily on security measures to safeguard the sensitive data it stores. However, the use of new technologies and services in cloud computing infrastructure means that many of these measures have not yet been thoroughly assessed for their effectiveness. This can lead to significant issues with trust, data security, performance, and regulatory compliance. One of the most significant concerns is the possibility of malicious insiders compromising the security of cloud services. As a result, companies are focusing on ways to mitigate these risks and ensure the trustworthiness of their cloud management. To address these concerns, various security issues must be taken into consideration and adequate measures put in place to mitigate them [4].

18.2.1 Security Challenge 1

The utilization of cloud computing entails a potential compromise of physical security as the computing resources are shared with other entities. This leads to a lack of knowledge and control over the resources being utilized.

ENSUE: Ensuring Secure Data Transfer [5].

18.2.2 Security Challenge 2

To ensure the data integrity during transfer, retrieval and storage. it is crucial to ensure that it only undergoes modifications following authorized transactions. However, there is currently no common standard in place to ensure integrity of data that includes ENSUE: Ensuring Secure Software Interfaces.

18.2.3 Security Challenge 3

The violation of privacy rights by cloud service providers can lead to potential lawsuits by customers, and damage to the reputation of the providers. People may become concerned when they are unsure about the reason for the gathering of their personal data or how it will be utilized or shared with other organizations.

ENSUE: Maintaining Data Separation [6]

18.2.4 Security Challenge 4

Ownership and control of encryption/decryption keys should ideally rest with the customer to ensure secure storage of data. This is important to ensure the privacy and security of the saved data.

18.2.5 Security Challenge 5

For Payment Card Industry – Data Security Standard (PCI – DSS) compliance, data logs must be made available to security managers and regulators. ENSUE: Ensuring User Access Control [7].

18.3 PROBLEM FORMULATION

Security is one of the major concerns in cloud computing, as it involves sensitive data. Other issues include privacy, segregation, storage, reliability, and capacity, among others. However, the security issue is the most crucial, and cloud service providers must ensure its maintenance. Cloud computing serves various customers like ordinary users, enterprises and academia each having different motivations for moving to the cloud. For instance, academia is concerned about the security impact on computing performance, while enterprises are primarily focused on security with a distinct perspective. This article focuses on user-cloud security in cloud computing, proposing a specific plan that utilizes encryption algorithms.

18.4 PROPOSED WORK PLAN

We suggested a set of security algorithms aimed at mitigating issues related to data privacy, loss, and segregation while interacting with web applications on cloud platforms. Our proposed algorithms included AES, DES, and RSA. Elliptic Curve Cryptography, TwoFish and Blowfish, and we presented a comparative study among them to ensure the security of data on the cloud. AES, DES, Elliptic Curve Cryptography, Blowfish and TwoFish are symmetric key algorithms that employ a sole key to perform both the encryption and decryption of messages. DES was developed by IBM in the early 1970s, while Blowfish in 1993 specifically for implementation in resource-limited settings, such as embedded systems. On the other hand, NIST designed AES in 2001. The Rivest-Shamir-Adleman (RSA) algorithm, introduced in 1978, is a type of public-key cryptography. It is also referred to as an asymmetric key algorithm because it utilizes distinct keys for encryption and decryption processes. Each algorithm has a unique key size that sets it apart from the others. The structure of a community cloud can be contributed by several agencies, may help a specific area, and facilitate team awareness of contributed problems. A community cloud is enabled by agencies and third parties. The key sizes employed by various algorithms are as follows: DES uses a 56-bit key, AES employs 128, 192, or 256-bit keys, Blowfish utilizes keys ranging from 128 to 448 bits, and RSA adopts a key size of 1024 bits. We implemented our idea using NetBeans IDE 7.3 and Java Runtime Environment by developing encryption or decryption algorithms based on the discussed algorithms and also compared them by evaluating their respective characteristics.

18.5 SECURITY ALGORITHM USED IN CLOUD COMPUTING

18.5.1 RSA Algorithm

The Public Key algorithm known as RSA, which stands for the last names of its inventors Rivest, Shamir, and Adelman, is the most widely used Encryption/Decryption algorithm, RSA, is asymmetric in nature. This means the public key, distributed to all, is used to encrypt the message, while the private key, which is kept secret and not shared with everyone, is used for decryption. Cloud computing relies on the RSA algorithm to protect sensitive data. To ensure that only authorized users can access the information, the data is encrypted using RSA and stored in the cloud. Whenever a user needs the data, they request it from the cloud provider, who then verifies the user's identity and delivers the data to them. In this proposed cloud setup, the public key is publicly available, while the private key is only accessible to the user who owns the data. The cloud service provider is

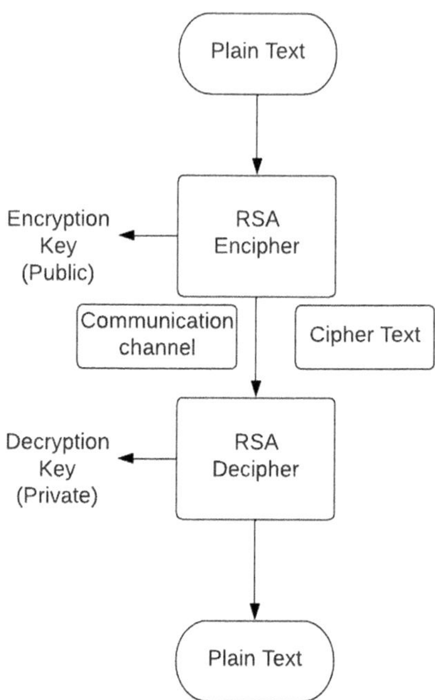

FIGURE 18.1 RSA algorithm flowchart.

responsible for encrypting the data with the public key, while the cloud user or consumer decrypts it using the corresponding private key. As RSA functions as a block cipher, each message is assigned an integer value [8] (Figure 18.1).

The time complexity of RSA encryption is O(k^3) where k is the size of the RSA key in bits. The time complexity of RSA decryption is O(k^3 * log(n)) where n is the modulus, and k is the size of the RSA key in bits.

```
// Generate public-private key pair
GenerateKeyPair(keySize)

// Encrypt user data using public key
encryptedData = Encrypt(data, publicKey)

// Send encrypted data to cloud
SendToCloud(encryptedData)

// Retrieve encrypted data from cloud
encryptedData = RetrieveFromCloud()

// Decrypt data using private key
decryptedData = Decrypt(encryptedData, privateKey)

// Use decrypted data
UseData(decryptedData)
```

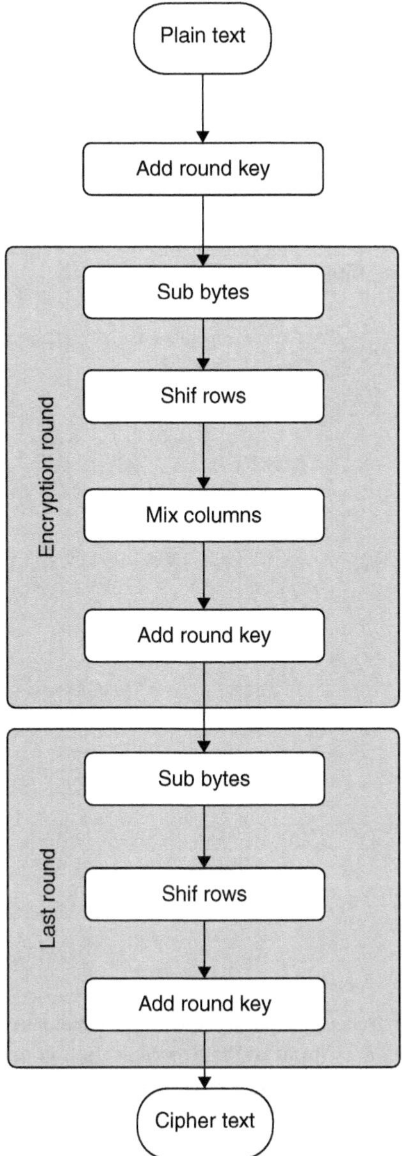

FIGURE 18.2 AES algorithm flowchart.

18.5.2 AES Algorithm

AES, or Advanced Encryption Standard, is a widely used symmetric block cipher that provides secure encryption of data in the cloud environment. In this proposed implementation, the user first decides to use cloud services and migrates their data to a Cloud Service Provider (CSP). The user submits their service requirements to the CSP and selects the best specified services offered by the provider. Before uploading data to the cloud, the application encrypts it using the AES algorithm with a key length of 128-bits. The encrypted data is then transmitted to the cloud provider. To read the data, it needs to be decrypted on the user's end, and only then can it be viewed in plain text format. It's worth noting that the plain text data is not stored anywhere on the cloud, ensuring its security. The encryption method in question can seamlessly integrate with the application without requiring any modifications. Additionally, to prevent key compromise, the key is not stored alongside the encrypted data. Instead, a dedicated physical server for key management can be set up on the user's premises. By utilizing this encryption method, the security of both data and keys is maintained, ensuring they are always under the user's control and never compromised during storage or transmission. For a variety of applications, AES has become the new standard, replacing DES [9] (Figure 18.2).

```
The time complexity is O(n), where n is the number of blocks of
input data and r is the number of rounds

// Generate random key and initialization vector (IV)
key = GenerateRandomKey(keySize)
iv = GenerateRandomIV()

// Encrypt user data using AES
encryptedData = AESEncrypt(data, key, iv)

// Send encrypted data, key and IV to the cloud
SendToCloud(encryptedData, key, iv)

// Retrieve encrypted data, key and IV from the cloud
encryptedData, key, iv = RetrieveFromCloud()

// Decrypt data using AES
decryptedData = AESDecrypt(encryptedData, key, iv)

// Use decrypted data
UseData(decryptedData)
```

18.5.3 DES Algorithm

DES is a type of block cipher that operates by encrypting data in chunks of 64 bits. This means that 64 bits of clear text are given as input to DES, which then generates 64 bits of cipher text. The algorithm employs a single key for both encryption and decryption, with slight variations. Despite having a key length of 56 bits, DES actually uses a 64-bit key as input. As a result, DES is classified as a symmetric key algorithm [10] (Figure 18.3).

```
The time complexity is O(n), where n is the number of blocks of
input data and r is the number of rounds.

// Step 1: Generate subkeys from the key using the key schedule
  subkeys = generate_subkeys(key)

// Step 2: Pad the plaintext to be a multiple of 64 bits
  plaintext = pad_plaintext(plaintext)

// Step 3: Divide the plaintext into 64-bit blocks
  blocks = divide_plaintext_into_blocks(plaintext)

// Step 4: Encrypt each block using the subkeys
```

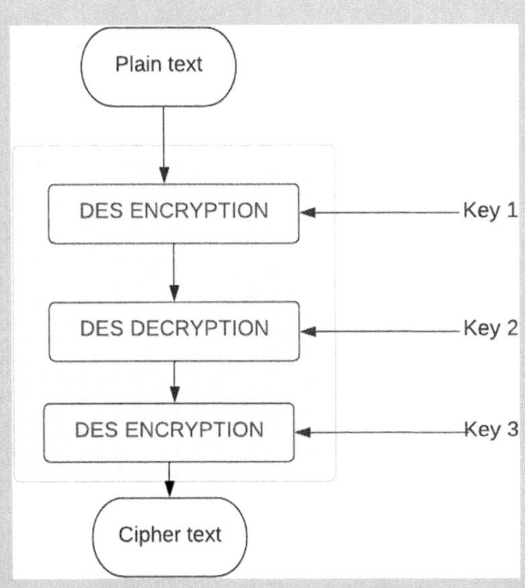

FIGURE 18.3 DES algorithm flowchart.

```
  encrypted_blocks = []
  for block in blocks:
    encrypted_block = initial_permutation(block)
    left = encrypted_block[0:32]
    right = encrypted_block[32:64]
    for i in range(16):
      temp = left
      left = right
    right = temp XOR feistel_function(right, subkeys[i])
    encrypted_block = final_permutation(right + left)
    encrypted_blocks.append(encrypted_block)

// Step 5: Combine the encrypted blocks into a single ciphertext
  ciphertext = combine_blocks_into_ciphertext(encrypted_blocks)
```

18.5.4 Blowfish Algorithm

Blowfish is considered secure and flexible due to its variable key length, which can be as long as 448 bits. Blowfish is suitable for applications that require a constant key over a long period, such as encryption of communication links; however, in situations where the key requires frequent changes, such as in packet switching, this approach may not be suitable [11] (Figure 18.4).

```
The time complexity is O(n), where n is the number of blocks of
input data and r is the number of rounds.

// Step 1: Initialize the Blowfish algorithm with the key
bf_state = initialize_blowfish(key)

// Step 2: Pad the plaintext to be a multiple of 64 bits
plaintext = pad_plaintext(plaintext)
```

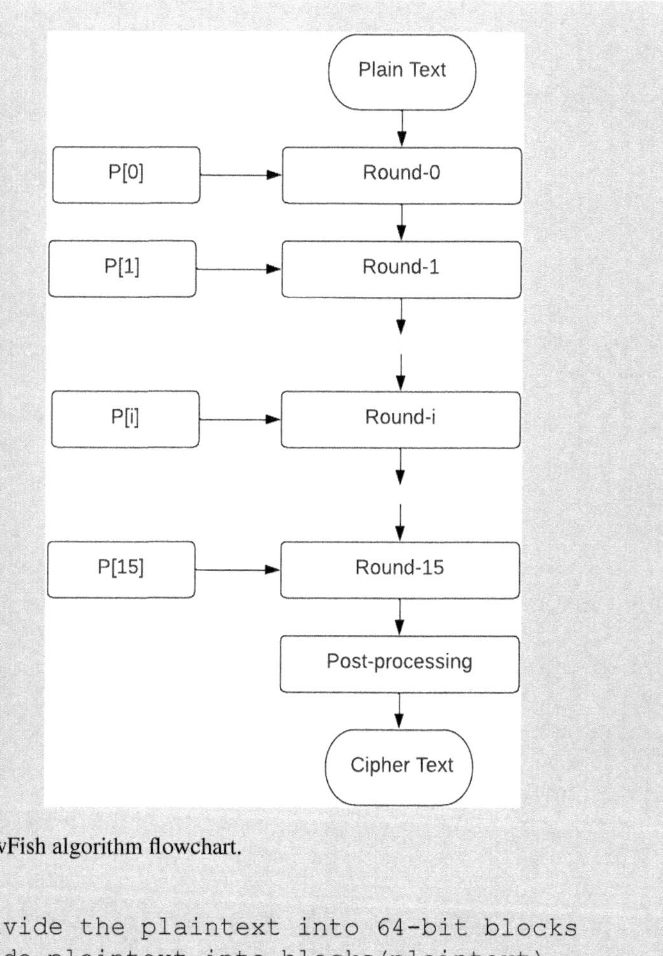

FIGURE 18.4 BlowFish algorithm flowchart.

```
// Step 3: Divide the plaintext into 64-bit blocks
blocks = divide_plaintext_into_blocks(plaintext)

// Step 4: Encrypt each block using the Blowfish algorithm
  encrypted_blocks = []
  for block in blocks:
    encrypted_block = bf_state.encrypt_block(block)
    encrypted_blocks.append(encrypted_block)

// Step 5: Combine the encrypted blocks into a single ciphertext
ciphertext = combine_blocks_into_ciphertext(encrypted_blocks)
```

18.5.5 Elliptic Curve Cryptography (ECC) Algorithm

Elliptic Curve Cryptography (ECC) is a type of encryption algorithm that involves the use of elliptic curves for the creation of both public and private keys. This cryptographic method is known as asymmetric encryption since it employs two distinct keys for the encryption and decryption processes. ECC was independently proposed by Neal Koblitz and Victor S. Miller in 1985. It is faster and more efficient than RSA for key exchange and digital signatures, and is becoming increasingly popular in cloud computing. ECC keys are much shorter than RSA keys for the same security level. As an illustration, if we consider a 256-bit ECC key and a 3072-bit RSA key, both would offer equivalent levels of security. This makes ECC more efficient in terms of key storage, transmission,

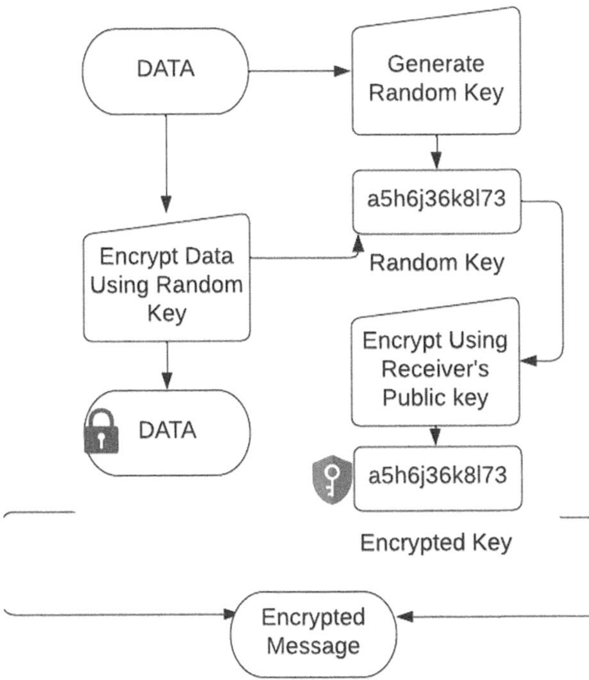

FIGURE 18.5 ECC algorithm flowchart.

and processing. ECC is generally faster than RSA for key generation, encryption, and decryption. This is because ECC operations involve fewer bits and computations than RSA operations. ECC and RSA are both considered secure when implemented correctly. Nonetheless, ECC is deemed to offer higher security than RSA when it comes to specific attack types, like brute force attacks and those based on factorizing large integers. ECC is a public domain algorithm that can be used freely (Figure 18.5).

```
The time complexity is O(k^3), where k is the key size in bits.

// Step 1: Generate a random private key
private_key = generate_private_key()

// Step 2: Calculate the shared secret
shared_secret = multiply_point_by_scalar(public_key, private_
key)

// Step 3: Generate a random ephemeral key
ephemeral_key = generate_private_key()

// Step 4: Calculate the encrypted point
encrypted_point = add_points(multiply_point_by_scalar(shared_
secret, ephemeral_key), plaintext)

// Step 5: Return the encrypted point and the ephemeral key
return (encrypted_point, ephemeral_key)
```

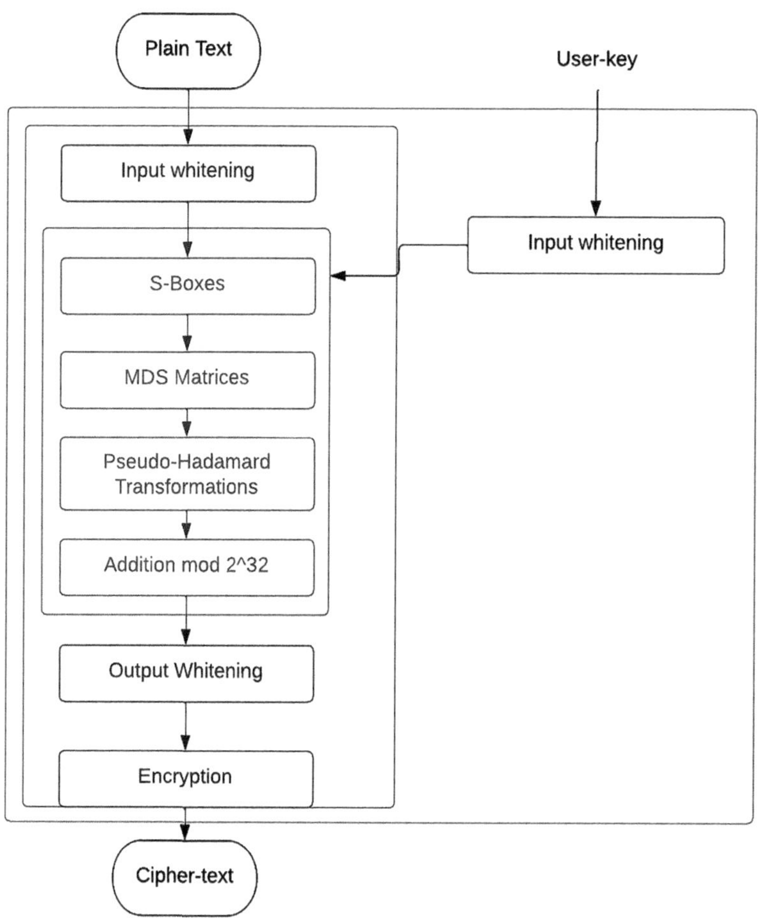

FIGURE 18.6 TwoFish algorithm flowchart.

18.5.6 Twofish Algorithm

Twofish is a symmetric encryption algorithm designed by Bruce Schneier, David Wagner, John Kelsey, Chris Hall, Doug Whiting, and Niels Ferguson. T-F uses a block cipher with a variable block size of max 256 bits and a variable-length key up to 256 bits [12]. It is known for its flexibility and high level of security, and is commonly used for data encryption and decryption in cloud computing. Twofish is a relatively fast encryption algorithm, and can be implemented efficiently on a wide range of computing platforms, including cloud computing environments. It is also a highly secure encryption algorithm, and has not been broken in practice. It is resistant to known attacks such as differential and linear cryptanalysis, and has been extensively analyzed and tested by the cryptographic community [13, 14]. Twofish is a flexible and secure encryption algorithm that is well-suited for cloud computing applications. Its support for variable key and block sizes, high speed, and open design make it a popular choice for data encryption and decryption in the cloud (Figure 18.6).

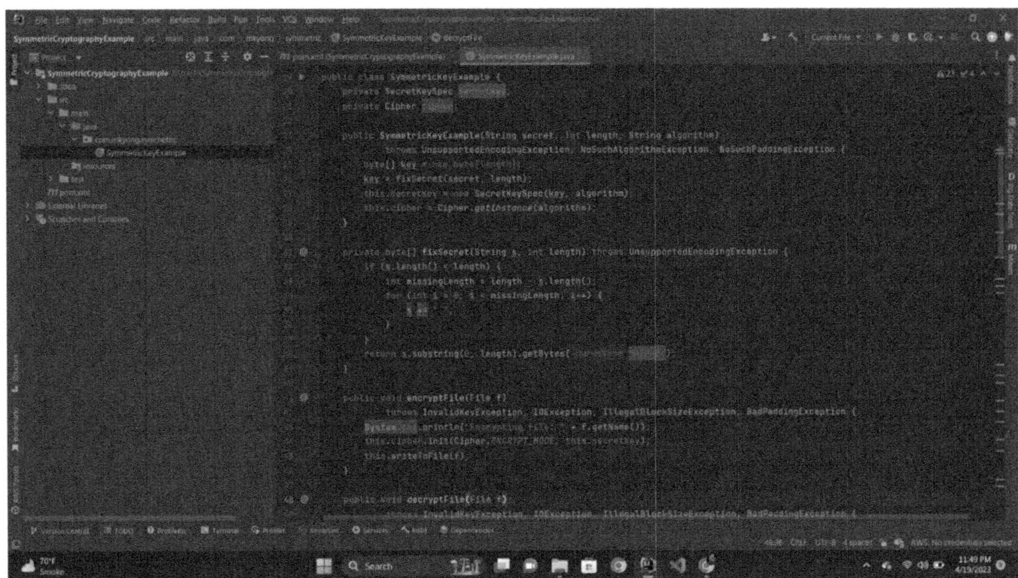

FIGURE 18.7 AES algorithm code.

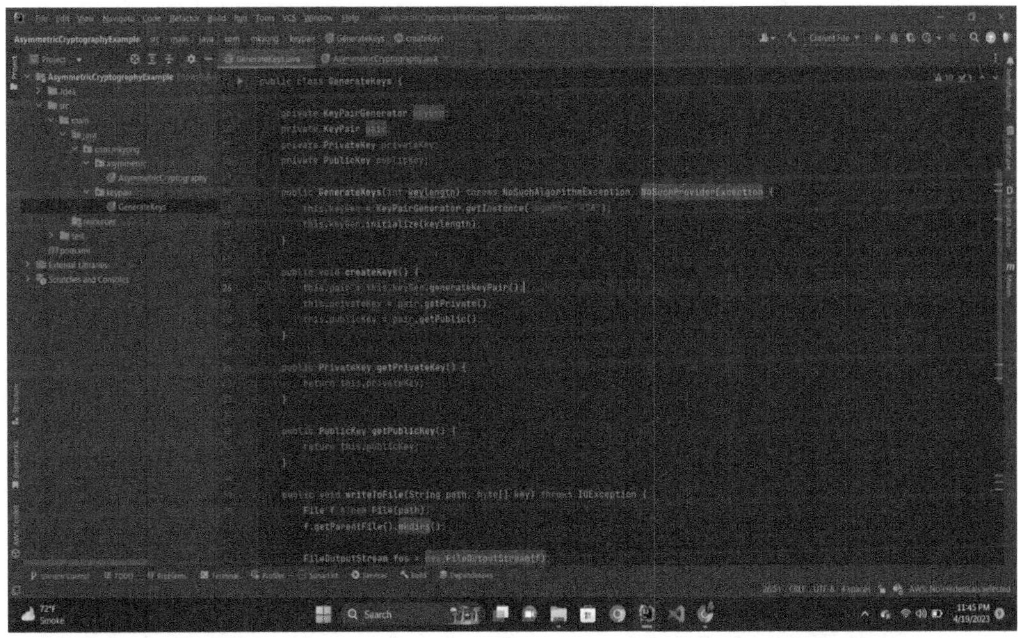

FIGURE 18.8 RSA algorithm code.

TABLE 18.1
Characteristics of Algorithm

Characteristics	AES	RSA	BLOWFISH	DES	ECC	TwoFish
Platform	Cloud Computing	Cloud Computing	Cloud Computing	Cloud Computing	Cloud Computing	Cloud Computing
Key Size	128,192,256 bits	1024 bits	32-448 bits	256 bits	128-521 bits	128, 256 bits
Key Used	The Same key is used to encrypt and decrypt blocks.	Public key is used for encryption where private key for decryption	Same key is used for encryption and decryption both.	For encryption and decryption - same key is used.	Public key is used for encryption where private key for decryption	For encryption and decryption - same key is used.
Scalability	Scalable	Not Scalable	Scalable	Scalable	Scalable	Scalable
Initial Vector Size	128 bits	1024 bits	64 bits	64 bits	N/A	128 bits
Security	Secure for both provider & user	Secure for only user	Secure for both client side and providers	Security applied to both user and providers	Secure for both user/client side and providers	Secure for both user/client side and providers
Data Encryption Capacity	Used for encrypt large amount of data	Used for encrypt small data	Less than AES	Less than AES	Less than AES	Less than AES
Authentication Type	Best authenticity provider	Robust authentic implementation	Comparable to AES	Less authentic than AES.	considered to be stronger than AES	More authentic than AES.
Memory Usage	Low RAM needed	Highest memory usage algorithm	Can execute in less than 5 kb	More than AES	ECC requires more memory than AES	AES is generally considered to be more memory-efficient than Twofish
Execution Time	Faster than others	Require a maximum time	Lesser time to execute	Equals to AES	The execution time of ECC is slower than AES	AES is considered to be slightly faster than Twofish

```
The time complexity is O(nr), where n is the number of blocks
of input data and r is the number of rounds.

// Step 1: Initialize the Twofish algorithm with the key
tf_state = initialize_twofish(key)

// Step 2: Pad the plaintext to be a multiple of 128 bits
plaintext = pad_plaintext(plaintext)

// Step 3: Divide the plaintext into 128-bit blocks
blocks = divide_plaintext_into_blocks(plaintext)

// Step 4: Encrypt each block using the Twofish algorithm
encrypted_blocks = []
for block in blocks:
encrypted_block = tf_state.encrypt_block(block)
encrypted_blocks.append(encrypted_block)

// Step 5: Combine the encrypted blocks into a single ciphertext
ciphertext = combine_blocks_into_ciphertext(encrypted_blocks)
```

18.5.5 Implementation and Results

The algorithms have been implemented using Java with the NetBeans IDE. The code used for the algorithms is shown in Figures 18.7 and 18.8.

18.5.6 Results

18.5.6.1 Characteristics and Comparison of Algorithms

Characteristics and comparison of the Algorithm is provided in Table 18.1.

18.6 CONCLUSION AND FUTURE PROSPECTS

This paper proposes encryption algorithms to enhance the security of cloud data, identify vulnerabilities, and address security issues and challenges. A comparison has been made among AES, DES, RSA, Elliptic Curve Cryptography, TwoFish, and Blowfish algorithms to decide the most effective security algorithm to be employed in cloud computing, with the aim of preventing cloud data from being compromised by attackers.

The encryption algorithms used in cloud computing play a crucial role in ensuring data security. By comparing various parameters used in the algorithms, it has been determined that AES algorithm is the fastest in executing cloud data, while the Blowfish algorithm requires the smallest space of memory. DES algorithm takes the least encryption time, while RSA requires the largest amount of memory and encryption time among the cryptographic algorithms. Through the utilization of an Integrated Development Environment (IDE) tool and JDK 1.7, it was possible to successfully attain the intended result for cloud data by implementing all relevant algorithms. As the demand for cloud computing increases, security concerns for both the cloud and its users are at the forefront. Therefore, the proposed algorithms are useful for meeting the current security requirements. It is possible to offer additional comparisons using diverse methods and outcomes in the future to showcase the efficacy of the suggested framework.

REFERENCES

[1] David, D.S., Anam, M., Kaliappan, C., Arun, S. and Sharma, D.K., 2022. Cloud security service for identifying unauthorized user behaviour. *CMC-Computers, Materials & Continua*, 70(2), pp.2581–2600.

[2] Kaliyamoorthy, P. and Ramalingam, A.C., 2022. QMLFD based RSA cryptosystem for enhancing data security in public cloud storage system. *Wireless Personal Communications*, 122(1), pp.755–782.

[3] Ghanmi, H., Hajlaoui, N., Touati, H., Hadded, M. and Muhlethaler, P., 2020. A Secure Data Storage in Multi-cloud Architecture Using Blowfish Encryption Algorithm. In *Advanced Information Networking and Applications: Proceedings of the 36th International Conference on Advanced Information Networking and Applications (AINA-2022)*, Volume 2, pp. 398–408. Cham: Springer International Publishing, 2022.

[4] Khan, N. and Anandaraj, S.P., 2022. A Survey on Preserving Data Confidentiality in Cloud Computing Using Different Schemes. In *Soft Computing and Signal Processing: Proceedings of 3rd ICSCSP 2020*, Volume 2 (pp. 211–219). Singapore: Springer.

[5] Sheik, S.A. and Muniyandi, A.P., 2023. Secure authentication schemes in cloud computing with glimpse of artificial neural networks: A review. *Cyber Security and Applications*, 1, p.100002.

[6] Bouchaala, M., Ghazel, C. and Saidane, L.A., 2022. Enhancing security and efficiency in cloud computing authentication and key agreement scheme based on smart card. *Journal of Supercomputing*, 78(1), pp.497–522.

[7] Samy, I.A.A. and Mary, M.S., 2022. Secure Data Transmission in Cloud Computing Using Std-rsa With Eslurnn Data Classification and Blockchain Based User Authentication System.

[8] Kadam, A.K.J., Varma, V., Patil, S., Patil, M. and Patil, M., 2022. Data storage security in cloud computing using Aes algorithm and Md5 algorithm. *SAMRIDDHI: A Journal of Physical Sciences, Engineering and Technology*, 14(Spl-2 issu), pp.296–300.

[9] Neela, K.L. and Kavitha, V., 2022. An improved RSA technique with efficient data integrity verification for outsourcing database in cloud. *Wireless Personal Communications*, 123(3), pp.2431–2448.

[10] Shyla, S.I. and Sujatha, S.S., 2022. Efficient secure data retrieval on cloud using multi-stage authentication and optimized blowfish algorithm. *Journal of Ambient Intelligence and Humanized Computing*, 13, pp.151–163.

[11] Mohamed, S.A. and Yousif, A., 2019, September. Bridging the gap between services oriented architecture and cloud software as a service. In *2019 International Conference on Computer, Control, Electrical, and Electronics Engineering (ICCCEEE)* (pp. 1–6). IEEE.

[12] Li, Y., Yu, Y., Min, G., Susilo, W., Ni, J. and Choo, K.K.R., 2017. Fuzzy identity-based data integrity auditing for reliable cloud storage systems. *IEEE Transactions on Dependable and Secure Computing*, 16(1), pp.72–83.

[13] Imran, M., Hlavacs, H., Haq, I.U., Jan, B., Khan, F.A. and Ahmad, A., 2017. Provenance based data integrity checking and verification in cloud environments. *PloS one*, 12(5), p.e0177576.

[14] Nagar, A., and Joshi, K.P., 2018. A semantically rich knowledge representation of PCI DSS for cloud services. In *6th International IBM Cloud Academy Conference ICACON 2018*, Japan.

19 Big Data Issues in Mobile Cloud Computing

Muhammad Jameel, Muhammad Waqas Nasir, Moodser Hussain, Naseer Ahmad, Mubasher Khalid, Taimoor Hassan Jabbar, and Dr. Muhammad Waseem

19.1 INTRODUCTION

Big Data can seen as the volume of data that is just beyond the capacity of technology to effectively manage, store, and process it. Big Data has become a target which is still far from being achieved. In other words, it is just beyond our instant grip – e.g. we need to work really hard to store it, manage it, process it, and access it. With the advancement of social media, multimedia and the Internet of Things (IoT), we have seen continuous growth of data in different flavors. Construction of data is rising at a record rate [1]. The progress in storage and data mining technologies which allows the conservation of gradually increasing quantity of data detained by organizations [2]. The rate of generation of data is staggering [1] This developing rate exceeds their ability to design suitable mobile cloud computing platform for big data examinations and updating exhaustive capacities is the need of the hour for practitioners and researchers.

MCC defines [3]: in Mobile Cloud Computing data processing and storage can be performed outside the infrastructure of mobile device. Applications of Mobile Clouds are placed away from devices into a cloud where large group of subscribers can access them [4]. Mobile applications have developed a new prototype in the form of MCC, where processing and storage can be stimulated from mobile devices to integrated environment in cloud. These shared cloud applications are then retrieved using the wireless connection or web browser on mobile devices. A combination of web and cloud computing is called MCC [5]. By using MCC, mobile devices are free from using CPU speed and memory capacity as all the work related to complex computing can be done in the clouds. Combined resources are the solution for using large amounts of computing power and storage at every device. Large scale and multifaceted computing can only be performed using MCC. Tackling Big Data and computing at a very large scale is the objective of MCC [6].

All types of resources using Big Data methods and computing infrastructure are liable to use Mobile Cloud computing [7]. This paper provides an overview of the status of Big Data in MCC, addressing the challenges of managing Big Data within this context. It also contributes to describe challenges of Big Data in MCC and suggests some solutions of the problems which we can face while incorporating Big Data in MCC.

Recent studies have landed a big hand in usage of both Big Data and MCC. Users got many benefits from these studies to use MCC for their personal and commercial use but still there is a need to incorporate Big Data in MCC. Our research will lead to a platform for others to merge Big Data and MCC. In this chapter, Section 19.2 covers the background of Big Data and MCC; Section 19.3 deals with the SCOPE of Big Data in MCC; Section 19.4 looks at the challenges of Big Data in MCC and taxonomy; Section 19.5 offers critical analysis; Section 19.6 covers current solutions; and Section 19.7 gives the conclusion (Figure 19.1).

DOI: 10.1201/9781003497851-19

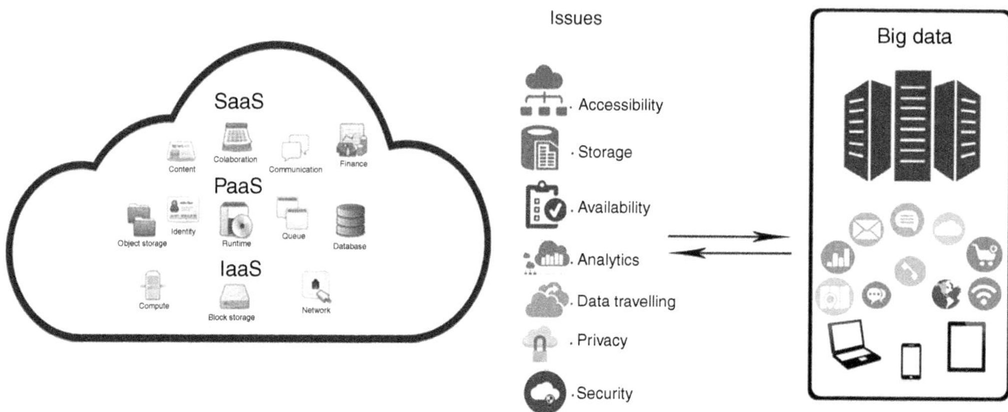

FIGURE 19.1 Architecture of big data issues in MCC.

19.2 BACKGROUND OF BIG DATA

We consider that Velocity, Volume and Variety were the 3Vs which initiated the concept of Big Data. Due to an increase in velocity, data is being generated fast and processed fast. Velocity of data generation is rising day by day. Data volume increased 0.8 zb to 35zb, which is a 44 times increase from 2009 to 2020. Data volume has increased exponentially [8]. Variety states types of data being generated in the form of text (web), semi-structured data (XML), semantic web (RDF) and real-time data. Only a single application can generate data of more than one type. Late processing of data means missing opportunities. Veracity is an updated addition to Big Data making it 4Vs approach. It deals with the uncertainty of data. It also refers to ambiguity, latency, deception, and inconsistency [9]. As there were only few companies like IBM, in the past that generated data and the rest of the world used it, now with the advancement of technology all of us are generating data and all of us are using it. For this reason, the vast amount of data we generate is now referred to as Big Data (Figure 19.2).

Background of MCC: Mobile cloud computing (MCC), which is a combination of Mobile Computing and Cloud Computing, has emerged as a discussion threat in the world of Information Technology since 2009 [24]. With the passage of time, there came a boom in cloud computing which resulted in a thorough change in storage, bandwidth, and processing. As a result, high purchase and management cost was charged by the users. To tackle high cost in managing data, there came an idea to take all the services, infrastructure, and resources on lease which result in cost effectiveness, less time to develop, reliability, and scalability. This phenomenon is known as Mobile Cloud Computing. The framework of MCC is based on three major layers, which are Infrastructure layer, platform layer and Application layer (Figure 19.3).

19.2.1 IaaS

Infrastructure as a Service (IaaS) is the delivery of software (operating systems virtualization technology, file system) and associated hardware (server, storage, and network). IaaS is an evolution of traditional hosting: it does not require long-term commitments and allows the users to occupy the resources on demands. Users manage the software services themselves as they would in their own data center. Example of IaaS are Amazon Web Services Elastic Compute Cloud and Secure Storage Service [8].

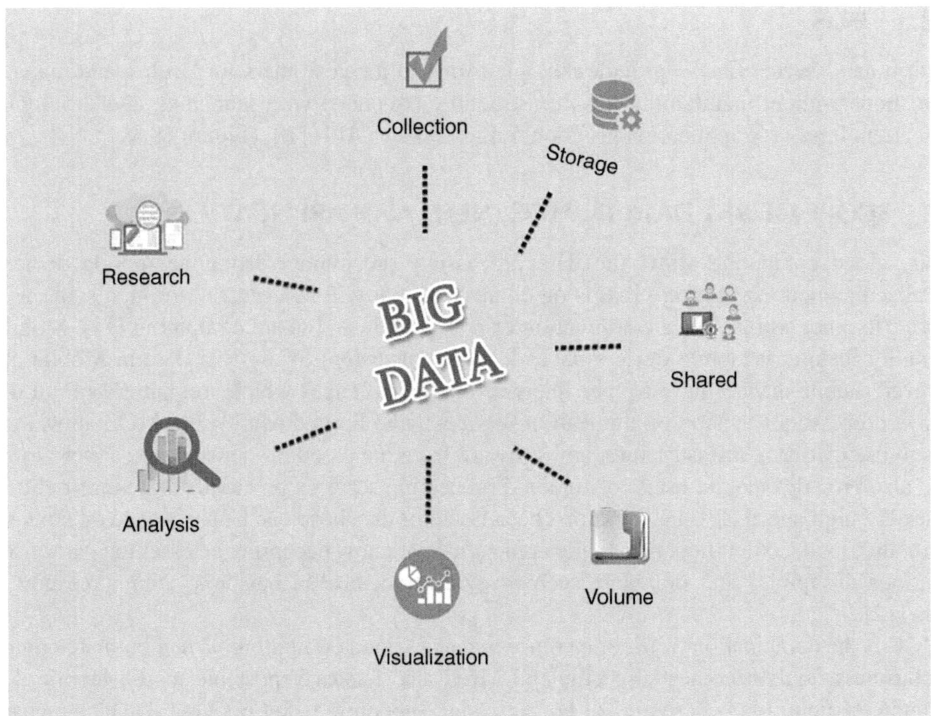

FIGURE 19.2 Big data architecture.

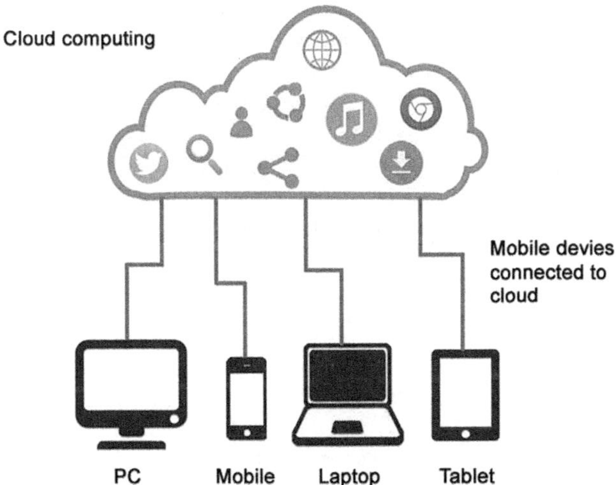

FIGURE 19.3 MCC architecture.

19.2.2 SaaS

Software as a service (SaaS) describes software application delivery as a service over internet. SaaS architecture based on multitenant architecture in which single configuration (hardware, network, operating system), of a single version of application, is used for all customers [9]. SaaS provides many benefits including cost reduction, elasticity in operations, fast upgrade, and easy implementation.

19.2.3 PaaS

Platform as a service (PaaS) provides the platforms to the customers to develop and manage the applications without maintaining the infrastructure associated with launching and developing an application. It provide application programming interface (API) [10] (Figure 19.4).

19.3 SCOPE OF BIG DATA IN MCC (NEED AND APPLICATIONS)

Mobile cloud computing shifts the data processing and storage from the mobile device into centralized computing platform that is on clouds and that will be accessed through some wireless internet. In other words, it is a combination of both mobile and cloud computing [11]. Mobility or portability feature of mobile device makes it more convenient to use Big Data in Mobile clouds. However, mobile devices have to face many resource challenges which are: battery life of device, storage, processing (division of application services), and limited bandwidth. MCC allows mobile users to use platform, infrastructure, and software from the cloud on-demand and at very low cost. MCC also provides mobile users to store and processing services on clouds that almost eliminates the need of high-speed CPU and memory, because all of the above can be performed on clouds [11].

Mobile Cloud Computing and Mobile applications are now becoming an essential part of society in various disciplines and domains, such as: education, health, business, commerce and many others [12].

MCC is the combination of Mobile computing and Cloud computing. When both are combined, then business fields become more efficient. MCC also makes it possible to use mobile devices in commerce field. Many activities of business are: Shopping, ticket booking, balance recharging, billing, advertising, recruitment process, and data sharing. Applications of mobiles also face many challenges which are: Low bandwidth, high complexity of devices and security [12].

Mobility and e-learning together generate mobile learning. Mobile learning is the need of the modern world because of rapid change of technology. It is the faster way to learn compared to traditional learning. Mobile learning overcomes many limitations of traditional learning, such as: high cost of device, high cost of network, low transmission resources, limited educational resources [12]. Use of MCC in healthcare provides convenient access of resources to mobile users, i.e. patient medical history. It also offers healthcare organizations and hospitals a variety of services on the demand of user from the cloud [13]. Games in mobile provide high potential revenue to service providers. Game engines require large computing resources. These resources cannot be offloaded to servers in the cloud. Mobile gaming is a high potential revenue-generating market for service providers.

Services of cloud computing

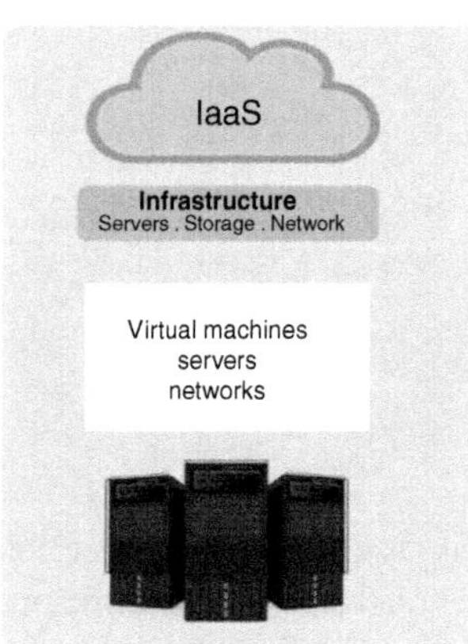

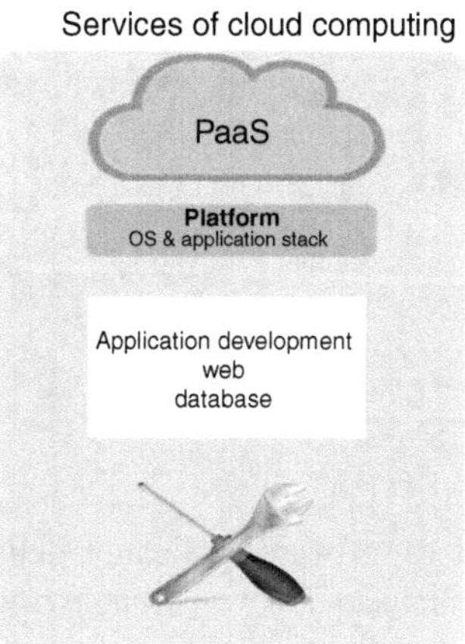

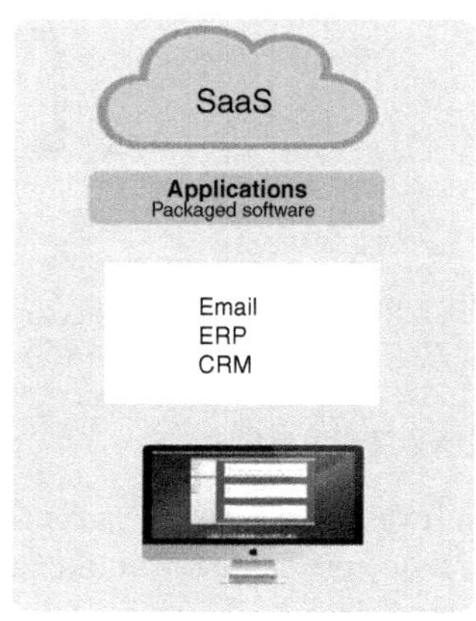

FIGURE 19.4 Services of cloud computing.

Game engines requiring large computing resources cannot be offloaded to servers in the cloud [13]. MCC also offers more than business applications. Other application includes: sharing of images, audio, video, voice, tag-based search, location search, house monitoring, keyword search and many more [11].

19.3.1 Challenges of Big Data in MCC

Despite all of the advantages there are several challenges and risks of big data and mobile cloud computing coming together – significantly, privacy, security and authority [14]. On a ground basis, Mobile devices, which are either a source of generating big data or the user of big data applications, are not that much powerful to handle such larger amount of computing, storing or processing (decision making) over the data which makes it more difficult.

Following is the more detailed list of issues of big data in MCC.

19.3.2 Privacy

Privacy is highly concerning issue in big data as by its genre big data is often valuable and containing sensitive data, therefore the risk of private data falling into the wrong hands because of a lost or stolen mobile device is much more than normal cases [14]. Moreover, there are privacy threats for individuals and organizational data to put on the cloud. As the network has grown and the ability to create, communicate, share, and access data on a very large scale that the world has become so small, chances to invade in to the privacy of individuals or the organization data have also grown [15]. Security agencies of different countries having a close observation on ones very private data, as a recent study mentioned the leakage of sensitive documentation of NASA has created a situation of privacy burst [16]. Big data itself high at the privacy concerns and when we discuss it regarding MCC, the concern is multiplied because security of mobile devices is much more critical.

19.3.3 Network Issues

There are certain network issues in mobile computing, which heightens when it comes to mobile computing over the cloud handling big data. In an article [17], Raj Jain has described the network issues as detection of collision and transmission latency which can be a very critical issue regarding time critical services, e.g. in health care virtual reality environments. Communication through mobile devices can be unpredictable sometimes because of the wireless network connections – not being able to access the cloud can completely disable the ability to perform any activity on such data extensive applications.

19.3.4 Security Concerns

A recent study by CSA and IEE pointed out the fact that organizations across the IT sectors and other businesses eagerly want to adopt cloud computing but ensuring security through the service models of cloud computing is difficult and yet has a large impact on the growth of the technology [10]. There are certain issues regarding service models of cloud computing which are also faced by the big data in MCC. As stated by a recent study, 74 percent of IT directors and CEOs referred to security as the main issue in adopting the service model of cloud computing [10]. Regarding SaaS, the client is dependent on the service provider to ensure security checks. Keeping multiple users from invading in each other's data is the responsibility of the service provider [18]. Regarding PaaS (Platform as a Service), the service provider can give some of the control over the application building. But security assurance below application level, e.g. hosting and network invasion prevention, would be the responsibility of the service provider. Unlikely SaaS and PaaS in the IaaS developer has the more control on security until unless there is a security laps in management of virtualization [10].

19.3.5 Resource Poverty of Mobile Devices

In MCC the mobile devices are the main concern as they are generating and using big data, yet they do not possess such power to handle the large data volume and its complexity because of the following facts about mobile devices: Limited Power and Low Computing Ability. In MCC large processing and storage is offloaded to the cloud but it is necessary and difficult as well to make a decision about what amount of processing will be done over the mobile devices by critically analyzing the computing power and life of a mobile device without charging. QoS of the Mobile devices can be more affected by the buildings, weather, and landforms because of the wireless environment. Taking all of the issues together, like hand-off delay, low computational ability and power limit issues, it certainly affects the quality of service, like the loss of data, late processing or analysis of data, decision makings on wrong times, and unauthorized invasions in the data. People and organizations do not trust on any service until they are assured of the service quality in each term discussed.

19.3.6 Data Analytics and Storage

Low computing ability of mobile devices can be a hindrance in the analysis of large data volumes. Moreover, the ownership of data and who is going to analyze the data is the big question mark due to privacy terms and security breaches threats. These factors directly affects the in-time analysis of data (real-time analysis). Large datasets of big data cannot be managed with present data mining techniques or methodologies due to their large size and complexity. Therefore the need to formulate new algorithms for Data Mining is becoming crucial. It is the large volume of data and the heterogeneity which make it big data. Storage of big data over mobile devices is not feasible due to resource poverty.

We have categorized the challenges of big data in MCC as privacy, network issues, security concerns, resource poverty, QoS, data analytics and storage, as shown in Figure 19.5. Further vulnerability and data breaches are discussed under privacy as mobile devices are more prone to get lost or stolen, which results in the data breaches and makes private data vulnerable. Mobile devices are connected through wireless networks which has their own limitations such as bandwidth which can cause the accessibility breaks to the cloud for such extensive data applications and affects the service quality. MCC has the one major issue of being poor in resource abilities like energy resource because it is bound to battery power, computing ability, storage problems which can be resolved somehow by offloading data and processing overhead to cloud.

19.5 CRITICAL ANALYSIS BIG DATA AND MOBILE CLOUD COMPUTING

In Table 19.1, we critically analyzed the features of Big Data compared to MCC. Tick mark shows the complete possession of the feature and means no feature is available and limited means the feature is available to some extent in the related field.

In Table 19.1 we take different features of Big Data and MCC, and then make comparative analysis of provided features. Some features are provided in both, and some are in one. There is more data than ever before in Big Data – that's why it's easy to make more accurate decisions [19].

Anywhere Access: In term of big data, access data is somehow limited because we cannot keep complete data with us any time at any place. However Mobile Cloud Computing makes it possible. MCC creates a convenient method to access and receive data remotely from the cloud on user request [20].

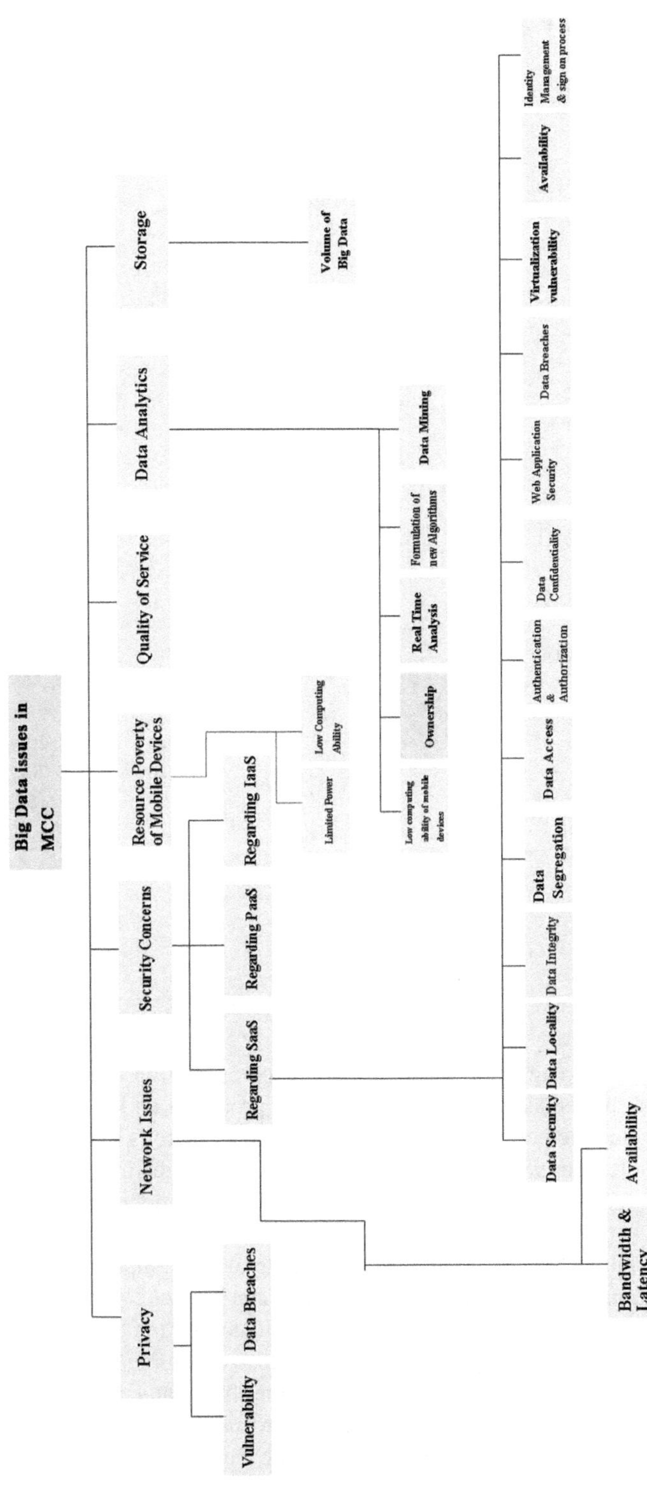

FIGURE 19.5 Taxonomy of challenges of big data in MCC.

TABLE 19.1
Analysis of the Features of Big Data Compared to MCC

	Big Data	MCC	References
Easy to use	✓	✓	[19]
Anywhere Access	✓	✓	[20]
Data Management/Tool	--	✓	[2, 21]
Consistency	--	--	[20]
Storage/Cost	✓	--	[2]
Content Representation	✓	--	[21–23]
Infrastructure representation	--	✓	[21–23]
IaaS	--	✓	[12, 22]
PaaS	--	✓	[12, 22]
SaaS	--	✓	[3, 12, 22]
Transfer Of Risk	✓	--	[24]
Data Replication	✓	✓	[20]
Virtualization	--	✓	[12, 25]
Authentication	--	✓	[20]
Processing	--	✓	[2]
Availability	✓	✓	[2]
Integrity	✓	✓	[2, 26]
Volume	✓	✓	[23]
Heterogeneity/Variety	✓	✓	[23]
Velocity	--	✓	[23]
Transformation	✓	✓	[2]
Scalability	✓	Limited	[12, 20]
Elasticity	✓	Limited	[12, 20]
Fault-tolerant	--	✓	[11, 12]
Internet Connectivity	✓	✓	[27]
Portable	--	✓	[19]
Large Scale	✓	✓	[22]
Autonomy	--	✓	[25]
Host	--	✓	[19]
Security/Privacy	??	??	[20]

Data Management Tool: Big Data is just a content representation and the nonstop rise in the size and detail of data that is taken by organizations has produced an irresistible flow of data that is either structured or unstructured format [2]. To manage data there is no tool. But MCC provides built in tools (Microsoft Azure/SQLDatabase, Clear DB) that makes the representation of data in some manners. The only tasks to do is make internet connection on mobile devices and use the cloud services [21].

Consistency is the major barrier in big data and mobile cloud computing. With the time data is increasing rapidly in terms of volume and variety so data can be replicated. Making consistent data is the key challenging issue in big data and in mobile cloud computing [20].

In big data we have to pay a very large amount of cost for storage of data. Data finding and accessing is another issue that we have to face in searching from large amounts of data. In Mobile Cloud computing, the only need is internet connection to use resources from the cloud on your mobile or laptop device. With Combing both, we overcome the storage issue. In Mobile cloud everything is a service [2].

Big Data is not only a description of the data when we say it is big. It is about how to organize data and how to label different kinds of it (structured, semi-structure, unstructured) and the technology to store and retrieve data [21] or in other words we can say that Big Data is the ability of mining useful information from these large streams of data [23].

Infrastructure representation: MCC is the paradigm of computing on the fly. In MCC everything is de-metalize and you don't need to worry a lot about the setup before starting. Like big data, you don't need to care about managing data or managing a farm of computers before using them. In MCC, the only need is internet connection to get access of clouds services [21]. Infrastructure AAS: In MCC, everything (Resources, Software and information) are provided to other computers and devices as a utility over the internet. IAAS is the main category of cloud computing services. It provides computation and storage facilities through virtualization using framework, i.e. Amazon EC2 [12]. It is well suited for those workloads that are temporary, experimental, and can change unexpectedly.

Platform AAS: The second main category of cloud computing service, provides application programming interface (API) and programming environment for developers without any need of building and maintaining the infrastructure that is associated with developing app (Google Application Engine (GAE), Amazon Web Services and Microsoft Azure) [12].

Software AAS: The third main category of Cloud Computing service is SAAS. It gives end users an approach to a distinct application, as Microsoft Office 365, Facebook, Gmail. Sometimes it is also called "on-demand-software" [25]. Transfer of Risk is the ubiquitous paradigm that makes MCC more feasible to use by user. As compared to Big Data, accessing data from various clouds, data is generated at a tremendous rate and chances of risk transfer also increases but with MCC chances to transfer risks are very few [27].

Consistency and replications is the major barrier to overcome in Big Data and MCC. Data is increasing at a tremendous rate. There is not any standard way to stop replicating data on the internet. With the generation of data, chances of replications possibilities also increase [20].Virtualization is not the requirement in big data, because we really don't need it. Root technology of Cloud computing that separate the resources (operating system, server, storage device) into dollop or entity that can be given when needed by applications [12]. In MCC Virtualization, the essential objective is to give a better involvement for mobile users whose device has defined system and scope (Computations speed, Storage and battery) [25].

There is lot of data in big data, anyone can make query and get data according to his need. In MCC authentication is the authorizations service to mobile user. When mobile user transmits file on cloud server for sharing with numerous users then there should be a system to authenticate the creator of file. With help of verification, we can verify the originator of file [20]. There is lot of data in big data to process, Cloud computing makes possible parallel processing. But parallel processing is not possible in MCC, and because mobile have limited processing capability, storage capacity. Mobile cloud permit users to outsource function to external service providers. Mobile Cloud utilization are drop box, iCloud and Gmail use to improve mobile cloud performance [2]. On the request of mobile users, data should be available within a short amount of time. Also, available data should deliver high-quality services to the users [2]. Integrity is the essential attitude of big data security. Integrity means that the data must be altered only by authorized or only by owner of the data to overcome misuse of the data [2]. In MCC, it becomes possible with the encryption and decryption keys [26].

There is more data than ever before and it's continuously increasing with time [23]. Heterogeneity: Big Data and MCC there are various types of data as graph, audio, image, text, video and more [23]. The Data is coming in flow. We are attentive only in that data that is relevant to our task and provided within a short processing time [23]. MCC provides more relevant data and within a short processing time. Data transformation is the transformation of data from one format to another. Transformation is usually from the format of a source system to a required format of a destination system [23].

Big Data scalability and elasticity is unlimited. In MCC Scalability and elasticity depend on the type of privacy and type of clouds. Local and private cloud has limited scalability and elasticity but is more secure in terms of privacy and provides high performance and efficient services on portable devices [12, 20]. Fault tolerance is the property that enables a system to continue operating in case of some failure or some component failure of the system. Increasing functionality of clouds increases the complexity of applications in mobiles and can leads to resource failures. It also reduces overall performance of MCC. By following Disease Resistance Approach (DRFT) we can identify the fault-tolerance machines mechanism [12].

Data is stored at a centralized place from where anyone can get access of data. However, to get access of data from that place, internet connectivity is required. Internet connectivity is required in both big data and in MCC. Portability: Portability is the major requirement of these days. Portable data is more valuable than a data that is at the same place. In Big Data terms you can't keep data with you at any place, but after the joining of cloud computing, the Big data become portable. Data Scale is very vast in both big data and in MCC [22]. Cloud System is automatic system. On the demand of clients, it automatically sets up and allocates the resources of software, hardware and storage. The management is transparent to end users [25]. Clouds are actually the platform that provide the sources (web access, storage, computational power) to the clients on demand. The services provided by clouds to different computing devices are called hosting services [19].

There is no privacy mechanism for data in both big data and in MCC. In the adoption of mobile cloud computing security threats are the major hurdles. It is the current topic in research that how to make our data secure in mobile cloud computing [20].

19.6 CURRENT SOLUTIONS

In Table 19.2 some solutions are presented for the big data issues in mobile cloud computing.

19.6.1 Solution for Privacy

Privacy is the major concern that hampers the users who upload their personal data on cloud storage. This has become more serious when a person's individual information is required for the relevant results in data mining and data analytics [2]. Privacy in the big data is more important when we are going to perform data mining in which we have to take care about data exposure [29].

Although PKI "Public key infrastructure" is a suitable method for this security issue, but the problem with the PKI is the additional computation cost. In addition, the connection among mobile devices in the cloud is extremely dynamic. In this scenario, at one time new devices can be connecting while the current connected devices may be disconnecting. In the result user verification needs will rise in to such extent that will cause the inadequate resources to do asymmetric key authentication and communicate heavy messages [30]. The solution for this issue is PKASSO, a Public key infrastructure based verification protocol [31]. To solve this resource problem, PKASSO offloads complex public key infrastructure procedures from the mobile, to the remote server. To solve this problem, cloudlet concept can also be used. Cloudlets can work as a local set-up in which the PKI operation can be offloaded [3 again used]. Techniques like anonymous routing in which onion routing is used can provide privacy for mobile clouds. The examples of this technique are in P2P domain [32].

In onion network messages are encrypted in layers like an onion. In this technique, encrypted data is transmitted in series of nodes that are named onion routers. In this technique the sender will remain anonymous, because of each middle way knows only location of directly previous and succeeding nodes [33].

Nevertheless, there are some risks and overheads with this anonymous routing protocol [34]. Furthermore, users of the mobile cloud have the ability to modify their confidential settings and the information that can be perceived. For example, the mobile cloud users do not want other devices to record their location [35].

TABLE 19.2
Solutions for Big Data in MCC

Challenges	Solutions
Privacy	• Public key infrastructure [30] • PKASSO [31] • Cloudlets [32] • Onion network [32] • Privacy rights management for mobile applications [3]
Network issues	• Femto cell [37–39] • 4G [37] • Cognitive radio [40, 41] • HTML5 [36] • CDNs
Security concerns	• Authentication code scheme [42] • SSL [43] • Cloud AV [36] • Web 3.0 servers [44] • Cryptographic techniques [11] • Digital certificates and signatures [46]
Resource poverty of mobile devices	• Applications operation offloaded on cloud [3] • Use such applications that consume less battery [47] • Non-displaying applications offload to the cloud [48]
Quality of service	• CloneCloud [50] • Cloudlets [49]
Data Analytics	• Offload the data to cloud [3, 48] • Data mining techniques
Storage	• Cloud storage servers [2]

The solution of this scenario is called "Privacy rights management for mobile applications" (PRiMMA). In this project the techniques were investigated by the authors, for protecting personal information that produced by ubiquitous computing apps from malicious and accidental misuse. The main objective of this development is to provide a tool to the user that that can control his privacy policy and can forecast the privacy requirements using the mechanism of monitoring [3].

19.6.2 Solution for Network Issues

Many network issues in the mobile cloud computing that increases when it occurs in mobile devices over cloud for handling the big data. The low bandwidth issue occurs because of increasing the mobile users and cloud users. Development of the 4G and Femtocell technologies are overcoming the limitation of bandwidth [36]. 4G network increase the bandwidth capacity for its users. 4G network can provide 100 Mbits to a28 Mbits to mobile users. It also provides area coverage and quick handoff [37].

Femtocell is cellular base station for small area. It is design for the home user and small business, also called "femtoAccess Point (AP)" [38]. Hay system limited (HSL) develops highly economical, secure and scalable network for mobile users [39]. The femtocell network expands and contract according to the user demand. The result will be highly cost-effective femtocell network with limited resources. The femtocell located in offices and homes of the users through internet to the cloud to increase the access of their network. Femtocell is highly efficient and cost-effective but it is only used for clouds [38].

Effective network access management provides advance link performance and high bandwidth usage to its users. Cognitive radio is a solution to have access management wirelessly in the mobile environment. It significantly increases efficiency of spectrum utilization. Cognitive radio allows the users who don't have the license to get access of the spectrum that is allocated for the licensed users. If we integrate this technology in MCC, spectrum will be used more affectively [40]. Cognitive radio is the radio technology that can be programmed or configured to use best wireless stations in its locality. This type of radio automatically senses accessible channels in the wireless spectrum [41].

Mobile users can detect the radio availability in the MCC while guaranteeing the traditional services will not have delayed. Latency for transferring big data in networks is also a major issue but different measures can improve this problem. If we keep the applications close to its users, it can decrease the latency significantly. It can save the bandwidths and remove transmission delay. It can be improved by the allowing service provider to logically re-route the traffic in internet, based on cache and location capabilities.

To get quicker internet access in mobile we can have used advance technologies like HTML5 to begin deployed. It provides the facility of the local caching. The researchers are also working on mobile web to have better access of internet [36]. Another solution to enhance network speed is to use CDN (Content delivery network), it is a distributed server based system that provide more speed and reliability. Through CDNs customer will have high speed on demand on internet all over the world.

19.6.3 Solution for Security Concerns

With growth of big data and clouding computing, mobile cloud computing users are still less than the expectations. According to a survey many IT executives are not involved in MCC due to risks and security purpose. The researchers proposed an integrity verification scheme for mobile users that can validate the integrity of stored files in cloud servers. They present a protocol that is energy-efficient and provide the integrity of the stored files in MCC. This scheme uses an authentication code that will be in the form of incremental message. It will less overhead jobs of mobile user by offloads many integrity authentication jobs on the cloud [42].

Secure socket layer (SSL) and digital certificate can provide external security [43]. For data security, organizations need to have operational security and information assurance policies. For big data concerns the cloud service provider have to build a trust for its users [42].

Security of the mobile device is very important, as the data in the device can be misused if it gets stolen. The data from misplaced device can be wiped remotely. This feature is mostly provided by wireless carriers and mobile manufacturers [44].

Many security threats to mobile devices have to face malevolent codes, viruses. Mobile devices have inadequate resources for protecting these types of threats, so we move sensing capabilities to the cloud. This technique is the extension of current cloud named Cloud AV, is used provide the service of detection in-cloud. It also let us to use many antivirus engines parallel by placing them in virtual containers. This approach also enhances the battery by 30% [36].

Networks may have malicious attacks that can be counter by implementing verifying security and control access all web 3.0 servers [44]. For malware exposure, we can used data mining in clouds to block this problem [45]. To secure big data in MCC you must have encrypted data that is going to store in cloud. Cryptographic techniques can be used to perform encryption in sensitive data [11].

In a cloud environment the customers did not know that's where they are going to store data. This is due to a privacy issue; a secure application can provide the reliability to its customers on the location where they are using the data [29]. Security of big data in mobile cloud computing MCC can be assure by using the services of the third party like digital certificates and signatures [46].

19.6.4 Solution for Resource Poverty of Mobile Devices

Service model of cloud computing is based on Platform as a Service (PaaS), Software as a Service (Saas), and Infrastructure as a Service (IaaS) but in MCC individual SaaS is applied because of limited resources like battery, storage, availability, and computing power [47]. As mobile as limited power computation capacity, so we have to move the execution of the application to the cloud to save the battery and computation cost. Basic operations like opening, inputting data and result display need to execute on device but application function is offloaded on cloud by the device [3]. For saving the battery of mobile devices, the user must have to use such applications that consume less battery. For smart phones, Wi-Fi consumes 23% less battery. On the other hand, GPRS consumes more battery in web browsing. For overcoming the limitation of the mobile resources problems, resources are added to the cloud; such as if the mobile has less storage capacity to store big data, storage is transferred to the cloud. Users can store their data on cloud and can access anytime remotely [47]. For having less consuming battery we divide the display of mobile into two categories that are: one is the displaying applications and the second is non-display applications. Displaying application can have consumed more power and computation cost and non-display have little display usage so they can be offloaded to the cloud [48].

19.6.5 Solution for Quality of Service

As big data is a collection of huge amount of data and IT manager focus on quality of data. Unstructured data is really difficult to maintain [3]. Mobile users have to face many problems like bottleneck due to the limited bandwidth and the signal attenuation produced by mobility. It causes delays that will reduce the quality of services QoS. Two techniques are used to improve the quality of service like Clone Cloud and Cloudlets. [49] Clone Cloud carries more power to your Smartphone's. CC used nearby data centers and computers to raise the speed of the running application of Smartphone [50]. Cloudlets is the trusted rich computer that is associated to internet and it is accessible for the use of the adjacent mobile device. When the users of mobile don't want to offload their data on cloud they can detect and can use the nearby Cloudlets [49].

19.6.6 Solution for Data Analytics

Processing quires in bid data in MCC is really difficult. Mobile cannot perform large quires and perform analytics. Analyzing huge amount of data by numerous applications in mobile is really difficult. We cannot perform analytics on mobiles [51]. For real-time analyzing the big data on mobiles, we have to first the offload the data to cloud then we can perform the data analyzing [3, 48]. However, in cloud it takes more time to process a query in big data [52]. Different data-mining techniques like classification, clustering, neural networks, genetic algorithms can be used for analyzing the data on the cloud.

19.6.7 Solution for Storage

Storage of big data in MCC is really difficult task. A mobile device has limited storage capacity. For storing huge amount of data, we need to design a hierarchal storage technique. The unstructured form of big data, we can storage cloud servers like Dropbox and Google drive to store these massive amounts of data [2].

19.7 CONCLUSION

As discussed in the paper, despite of several advantages and dire need of putting together the Big Data and MCC there are several challenges and issues in assuring feasibility of the big data in

MCC. We can infer from this survey that PRIVACY, SECURITY and AUTHORITY are the main challenges regarding mobile cloud computing (MCC), which can affect the credibility of big data on larger scale. Moreover, cloud computing service models (PaaS, IaaS and SaaS) are the main factors responsible for assuring the security concerns, Security is ensured on different levels in each case SaaS, PaaS or IaaS. We have discussed in the paper that the resource poverty of mobile devices also have a great impact over the ability to compute, process, analyze and store the Big Data in MCC. Network latency can be hindrance in real time decision making but it can be solved through the offloading technique by putting load over the cloud which is also a difficult decision about what part of application can be offloaded to cloud and what should be run on the mobile device.

REFERENCES

1. Kaisler, S., et al., Big data: Issues and challenges moving forward. in *System Sciences (HICSS), 2013 46th Hawaii International Conference on*. 2013. IEEE.
2. Hashem, I.A.T., et al., The rise of "big data" on cloud computing: Review and open research issues. *Information Systems*, 2015. **47**: pp. 98–115.
3. Dinh, H.T., et al., A survey of mobile cloud computing: Architecture, applications, and approaches. *Wireless Communications and Mobile Computing*, 2013. **13**(18): pp. 1587–1611.
4. Christensen, J.H., Using RESTful web-services and cloud computing to create next generation mobile applications. in *Proceedings of the 24th ACM SIGPLAN conference companion on Object Oriented Programming Systems Languages and Applications*. 2009. ACM.
5. Liu, L., R. Moulic, and D. Shea, Cloud service portal for mobile device management. in *e-Business Engineering (ICEBE), 2010 IEEE 7th International Conference on*. 2010. IEEE.
6. Ghemawat, S., H. Gobioff, and S.-T. Leung, The Google file system. in *ACM SIGOPS Operating Systems Review*. 2003. ACM.
7. Dean, J., and S. Ghemawat, MapReduce: Simplified data processing on large clusters. *Communications of the ACM*, 2008. **51**(1): pp. 107–113.
8. Normandeau, K., Beyond volume, variety and velocity is the issue of big data veracity. Inside Big Data, 2013.
9. Bhardwaj, S., L. Jain, and S. Jain, Cloud computing: A study of infrastructure as a service (IAAS). *International Journal of engineering and information Technology*, 2010. **2**(1): pp. 60–63.
10. Cusumano, M., Cloud computing and SaaS as new computing platforms. *Communications of the ACM*, 2010. **53**(4): pp. 27–29.
11. Subashini, S., and V. Kavitha, A survey on security issues in service delivery models of cloud computing. *Journal of Network and Computer Applications*, 2011. **34**(1): pp. 1–11.
12. Kaur, A., A review on Mobile Cloud Computing (MCC) and big data convergence. *International Journal of Advanced Research in Computer and Communication Engineering,* 2015. **4**(6): pp. 29–34.
13. Rahimi, M.R., et al., Mobile cloud computing: A survey, state of art and future directions. *Mobile Networks and Applications*, 2014. **19**(2): pp. 133–143.
14. Wu, J.-H., S.-C. Wang, and L.-M. Lin, Mobile computing acceptance factors in the healthcare industry: A structural equation model. *International Journal of Medical Informatics*, 2007. **76**(1): pp. 66–77.
15. Tene, O., and J. Polonetsky, Privacy in the age of big data: a time for big decisions. *Stanford Law Review*, Online, 2011. 64: p. 63.
16. Polonetsky, J., and O. Tene, Privacy and big data: making ends meet, *Stanford Law Review*, 2013. 66: p. 26.
17. Kasabian, A., Litigating in the 21st century: Amending challenges for cause in light of Big Data. *Pepperdine Law Review*, 2015. **43**: p. iv.
18. Jain, R., Networking issues for mobile computing. in *Recent Advances in Networking and Telecommunications Seminars in Columbus,* 1(1), 1999, Ohio. Citeseer.
19. Vidyanand, C., Software as a service: Implications for investment in software development, in *Proceedings of the 40th Hawaii International Conference on System Sciences (HICSS), Waikoloa, Hawaii*. 2007.
20. Manakshe, A., R.S. Satakolla, and N. Tiwari, *Cloud computing in mobile application.*

21. Khan, A.N., et al., Towards secure mobile cloud computing: A survey. *Future Generation Computer Systems*, 2013. **29**(5): pp. 1278–1299.
22. Ranjan, R., Streaming big data processing in datacenter clouds. *IEEE Cloud Computing*, 2014. **1**(1): pp. 78–83.
23. Assunção, M.D., et al., Big Data computing and clouds: Trends and future directions. *Journal of Parallel and Distributed Computing*, 2015. **79**: pp. 3–15.
24. Fan, W., and A. Bifet, Mining big data: Current status, and forecast to the future. *ACM sIGKDD Explorations Newsletter*, 2013. **14**(2): pp. 1–5.
25. Cuzzocrea, A., I.-Y. Song, and K.C. Davis, Analytics over large-scale multidimensional data: The big data revolution!, in *Proceedings of the ACM 14th International Workshop on Data Warehousing and OLAP*. 2011. ACM.
26. Qi, H., and A. Gani, Research on mobile cloud computing: Review, trend and perspectives. in *Digital Information and Communication Technology and it's Applications (DICTAP), 2012 Second International Conference on*. 2012. IEEE.
27. Zissis, D. and D. Lekkas, Addressing cloud computing security issues. *Future Generation Computer Systems*, 2012. **28**(3): pp. 583–592.
28. Agrawal, D., S. Das, and A. El Abbadi, Big data and cloud computing: Current state and future opportunities. in *Proceedings of the 14th International Conference on Extending Database Technology*. 2011. ACM.
29. Ji, C., et al., Big data processing in cloud computing environments. in *2012 12th International Symposium on Pervasive Systems, Algorithms and Networks*. 2012. IEEE.
30. Fernando, N., S.W. Loke, and W. Rahayu, Mobile cloud computing: A survey. *Future Generation Computer Systems*, 2013. **29**(1): pp. 84–106.
31. Han, J. and Y. Liu, Mutual anonymity for mobile p2p systems. *IEEE Transactions on Parallel and Distributed Systems*, 2008. **19**(8): pp. 1009–1019.
32. Tapia, M.G. and J. Shorter, Into the depths of the internet: The deep web. *Issues in Information Systems*, 2015. **16**(3): pp. 230–237.
33. Han, J., et al., Provide privacy for mobile p2p systems. in *25th IEEE International Conference on Distributed Computing Systems Workshops*. 2005. IEEE.
34. Corapi, D., et al., Learning rules from user behaviour. in *IFIP International Conference on Artificial Intelligence Applications and Innovations*. 2009. Springer.
35. Park, K.-W., S.S. Lim, and K.H. Park, Computationally efficient pki-based single sign-on protocol, PKASSO for mobile devices. *IEEE Transactions on Computers*, 2008. **57**(6): pp. 821–834.
36. Zheng, J., et al., E-shaped patch antenna using L-probe for 4G communication systems. *Sensors & Transducers*, 2014. **182**(11): p. 105.
37. Yeboah, P.N., Proposal and implementation of an IDS for potential SMS spam signaling messages on SS7, 2016, NTNU.
38. Andrews, J.G., et al., Femtocells: Past, present, and future. *IEEE Journal on Selected Areas in Communications*, 2012. **30**(3): pp. 497–508.
39. Yucek, T. and H. Arslan, A survey of spectrum sensing algorithms for cognitive radio applications. *IEEE Communications Surveys & Tutorials*, 2009. **11**(1): pp. 116–130.
40. Priyadharshini, R.A. and K.U. Haimavathi, Detection of attacks and countermeasures in Cognitive Radio Network. in *Wireless Communications, Signal Processing and Networking (WiSPNET), International Conference on*. 2016. IEEE.
41. Itani, W., A. Kayssi, and A. Chehab, Energy-efficient incremental integrity for securing storage in mobile cloud computing. in *2010 International Conference on Energy Aware Computing*. 2010. IEEE.
42. Sahu, D., et al., Cloud computing in mobile applications. *International Journal of Scientific and Research Publications*, 2012. **2**(8): pp. 1–9.
43. Nafi, K.W., et al., A newer user authentication, file encryption and distributed server based cloud computing security architecture. arXiv preprint arXiv:1303.0598, 2013.
44. S O, K. and O. F Y, Big data and cloud computing issues. *International Journal of Computer Applications*, 2016. **133**(12): pp. 14–19.
45. Deng, R., Y. Xiang, and M.H. Au, *Cryptography in Cloud Computing*. 2014. **1**(5): pp. 20–37.

46. Qureshi, S.S., T. Ahmad, and K. Rafique, Mobile cloud computing as future for mobile applications-Implementation methods and challenging issues. in *2011 IEEE International Conference on Cloud Computing and Intelligence Systems*. 2011. IEEE.
47. Maan, J., Extending the Principles of Cloud Computing in Mobile Domain, in *Trends in Network and Communications*. 2011, Springer. pp. 197–203.
48. Katal, A., M. Wazid, and R. Goudar, Big data: issues, challenges, tools and good practices. in *Contemporary Computing (IC3), 2013 Sixth International Conference on*. 2013. IEEE.
49. Chun, B.-G., et al., Clonecloud: elastic execution between mobile device and cloud. in *Proceedings of the Sixth Conference on Computer Systems*. 2011. ACM.
50. Bilal, M., et al., Big Data in the construction industry: A review of present status, opportunities, and future trends. *Advanced Engineering Informatics*, 2016. **30**(3): pp. 500–521.
51. Chen, H., R.H. Chiang, and V.C. Storey, Business intelligence and analytics: From big data to big impact. *MIS Quarterly*, 2012. **36**(4): pp. 1165–1188.
52. Chen, Q. and Q. Deng, Cloud computing and its key techniques. *Journal of Computer Applications*, 2009. **29**(9): p. 2565.

20 A Comprehensive Review on the Role of Blockchain in Reinforcing Security for Industrial IoT

Shehla Afzal, Muhammad Ameer Hamza, Danish Shehzad, Irfan Ud Din, Muhammad Arif, and Arfan Arshad

20.1 INTRODUCTION

The Internet of Things (IoT) refers to the interconnectedness of physical devices, vehicles, buildings, and other items embedded with sensors, software, and network connectivity. IoT can support a wide range of industrial applications, such as manufacturing, logistics, food industry, and utilities. The goal of IoT is to improve operational efficiency and production throughput, reduce machine downtime, and improve product quality with increasing security. In particular, IoT has the following characteristics: 1) decentralization of IoT systems, 2) diversity of IoT devices and systems, 3) heterogeneity of IoT data, and 4) network complexity. All of them lead to challenges, including IoT system heterogeneity, poor interoperability, IoT device resource limitations, and privacy and security vulnerabilities. The current system is a centralized cloud dependent on trust based on a model that limits interoperability due to different configuration of multiple clouds and devices.

Also, the security mechanism in the current infrastructure is limited to only discrete security leaves many vulnerable entry/exit points IoT systems [1]. Due to the involvement of multiple stakeholders in the IoT, it is essential to build trustless, preventive, decentralized and interoperable systems for creating IoT systems practical. The Industrial Internet of Things (IIoT) comprises extensive awareness from industries and academics, which could be a significant factor in the upcoming conversion of industrial schemes [2]. The significant difference between IoT and IIoT is an automation equipment and industrial devices in an IoT environment; IIoT is about utilization of applications usage in smart factories and smart manufacturing [3]. Using actuators and sensing tools, IIoT presents intelligence and interconnection to industrial schemes including ubiquitous network and processing capabilities. To enhance business competence and manufacture throughput, decrease the device's downtime and improve the excellence of the produced product is the main goal of implementing IIoT. Precisely, IIoT consists of the following characteristics: 1) IIoT systems decentralization, 2) IIoT applications and schemes, 3) IIoT information heterogeneity with 4) strain on the network. Blockchain is a peer-to-peer (P2P) distributed ledger.

Safe transfer encryption algorithm based on hash chain and time stamp the certificate value mechanism ensures data traceability and irreversible adjustment. Consistency algorithms are used ensure the relationship between node and block data. Programmable smart contracts are guaranteed based on automated script and virtual Turing code consistency machine. Blockchain is revolutionizing the IoT system in adversity of aspects. Unique features of blockchain – its P2P decentralized network, its open and transparent multilateral consensus and its unaffected data [4, 5]. As Blockchain

DOI: 10.1201/9781003497851-20

is a decentralized system that allows for secure and transparent record-keeping [6, 7]. Transactions are verified through consensus algorithms, making it resistant to tampering or central control [8, 9]. Moreover, blockchain is different from other distributed systems based on consensus and the following properties [21]:

- No trust: The entities involved in the network do not know each other. However, they can communicate, collaborate, and cooperate with each other without knowing each other, which means that there is no requirement for certified digital identity to perform any transaction between entities.
- Permission less: There is no restriction on who can or cannot work within the network, i.e. there is no species authorization.
- Censorship-proof: Since it is a controller less network, anyone can communicate or transact on the blockchain. In addition, no confirmed transaction can be edited or censored.

20.1.1 Blockchain Classification

Blockchain technology can be classified into several categories:

- Public vs. Private Blockchains: Public blockchains are open and decentralized, whereas private blockchains are permissioned and centralized.
- Permissioned vs. Permission less Blockchain: Permissioned blockchain allow only authorized participants to access and validate transactions, while permission less blockchain are open to everyone.

20.1.1.1 Consortium vs. Federated Blockchains

Consortium blockchain are governed by a group of organizations, while federated blockchains are governed by a specific group of participants.

20.1.1.2 Proof-of-Work (PoW) vs. Proof-of-Stake (PoS) Blockchains

PoW blockchain use computational power to validate transactions and create new blocks, while PoS blockchains use stakeholders' stakes to validate transactions and create new blocks (Figure 20.1).

20.1.2 Blockchain Characteristics

Blockchain technology has several key characteristics. Table 20.1 summarizes the basic characteristics of blockchain.

Table 20.2 shows comparison of characteristics between three classifications of blockchain technology: private, public, and consortium blockchain.

20.1.3 Industrial Internet of Things

New standards, innovative business methods, competitive pressure, and the need to transport goods on time are challenges in starting a new business in today's world [10, 11]. As a result, many businesses rely on the Industrial Internet of Things (IIoT) [24], which refers to all or any performance performed by enterprises to model, monitor and improve their business processes during statistics obtained from thousands connected machines, things and computers to help them achieving economic profit [21]. Edge Appliances for monitoring, control, and optimization of manufacturing processes and industrial facilities are described in Table 20.3.This table consists of IIOT

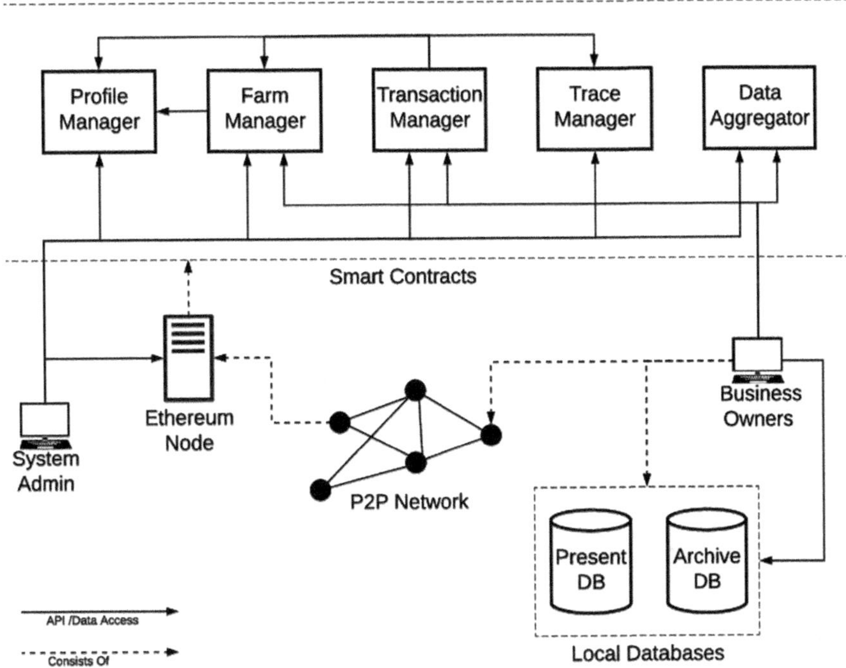

FIGURE 20.1 Basic structure of blockchain.

TABLE 20.1
Characteristics of Blockchain

Key points	Explanation
Decentralization	Blockchain decentralization refers to the distributed nature of the technology, where the data is stored across a network of nodes rather than in a central repository. This decentralized architecture provides several benefits, such as increased security, transparency, and resilience, as no single point of failure can compromise the network.
Transparency	It refers to the open and publicly accessible nature of the data stored on the blockchain network. This means that all transactions and information recorded on the blockchain can be viewed and verified by anyone with access to the network. The use of cryptographic techniques, such as hash functions and digital signatures, helps ensure the authenticity and integrity of the data, making it possible to track and verify every transaction on the blockchain.
Non-Repudiation	Non-repudiation in blockchain refers to the ability to prevent a participant from denying a previous transaction or action.
Persistency	It provides a high degree of persistence, as the data on a blockchain is maintained and can be accessed even if individual nodes or participants in the network fail or leave. This helps to ensure the durability and long-term preservation of the data on the blockchain.
Auditability	Audibility is achieved through the transparent and decentralized nature of the technology, which provides a clear and verifiable record of all transactions. Every node in the network has a copy of the blockchain, allowing for the easy tracking and verifying of transactions.
Anonymity	Anonymity in blockchain technology refers to the ability of users to conduct transactions and store data without revealing their identities.

TABLE 20.2
Comparison Classifications of Blockchain Technology

Key characteristics	Private Blockchain	Public Blockchain	Consortium Blockchain
Immutability	**Strong**	**Weak**	**Medium**
Traceability	**Strong**	**Strong**	**Medium**
Non-Repudiation	**Strong**	**Medium**	**Strong**
Decentralization	**Strong**	**Strong**	**Strong**
Flexibility	**Strong**	**Weak**	**Medium**
Consensus	**Ripple**	**PoW,PoS**	**PoA,PBFT**

TABLE 20.3
Features of Industrial Internet of Things Applications

Features	Explanation
Real-time data collection and analysis	*IoT sensors and devices are used to collect data in real-time from various industrial systems and machines, which is then analyzed to make informed decisions.*
Predictive Improved efficiency and productivity	*Industrial IoT needs to provide efficiency and productivity across various industrial operations as its real time data processing.*
Remote monitoring and control	*Real time monitoring and control is a main feature for industrial IoT of industrial processes and systems, enabling operators to make adjustments and respond to changes in real-time.*
Interoperability	*Industrial IoT enables interoperability between different systems and devices, allowing for seamless communication and data exchange between different components of an industrial operation.*
Enhanced Security	*Security is a crucial aspect of Industrial IoT (IIoT), as industrial systems and processes can be vulnerable to cyber-attacks that can result in significant harm to operations and safety.*

features necessary to perform actions as it works real-time data and collection of data via sensors and devices.

Industry 4.0 uses smart devices to manage data and activities, however corrupted sensors/IoT devices could exponentially reduce economic growth. IIoT mainly applies instrumentation, connected sensors, and other devices to machinery, vehicles in the transport, and the energy and industrial sectors. Traditional IoT applications such as Internet-connected refrigerators are a subset of IIoT. It has four design principles [25]:

1. Interoperability: the possibility that machines and related components connect and communicate with people.
2. Information transparency: the necessity of creating virtual copies of the physical world.
3. Technical assistance: essential comprehensive aggregation and information visualization to support human capabilities.
4. Decentralization of decisions: the ability of a network-enabled system to independently make decisions and perform its own specialized functions.

20.1.4 Merging Blockchain with IIoT

Without a dealer-generated system, middlemen cost. Sensor users distrust technical suppliers who, like manufacturers, governments, and service partners, give device control authority to acquire and analyze user data. This research presents different ways that are prone to feat and bugs but less susceptible to nasty embedment from third parties because they may be evaluated by several consumers.

RQ1: What are the limitations of using blockchain technology in IIOT for security purposes? Blockchain-related applications for IoT and IIoT systems:

a. Electric vehicle cloud and edge.
b. Mobile commerce.
c. Authentication and access management.
d. Cloud storage.
e. Big data.
f. Smart home.
g. Smart cities.

Blockchain improves security when it provided the encryption and validation of data. Using decentralization in data storage, block chain's data sensitivity has been protected. Some problems identified from peer articles are as follows:

a. Cost of technology:
Blockchain technology may be capable of saving the cost of the transaction (transaction fee) but "Proof of work system" requires more computational power to validate the transactions.

b. Inadequate Performance:
Blockchain-based solutions usually perform with some inadequacies when it comes to centralized databases. The main reason behind these issues is the scalability of their consensus mechanism that reduces the speed of transaction throughput.

c. Susceptibility:
Crypto currencies may be susceptible to 51% attacks. These attacks may not be easy to execute because they require high-end computational power.

20.2 LITERATURE REVIEW

Preexisting surveys explored from different aspects of blockchain with IoT. Some of these are discussed as follows. M. Crosby et al. explained blockchain and about Bitcoin technology. Khan et al. [20] focused on the security requirements of IoT using blockchain technology. Florea et al. [21] described the use of blockchain technology. Dorri et al. [22] projected a secure and lightweight architecture for IoT; based on blockchain technology. Atlam et al. [23] provided an overview of the integration of IoT and blockchain, while highlighting its benefits and challenges. S. Sicari et al. [1] discussed security and privacy issues in IoT. Guizi Chen et al. [2] proposed an efficient framework securing IoT devices. Tiago et al. [3] discussed applications that are using blockchain in cyber security systems. But, to the best of our knowledge, these had been focused on how to use blockchain with IoT applications on decentralized storage purpose or cyber-attacks only.

Blockchain with IIOT has been focusing now frequently so this review considered the potential of blockchain with IIOT; how it can help for improving security in industrial applications and what will be its future. Therefore, this comprehensive survey has covered the span of last five years (2018–till date), mainly in the perspective of security and as well as challenges and issues linked with it. Table 20.4 shows a few of the existing surveys in industrial internet of things with blockchain.

RQ2: **How can blockchain technology can be used to enhance security in IIOT systems?**
Blockchain utilizes encryption and hashing to store immutable records and many of the existing cyber security solutions utilize very similar technology as well. As outlined in other sections of

TABLE 20.4
Related Surveys on Blockchain for IIOT

Year	Author	Focus
2018	Ferrag et al. [12]	Blockchain based security and privacy solutions for IoT
2019	Lu et al. [13]	Review of blockchain in oil and gas industry
2019	H. Juma et al. [14]	Survey of blockchain in Trade Supply Chain industry
2019	Xie et al. [15]	Blockchain applicability to smart cities
2019	Fraga et al. [16]	Review of blockchain in automotive industry
2020	Megha S. et al. [17]	Survey of blockchain in the Energy industry
2021	U Majeed et al. [18]	Blockchain based IOT for smart cities
2022	Kumar et al. [19]	Integrated blockchain and IoT in food supply chain

this paper, this is achieved in many different ways, but the bottom line is this: many members of a group who have access to the same information will be able to secure that group far better than a group made up of one leader and a host of members who rely on the leader for their information, particularly when bad actors could come in the form of group members or even as the leaders themselves. The existing solutions and Key qualitative and quantitative Data used previously using blockchain with IIoT have been described in Table 20.5. Proposed solutions, a few of them from peer reviewed articles, are described in Table 20.6.

RQ3: **What are the existing solutions and Key qualitative and quantitative Data used previously using blockchain with IIoT?**

20.3 ORGANIZATION AND RESEARCH METHODOLOGY

To achieve research objectives we have selected these strategies (Figure 20.2).

20.3.1 Selection of Primary Studies

Primary studies were highlighted by passing keywords to the search facility of a particular publication or search engine. The keywords were selected to promote the emergence of research results that would assist in answering the research questions. The Boolean operators were restricted to AND and OR. The search strings were: ("blockchain" OR "block-chain") AND "security" AND "IIOT".

The platforms searched were:

- IEEE Xplore
- Science Direct
- Springer Link
- ACM Digital Library

The searches were run against the title, keywords or abstract, depending on the search platforms. The searches were conducted on 15 January 2023, and we processed all studies that had been published up to this date. The results from these searches were filtered through the inclusion/exclusion criteria. Any results from Google Scholar will be checked for compliance with these criteria. Only the most recent version of a study will be included this SLR. The key inclusion and exclusion criteria are shown below in Table 20.7.

TABLE 20.5
Main Findings and Themes of the Primary Studies

Primary Study	Key qualitative and quantitative Data	Applied security mechanism
[28]	Proof of concept pseudonymous protocol for secure communications between IoT devices using bitcoin blockchain for case study.	**IoT**
[29]	Blockchain. For example, IoT devices from one manufacturer are on the same blockchain and then distribute firmware upgrades peer to peer rather than pushing from the center. Recognition of requirement for token. Possible solutions offered.	**Industrial IoT**
[30]	Blockchain-based system for providing authenticity for Docker images	**IoT devices**
[31]	Bitcoin blockchain-based proposal for securing smart home IoT devices on a local blockchain. Assessment of network overheads when utilizing blockchain.	**Smart Home**
[32]	Multi-level network of IoT devices utilizing blockchain. Managing security of the blockchain through communication between layers rather than fully decentralized nodes and miners.	**IIoT**
[33]	Blockchain is an ideal solution for IIoT security. The work proposed in this paper is a blockchain platform for IIoT. This platform, with smart contracts deployed, enables development of different distributed applications for manufacturing using a decentralized, trustless, peer-to-peer network for IIoT applications.	**IIoT**
[34]	Position paper highlights increasing importance of blockchain application to IoT in homes, battlefields and healthcare. Conceiving a way for IoT to install secure firmware updates	**IoT**
[35]	Discussion on strengths of blockchain in improving security, particularly with IoT. Highlighting the security benefits of IoT supply chain from manufacturer to end user.	**IoT**
[36]	Provides an application of blockchain in the form of securing historic IoT connections and sessions and detecting malicious behavior.	**IoT**
[37]	Suggestion for how low-power IoT devices could communicate with a sufficient gateway to enable node communication on the Ethereum blockchain	**IoT**
[38]	Proposing "ControlChain", a blockchainbased solution for IoT device access control.	**IoT**
[39]	Substantial review of IoT security and how blockchain could meet the challenges of reducing the existing security threats against such devices.	**IoT**
[40]	A thorough review of how blockchain works, current Proof-of-X concepts and their advantages and disadvantages.	**IoT**
[41]	A blockchain-enabled IIoT framework is introduced and involved fundamental techniques are presented.	**IIoT**

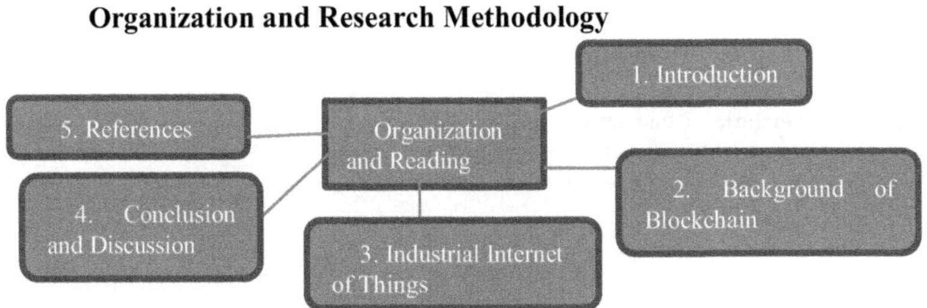

FIGURE 20.2 Organization and reading map to be followed in this research paper.

TABLE 20.6
Proposed Solution and Detail

Year	Author	Proposed Solution	Blockchain approach	Problem addressed
2020	Geetanjali et al. [42]	A secure IoT sensors communication in industry 4.0 using blockchain technology in order to trace and manage every single activity.	Consensus Algorithm	Authentication, transparency and Provability
	Meng et al. [43]	An efficient blockchain-assisted secure device authentication mechanism BASA for cross-domain IIoT.	Consortium blockchain	To construct trust among different devices
	Lu et al. [26]	Secure data sharing architecture using blockchain	Permissioned	Secure data sharing
	Ahmed et al. [44]	Applying blockchain technology to the healthcare industry could improve information security management.	Consensus Algorithm	Improving diagnosis accuracy, privacy and security preservation
2021	Latif et al. [28]	Blockchain-based architecture that ensures secure and trustworthy industrial operations.	Consensus Algorithm	security, immutability, trust, decentralization
2022	Ali et al [27]	Secure patient healthcare data access scheme is devised, which integrates blockchain and trust chain to fulfill the efficiency and security issues.	Blockchain homomorphic encryption technique	Efficiency and security issues
	Prabhat et al. [45]	Blockchain and Deep Learning (DL) integrated framework for delivering decentralized data pro-cessing and learning in IIoT network.	Smart contracts	Ensure integrity and authenticity of data.

TABLE 20.7
Inclusion Criteria and Exclusion Criteria

Inclusion criteria	Exclusion criteria
C1: A List of papers between 2017-2023	E1: Papers focusing on economic, business or legal impacts of blockchain applications E2: Non-English papers.
C2: Papers included in this review are from selected journals of Q1, Q2, Q3	E3: Older papers published before 2017.
C3: Papers comparing types of blockchain for IIoT	E4: Non peered review papers such as news, articles
C4: Papers discussing challenges to embed blockchain with IIoT	
C5: Papers proposed solutions for IIoT in security perspectives using blockchain	
C6: The paper must present empirical data related to the application and the use of blockchain in IIoT	

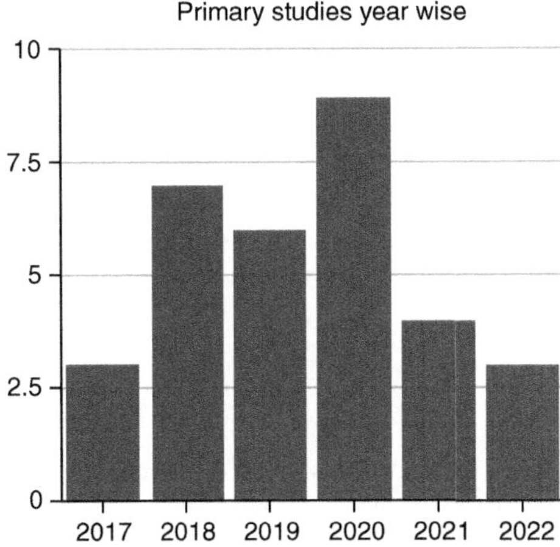

FIGURE 20.3 Number of primary studies published over time selected in this review.

20.3.2 Research Goals

The purpose of this research is to analyze existing studies and their findings and to summarize the efforts of research in blockchain applications to improve cyber security in Industrial IoT. To make the work more focused, we developed three research questions, as mentioned below:

1. What are the limitations of using blockchain technology in IIOT for security purposes?
2. How blockchain technology can be used to enhance security in IIOT systems?
3. What are the existing solutions and key qualitative and quantitative data used previously using blockchain with IIoT?

There were a total of 75 studies identified from the initial keyword searches on the selected platforms. This was reduced to 70 after removing duplicate studies. After checking the studies under the inclusion/exclusion criteria, the number of papers remaining for reading was 60. The 40 papers were read in full with the inclusion/exclusion criteria being re-applied, and 35 papers remained.

As shown in Figure 20.3, there is an upward trend in the usage of blockchain in the cyber security for IIOT context. We envisage that in the future we will see a significant number of research studies regarding the adoption of blockchain in real-world applications.

20.4 CONCLUSION

The combination of IIoT (Industrial Internet of Things) and blockchain technology offers a promising solution for enhancing security in industrial systems. This paper focuses on the use of blockchain technology to enhance the security of IIoT systems in areas such as data sharing, immutable records, identity management, and decentralization. While many applications are already utilizing this technology, implementing IIoT and blockchain for security enhancement requires careful planning and consideration of potential challenges and limitations. These challenges include scalability and performance issues, ensuring compatibility with existing systems, and addressing regulatory and compliance requirements. The paper comprehensively analyzes proposed solutions and models for IIoT using blockchain technology to enhance security. These solutions include blockchain-based access control, blockchain-based authentication, blockchain-based data sharing, and blockchain-based data provenance on industrial IIoT. These solutions ensure that data is securely stored and transferred, and that only authorized parties can access and modify the data.

Overall, the integration of IIoT and blockchain technology has the potential to significantly enhance the security of industrial systems. This paper provides a comprehensive analysis of the current state of IIoT and blockchain technology, and highlights potential areas for future research and development. In conclusion, the integration of blockchain and IIoT is a promising development in the field of computational communication systems. Blockchain-based solutions provide a decentralized, trustless, and peer-to-peer approach to securing data storage and transfer in IIoT applications.

REFERENCES

1. Sicari, Sabrina, Alessandra Rizzardi, Luigi Alfredo Grieco, and Alberto CoenPorisini. "Security, privacy and trust in Internet of Things: The road ahead." *Computer Networks*, 76.Supplement C (2015): 146–164, 2015.
2. Chen, Guizi, Wee Siong Ng. "An efficient authorization framework for securing industrial internet of things." *Proc. of the 2017 IEEE Region 10 Conference (TENCON)*, Malaysia, November 5–8, 2017.
3. Ferna´ndez-Carame´s, Tiago M, Paula Fraga-Lamas. "A review on the application of blockchain to the n generation of cyber secure industry 4.0 smart factories." *Access IEEE* 7 (2019): 45201–45218.
4. Lu, Yang. "Artificial intelligence: A survey on evolution, models, applications and future trends." *Journal of Management Analytics* 6.1 (2019): 1–29.
5. Crosby, Michael, Pradan Pattanayak, Sanjeev Verma, and Vignesh Kalyanaraman, "Blockchain technology: Beyond bitcoin." *Application Innovations* 2 (2016): 6–10.
6. Nakamoto, Satoshi. "Bitcoin: A peer-to-peer electronic cash system." (2008): 17–23.
7. Larimer, Daniel. "Transactions as proof-of-stake," Nov-2013 (2013): 1–8.
8. "Delegated proof-of-stake (dpos)." Bitshare whitepaper (2014): 14–22.
9. Castro, Miguel, Barbara Liskov. "Practical byzantine fault tolerance." *OSDI* 99 (1999): 173–186.
10. Parizi, Reza M., Amritraj, and Ali Dehghantanha. "Smart contract programming languages on blockchains: An empirical evaluation of usability and security." *International Conference on Blockchain (ICBC 2018)*. Springer, 2018, pp. 75–91.
11. Parizi, Reza M., Ali Dehghantanha, Kim-Kwang Raymond Choo, and Amritraj Singh. "Empirical vulnerability analysis of automated smart contracts security testing on blockchains." *Proceedings of the 28th Annual International Conference on Computer Science and Software Engineering (CASCON 2018)*. IBM Corp., 2018, pp. 103–113.
12. Ferrag, Mohamed Amine, Makhlouf Derdour, Mithun Mukherjee, Abdelouahid Derhab, Leandors Maglaras, and Helge Janicke. "Blockchain technologies for the internet of things: Research issues and challenges." *IEEE Internet of Things Journal*, 6 (2018): 2188–2204.
13. Lu, Hongfang, Kun Huang, Mohammadamin Azimi, and Lijun Guo. "Blockchain technology in the oil and gas industry: A review of applications, opportunities, challenges, and risks." *IEEE Access* 7 (2019): 41 426–41 444.
14. Juma, Hussam, Khaled Shaalan, and Ibrahim Kamel. "A survey on using blockchain in trade supply chain solutions." *IEEE Access* 7 (2019): 184115–184132.
15. Xie, Junfeng, Helen Tang, Tao Huang, F. Richard Yu, Renchao Xie, Jiang Liu, and Yunjie Liu. "A survey of blockchain technology applied to smart cities: Research issues and challenges." *IEEE Communications Surveys & Tutorials* 21 (2019): 2794–2830.
16. Fraga-Lamas, Paula, and Tiago M. Fernández-Caramés. "A review on blockchain technologies for an advanced and cyber-resilient automotive industry." *IEEE Access* 7 (2019): 17 578–17 598, 2019.
17. Megha, Swati, Joseph Lamptey, Hamza Mohamed Salem, and Manuel Mazzara. (2020) "A survey of blockchain-based solutions for energy industry." *Artificial Intelligence and Network Applications, WAINA 2020, Advances in Intelligent Systems and Computing*, vol 1150, Springer, Cham.
18. Majeed, Umer, Latif U. Khan, Ibrar Yaqoob, SM Ahsan Kazmi, Khaled Salah, and Choong Seon Hong. "Blockchain for IoT-based smart cities: Recent advances, requirements, and future challenges." *Journal of Network and Computer Applications* 181 (2021): 103007.

19. Kumar, Shashank, Rakesh D. Raut, Nishant Agrawal, Naoufel Cheikhrouhou, Mahak Sharma, and Tugrul Daim. "Integrated blockchain and internet of things in the food supply chain: Adoption barriers." *Technovation* 118 (2022): 102589.
20. Khan, Sumaiyya Z, Snehal R. Kamble, Ambarish R. Bhuyar. "A review on BIoT: Blockchain IoT." *IJREAM* 04.03 (2018): 808–812.
21. Florea, Bogdan Cristian. "Blockchain and Internet of Things data provider for smart applications." *2018 7th Mediterranean Conference on Embedded Computing (MECO)*, IEEE, Budva, Montenegro, July 2018, pp. 1–4. https://doi.org/10.1109/ MECO.2018.8406041
22. Dorri, A, Salil S. Kanhere, and Raja Jurdak. "Blockchain in internet of things: Challenges and solutions." 2016. arXiv:1608.05187.
23. Atlam, Hany F., Ahmed Alenezi, Madini O. Alassafi, and Gary Wills. "Blockchain with internet of things: Benefits, challenges, and future directions." *International Journal of Intelligent Systems and Applications* 10.6 (2018): 40–48.
24. Lade, Prasanth, Rumi Ghosh, and Soundar Srinivasan. "Manufacturing analytics and industrial Internet of Things." *IEEE Intelliegnt System* 32.3 (2017): 74–79.
25. Underwood. Industry 4.0: Key design principles. https://kingstar.com/industry-4-0-keydesign-principles/. April 24, 2017.
26. Yunlong Lu, Xiaohong Huang, Yueyue Dai, Sabita Maharjan, and Yan Zhang, "Blockchain and federated learning for privacy-preserved data sharing in industrial IoT." *IEEE Transactions on Industrial Informatics* 16.6 (June 2020): 4177–4186.
27. Ali, Aitizaz, Mohammed Amin Almaiah, Fahima Hajjej, Muhammad Fermi Pasha, Ong Huey Fang, Rahim Khan, Jason Teo, and Muhammad Zakarya. "An industrial IoT-based blockchain-enabled secure searchable encryption approach for healthcare systems using neural network." *Sensors* 22 (2022): 572. https://doi.org/10.3390/s22020572
28. Zyskind, Guy, and Oz Nathan. "Decentralizing privacy: Using blockchain to protect personal data." *2015 IEEE Security and Privacy Workshops*. IEEE, 2015.
29. Christidis, Konstantinos, and Michael Devetsikiotis. "Blockchains and smart contracts for the Internet of Things." *IEEE Access* 4 (2016): 1–1. https://doi.org/10.1109/ACCESS.2016.2566339
30. Xu, Quanqing, Chao Jin, Mohamed Faruq Bin Mohamed Rasid, Bharadwaj Veeravalli, and Khin Mi Mi Aung. "Blockchainbased decentralized content trust for docker images." *Multimedia Tools and Applications* 77 (2018): 18223–18248.
31. Dorri, Ali, Salil S. Kanhere, Raja Jurdak, and Praveen Gauravaram. "Blockchain for IoT security and privacy: The case study of a smart home." *2017 IEEE international conference on pervasive computing and communications workshops (PerCom workshops)*. IEEE, 2017.
32. Li, Cheng, and Liang-Jie Zhang. "A blockchain based new secure multi-layer network model for internet of things." *2017 IEEE international congress on internet of things (ICIOT)*. IEEE, 2017.
33. Bahga, Arshdeep, and Vijay K. Madisetti. "Blockchain platform for industrial internet of things." *Journal of Software Engineering and Applications* 9 (2016): 533–546.
34. Banerjee, Mandrita, Junghee Lee, and Kim-Kwang Raymond Choo. "A blockchain future for internet of things security: A position paper." *Digital Communications and Networks* 4.3 (2018): 149–160.
35. Kshetri, Nir. "Blockchain's roles in strengthening cybersecurity and protecting privacy." *Telecommunications Policy* 41.10 (2017): 1027–1038.
36. Gupta, Yash, Rajeev Shorey, Devadatta Kulkarni, and Jeffrey Tew. "The applicability of blockchain in the Internet of Things." *2018 10th International Conference on Communication Systems & Networks (COMSNETS)*. IEEE, 2018.
37. Özyılmaz, Kazım Rıfat, and Arda Yurdakul. "Integrating low-power IoT devices to a blockchain-based infrastructure: Work-in-progress." *Proceedings of the Thirteenth ACM International Conference on Embedded Software 2017 Companion*. 2017.
38. Pinno, Otto Julio Ahlert, Andre Ricardo Abed Gregio, and Luis CE De Bona. "Controlchain: Blockchain as a central enabler for access control authorizations in the iot." *GLOBECOM 2017-2017 IEEE Global Communications Conference*. IEEE, 2017.
39. Khan, Minhaj Ahmad, and Khaled Salah. "IoT security: Review, blockchain solutions, and open challenges." *Future Generation Computer Aystems* 82 (2018): 395–411.

40. Jesus, Emanuel Ferreira, Vanessa RL. Chicarino, Célio VN. De Albuquerque, and Antônio A. de A. Rocha. "A survey of how to use blockchain to secure internet of things and the stalker attack." *Security and Communication Networks* 2018 (2018): 9675050.
41. Zhao, Shanshan, Shancang Li, and Yufeng Yao. "Blockchain enabled industrial Internet of Things technology." *IEEE Transactions on Computational Social Systems* 6.6 (2019): 1442–1453.
42. Rathee, Geetanjali, et al. "A secure IoT sensors communication in industry 4.0 using blockchain technology." *Journal of Ambient Intelligence and Humanized Computing* 12 (2021): 533–545.
43. Shen, Meng, Huisen Liu, Liehuang Zhu, Ke Xu, Hongbo Yu, Xiaojiang Du, and Mohsen Guizani. "Blockchain-assisted secure device authentication for cross-domain industrial IoT." *IEEE Journal on Selected Areas in Communications* 38.5 (2020): 942–954.
44. Farouk, Ahmed, Amal Alahmadi, Shohini Ghose, and Atefeh Mashatan. "Blockchain platform for industrial healthcare: Vision and future opportunities." *Computer Communications* 154 (2020): 223–235.
45. Kumar, Prabhat, Randhir Kumar, Abhinav Kumar, A. Antony Franklin, Sahil Garg, and Satinder Singh. "Blockchain and deep learning for secure communication in digital twin empowered industrial IoT network." *IEEE Transactions on Network Science and Engineering* 10 (2022): 2802–2813.

21 Design and Implementation Role of Middleware in Shared Network Environments

A Systematic Review

Muddassar Ali, Hamayun Khan, Irfan Ud Din, Muhammad Imran Tariq, and Aetsam Javed

21.1 INTRODUCTION

A shared network environment is a complex system that involves multiple applications and systems that need to communicate with each other to achieve a common goal. In such an environment, middleware serves as a bridge between applications, allowing them to exchange data and communicate with each other. Middleware is defined as software that provides a layer of abstraction between applications and the underlying operating system or network infrastructure. Middleware can be thought of as a layer of software that provides a common set of services and protocols that enable communication and data exchange between disparate applications and systems. Examples of middleware in a shared network include application servers, message brokers, and API gateways [1]. These components provide a layer of abstraction that shields applications from the underlying network infrastructure, making it easier to build and manage complex distributed systems.

21.1.1 Architecture

Middleware refers to software that acts as a bridge or intermediary between different applications or systems, allowing them to communicate and interact with each other. It essentially sits in the middle of the communication between two or more systems, hence the name middleware. Middleware can be used to integrate disparate systems, enable communication [1] between systems with different protocols, and provide common services to applications. The architecture of middleware typically involves three layers: the transport layer, the middleware layer, and the application layer. Each of these layers plays a specific role in enabling communication and interaction between different systems. The transport layer [2] is responsible for the physical transfer of data between systems. It includes protocols such as TCP/IP [3], HTTP, and FTP, which provide the foundation for data transfer across networks. The transport layer is responsible for ensuring that data is transmitted reliably, securely, and efficiently. The middleware layer [4] is the heart of the middleware architecture. It is responsible for providing the necessary services to enable communication between different systems. This layer includes a range of middleware technologies such as message queues, service buses, and API gateways. These technologies provide the necessary routing, transformation, and orchestration services to enable systems to communicate with each other. Message queues are used to decouple systems and enable asynchronous communication. They provide a way for systems to send messages to each other without the need for real-time interaction. This can be useful for systems that are distributed across different geographies or have different processing speeds [5]. Service buses are used to provide a common communication platform for different systems. They

DOI: 10.1201/9781003497851-21

provide a way for systems to publish and subscribe to events, enabling them to interact in real time. Service buses can also provide transformation services to ensure that messages are delivered in the correct format. API gateways are used to provide a unified interface for different systems to interact with each other. They provide a way for systems to interact through a set of well-defined APIs, which can be used to abstract the underlying complexity of different systems. The application layer [6] is the layer where the actual business logic resides. It includes the applications that are interacting with each other through the middleware layer. These applications can be either custom-built or off-the-shelf software, depending on the requirements of the system.

Middleware architecture [7] can be implemented using different approaches, including: Message-Oriented Middleware (MOM) – MOM [8] is a type of middleware that provides a platform for the exchange of messages between distributed software applications. In MOM, applications communicate by sending and receiving messages, rather than calling each other's functions or methods directly. MOM is used in a wide range of applications, including financial trading systems, supply chain management, and real-time data processing. Some popular MOM [9] implementations include Apache ActiveMQ, IBM MQ, and RabbitMQ.

Remote Procedure Call (RPC) – RPC [10] middleware is a type of middleware that allows software applications to communicate with each other by invoking procedures or functions in a remote system, as if they were local functions. RPC [11] provides a way for distributed applications to work together, without the need for developers to worry about the details of network communication. In RPC, a client application sends a request to a server application, specifying the name of the function to be invoked, along with any input parameters. The server application then executes the function and sends back the results to the client. The communication between the client and server is typically done using a protocol such as HTTP, TCP/IP, or UDP.RPC can be used in a variety of scenarios, including client–server applications, distributed systems, and micro services architectures. Some popular implementations of RPC middleware include Apache Thrift, gRPC, and CORBA.

(CORBA) – Common Object Request Broker Architecture [12] is a middleware technology that enables distributed software applications to communicate with each other over a network. CORBA provides a standard way for objects to interact with each other, regardless of the programming languages, operating systems, or hardware platforms used. In CORBA, objects are defined using an Interface Definition Language (IDL), which specifies the methods and properties of the object's interface. Clients can then use a CORBA [13] object request broker (ORB) to invoke methods on the object, regardless of where it is located in the network CORBA has been widely used in enterprise environments for building complex, distributed systems, such as telecommunications, finance, and defense. One of the main advantages of CORBA is its support for interoperability between different programming languages and platforms. CORBA also provides a rich set of features, such as security, transaction management, and persistence. However, CORBA can be complex to use, and requires developers to learn a new programming model and IDL syntax [14]. Additionally, CORBA has been largely superseded by newer middleware technologies such as web services and message-oriented middleware.

Simple Object Access Protocol (SOAP) – SOAP [15] is a messaging protocol used for exchanging structured data between different applications over the internet. SOAP is a protocol that was designed to be platform-independent and language-independent, making it a popular choice for web services. SOAP uses XML (Extensible Markup Language) as its message format, which is a text-based markup language that is easy to read and understand by both humans and machines. SOAP messages are typically sent over HTTP orHTTPS, and can be transmitted using different transport protocols. One of the key advantages of SOAP is its support for WSDL (Web Services Description Language) [16], which provides a standard way of describing web services and their interfaces. This makes it easier for developers to discover and consume web services. However, SOAP is a relatively heavyweight protocol and can be slower and more complex to use than other messaging protocols like REST. Additionally, SOAP messages can be larger than equivalent messages in other formats, which can impact performance in low-bandwidth environments. Overall, SOAP is a useful protocol for certain types of web services, but its complexity and overhead makes it less suitable for others.

Table 21.1 provided a comparison [17] of middleware models that clearly shows the features and implementation as well as disadvantages of each model. In summary, middleware architecture provides a layered approach to implementing middleware software that enables communication and data exchange between disparate applications and systems. The architecture consists of the presentation layer, application layer, and communication layer, each with specific components and services. Implementing middleware architecture requires careful consideration of the specific requirements of the system and the capabilities of the middleware technology being used.

There are several key features of middleware shown in Table 21.1, which make it essential in a shared network environment. These include: Interoperability – Middleware enables disparate applications [18] and systems to communicate with each other, regardless of their underlying technologies and platforms.

Scalability – Middleware can handle large volumes of data and transactions [19], making it suitable for use in enterprise-level systems. Security – Middleware provides security features that protect data and transactions from unauthorized [20] access and ensure the integrity of the system. Reliability – Middleware provides features that ensure the reliability [5] and availability of the system, including fault tolerance, load balancing, and failover.

There are several benefits of middleware that make it an essential component of shared network environments. These include: Reduced Complexity – Middleware simplifies the integration process by providing a common set of services and protocols that enable communication [5] and data exchange between disparate applications and systems. Increased Agility – Middleware enables organizations to respond quickly to changing business requirements by providing a flexible [21] and adaptable architecture. Improved Efficiency – Middleware can automate many of the manual processes that are required to integrate applications [22], reducing the time and cost associated with integration. Enhanced Visibility – Middleware provides a centralized view of the system, enabling organizations to monitor and manage their applications [23] and systems more effectively.

21.2 METHODOLOGY

Methodology refers to the systematic and transparent process used to conduct research or analysis. In the context of a systematic literature review (SLR), methodology involves a rigorous and standardized approach to identifying, selecting, evaluating, and synthesizing relevant studies. This involves defining a clear research question or objective, developing a comprehensive and reproducible search strategy, applying specific inclusion and exclusion criteria to identify relevant studies, evaluating the quality and risk of bias of included studies, and synthesizing the findings of the selected studies. A well-defined and transparent methodology is crucial in ensuring the reliability and validity of the review process and findings. We conducted a systematic literature (SLR) review that is divided in three phases and each phase is further divided in activities which are performed to conduct the SLR. The detail of phases involved in SLR is as follows: Planning is a critical first step in conducting a systematic literature review (SLR), as it lays the foundation for a well-organized and comprehensive review. Overall, effective planning is essential for a successful SLR, as it helps to ensure that the review is comprehensive, rigorous, and well-organized. The primary goal of the review was established at this phase, and the activities that followed thoroughly described each stage.

21.2.1 Need Identification of Review

The lack of systematic literature review (SLR) [8] in the field of middleware can be attributed to several factors. First, middleware is a broad and diverse field, encompassing a wide range of technologies and applications. This makes it difficult to define a specific research question or scope for

TABLE 21.1
Comparison of Middleware Models

Sr. #	Middleware Model	Features	Implementation	Disadvantages
1.	• Message-Oriented Middleware (MOM)	• Asynchronous communication. • High message throughput. • Guaranteed delivery and transactions. • Apache ActiveMQ, IBM MQ, and RabbitMQ.	Implements a message queue to buffer messages	• Can be complex to set up and manage • Can introduce latency due to message queuing
2.	• Remote Procedure Call (RPC)	• Remote procedure calls • Request-response model. • Multiple languages. • gRPC, Apache Thrift, and CORB	Implements a protocol for encoding data and invoking procedures remotely	• Tightly coupled, which can make changes difficult • Can have issues with versioning and compatibility
3.	• Web Services	• Standardized way for communication. • Open protocols like SOAP and REST. • Multiple programming languages. • Apache CXF, Microsoft WCF, and Oracle WebLogic.	Implements a set of standardized protocols for communication and data exchange, such as SOAP and REST	• Can be complex to implement and manage • Requires a lot of upfront planning
4.	• Event-Driven Middleware	• Asynchronous processing, • Reactive, Real-time processing • Publish-Subscribe Model. • Highly scalable. • Apache Kafka, Apache NiFi, RabbitMQ, Apache Flink	Implements an event loop and event handlers to handle and react to events	• Can be difficult to design and implement • Requires careful consideration of event handling and messaging patterns
5.	• Distributed Data Management Systems	• Data Distribution and Replication. • Consistency and security to access data. • Apache Cassandra, Hadoop, Kafka, MongoDB, Riak	Implements a distributed database and replication protocols to ensure consistency	• DDMS middleware can introduce additional latency due to data replication and synchronization, which can impact application performance • not be compatible and also costly
6.	• Common Object Request Broker Architecture (CORBA)	• Service discovery and load balancing. • Middleware chaining and transport independence • Context propagation and plug-in architecture • Language and platform independence • Location Transparency and object management.	Implements an object interface and a protocol for object invocation and marshaling	• Vendor lock-in: CORBA implementations can vary between vendors, potentially leading to vendor lock-in or compatibility issues. • Integration challenges and performance overhead

an SLR. Second, the fast-paced nature of the field means that new technologies and approaches are constantly emerging, making it difficult to keep up with the latest developments. Finally, there is a lack of standardization in the terminology and concepts used in middleware research, which makes it difficult to compare and synthesize results from different studies.

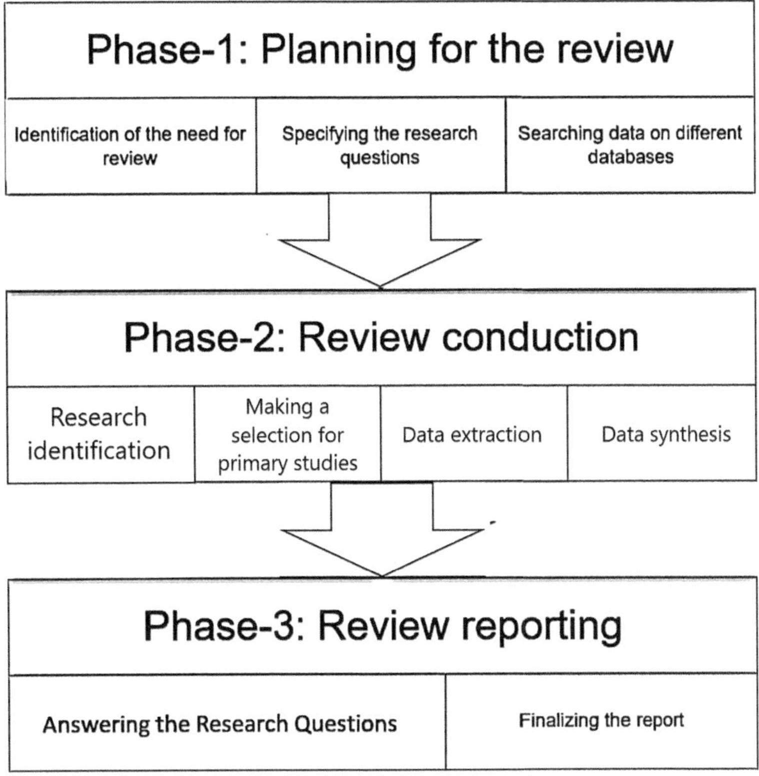

FIGURE 21.1 SLR phases.

21.2.1.1 Specifying the Research Question(S)

In order to gain a comprehensive understanding of the topic, this study aims to identify and analyze research related to role of middleware in shared network environment that have been published from 2000 to the present day. The primary objective has been divided into several research questions.

- RQ1: What is the role of middleware in a shared network environment?
- RQ2: Which facilities are provided by the middleware in the context of a shared network environment?
- RQ3: How much middleware is needy for a shared network?

21.2.1.2 Identifying the Relevant Bibliographic Databases

Initial Search: At this juncture, we are conducting an automated search using Google Scholar, which is currently one of the largest and most exhaustive databases and indexing systems for scientific literature. Our rationale for utilizing this data source is based on two main factors. Firstly, according to the literature, this indexing tool has been a wise selection in terms of identifying the initial set of literature studies for the snowballing process. Secondly, Google Scholar offers the highest number of potentially relevant studies in comparison to other notable libraries such as Scopus, ACM Digital Library, IEEE Explore, and Web of Science, and furthermore, the query results can be easily extracted from the indexer through automation. Figure 21.2 shows the database records according to frequency and percentage of articles.

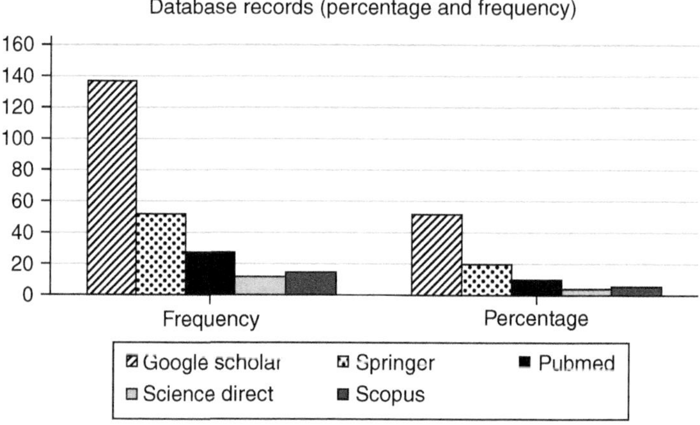

FIGURE 21.2 Database records.

21.2.2 Phase 2: Conducting the Review

To create a suitable search string, it is necessary to strike a balance between the scope of the search and its manageability. Therefore, we extracted the main phrases that matched the study questions and looked for synonyms of important terms to include in the search string. We used the Boolean AND to connect the main components and the Boolean OR to link alternate phrases. In this way, we compiled a comprehensive list of search terms:

{(middleware "OR" adapter "OR" intermediate layer) AND (shared "OR" distributed "OR" divided) AND (network "OR" system) AND (role "OR" importance)}

Listing 1 provides the search term that is utilized for the initial search. The search phrase is deliberately kept general to encompass a wide range of relevant papers related to the subject of the research. It consists of four primary parts, each emphasizing role, features, and importance of middleware in a shared network. To ensure the initial search results are as targeted as possible, the query is applied only to the titles of the studies. The search string is tested through pilot searches on Google Scholar. The time frame for the search spans from 2007 until the date the query is executed, which is March 2023.

21.2.2.1 Application of Selection Criteria

We are currently considering all 265 papers that were obtained during the initial search and applying a set of specific criteria for inclusion and exclusion to filter them. In order to make unbiased and cost-effective research choices, we employ an adaptive reading approach, which involves selectively reading only the relevant portions of the studies that meet our criteria, rather than reading them in their entirety. The inclusion and exclusion criteria that we are using for this study are outlined in the following sections.

21.2.2.1.1 Exclusion Criteria

After applying the exclusion criteria [5], Figure 21.3 indicates that 115 papers were eliminated. The criteria for deletion are as follows: Studies that aren't in English as well as those that were published before 2000 and after 2023. There was no peer review mechanism for the other materials, which included theses, technical reports, and other documents. Secondary or postsecondary education (e.g., systematic literature reviews, surveys, etc.). Editorials, tutorials, and poster papers for studies, as they don't offer adequate details. Papers that are expansions of previously included papers or

TABLE 21.2
Description of SLR

Sr. #	Description of indicator	Quantity of papers
1.	Articles identified through database searching	265
2.	Paper selected after application of exclusion criteria	115
3.	Paper selected after application of inclusion criteria	100
4.	Papers retained after application of quality assessment indicator	47

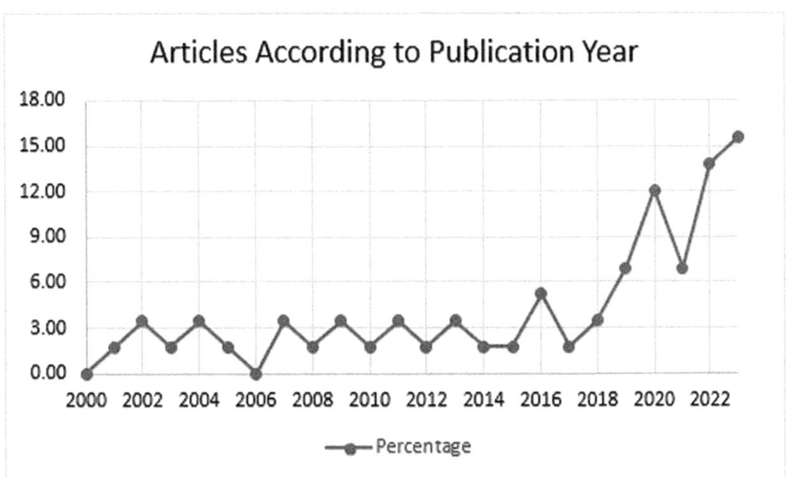

FIGURE 21.3 Percentage of articles according to publication year.

duplicates. Research that hasn't undergone peer review. We are unable to inspect certain documents, which makes them unavailable.

If a publication meets all the inclusion requirements, it is classified as a primary study; however, if it fails to meet any of these criteria, it is rejected. Two researchers have refined and evaluated the criteria through several pilot investigations. It's important to mention that secondary research is excluded due to the initial exclusion criterion, but it is still addressed in the related work section.

21.2.2.1.2 Inclusion Criteria

Table 21.3 shows that 100 papers were retained after the application of inclusion criteria. We examined the summary [24], opening, and closing sections of the papers. From the collected papers, we picked the ones that fulfilled at least one of the subsequent standards: The research is centered on the concept of role of middleware. The research concentrates on shared network, as described in Section II. Studies published in English languages. Studies published between the period of 2000 and 2023

21.2.2.2 Quality Assessment

After employing specific guidelines for selection and rejection, the full contents of the selected articles were thoroughly assessed for their standard [25], the remaining papers were subjected to the quality evaluation criteria, and low-quality papers were eliminated. In the end, 47 publications were used to extract the data and offer responses to the SLR's study questions.

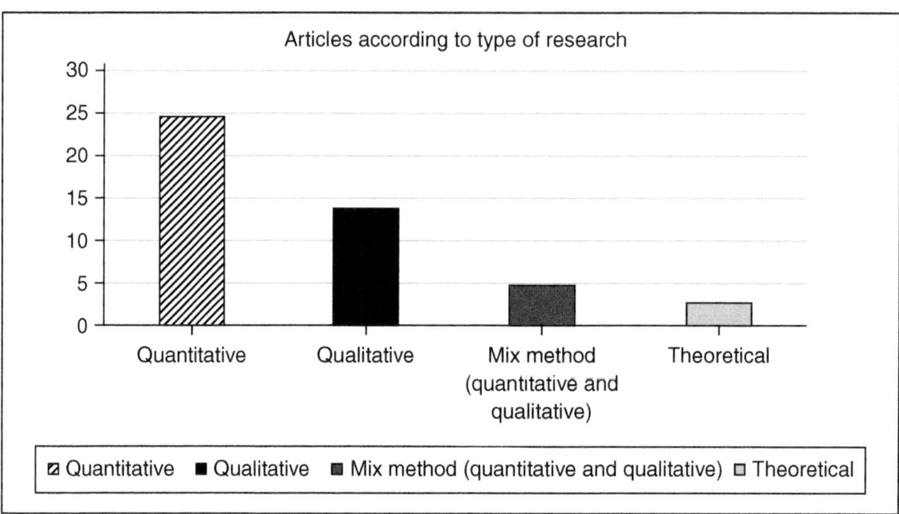

FIGURE 21.4 Articles according to type of research.

The goal of quality evaluation is typically to identify the impartial and pertinent research. So that we could refine our search results and evaluate the suitability and rigor of the candidate articles, we came up with several quality evaluation markers. The following SLRs served as the foundation for the quality evaluation questionnaires. A paper that includes evidence that has been quantitatively or qualitatively assessed. Figure 20.4 shows the strength of papers accordingly. The paper has an intricate experimental plan. The applicable approaches' accuracy rates are quantified and documented in papers. In articles, the comparison of various methodologies is offered. The article outlines the pros and cons of various techniques.

21.2.3 Data Extraction and Synthesis

The study's queries were addressed by employing a data extraction form to collect the required information from the selected papers.

RQ1: The aim of this question is to understand the importance and function of middleware in a shared network environment. The question is asking about the specific role that middleware plays in facilitating communication and data exchange between different applications and systems within a shared network environment. The answer to this question will provide insights into the key benefits of using middleware, such as improved interoperability [18], scalability [19], and flexibility in networked systems. It will also help to illustrate the various types of middleware that are commonly used in shared network environments [23] and how they are deployed to enhance communication and integration between different components of a networked system. Overall, the aim of this question is to provide a deeper understanding of the critical role that middleware plays in supporting the functionality and performance of shared network environments.

RQ2: The aim of this question is to inquire about the specific facilities or functionalities provided by middleware in a shared network environment. The question seeks to understand the role of middleware in enabling communication and coordination between different software applications, services, or systems that operate within a shared network environment. The answer to this question may include a discussion of the various types of middleware and their functionalities, such as message passing, remote procedure calls, object-oriented middleware, transaction processing, and security services, among others. The response may also describe how middleware helps to simplify the development, deployment, and maintenance of distributed systems in a shared network environment.

RQ3: The aim of the question appears to be to determine the extent to which middleware is required for a shared network. Middleware is a type of software that provides a platform for different applications to communicate [20] with each other and share data in a networked environment. The question seeks to understand whether middleware is a necessary component for a shared network, or if there are alternative approaches that can achieve the same objectives without middleware. Answering this question would require an understanding of the specific requirements and characteristics of the shared network, as well as the capabilities and limitations of different middleware solutions.

21.3 RESULTS

This section presents the findings of this review (Phase 3: reporting the review). The outcomes of the selection process are first presented in their entirety, followed by individual reports of each research question's findings.

21.3.1 Overview of the Selected Studies

We discovered the 265-candidate document as seen in Figure 21.2 by conducting searches across five digital databases. 115 papers out of the 265 total papers were eliminated using exclusion criteria. To identify the most significant studies, a thorough analysis of the remaining 150 publications was conducted. After examining the titles, abstracts, and keywords, only papers that met at least one of the inclusion criteria were considered. Out of the initial pool, 100 papers remained. The quality evaluation criteria were subsequently applied to these papers, and those that were deemed to be of low quality were eliminated. This process enabled the assessment of the overall quality of the selected articles based on their entire texts. Eventually, 47 publications were utilized to extract data and provide answers to the study questions posed in the systematic literature review (SLR) [22].

21.3.2 Phase 3: Reporting the Review

In this phase, we have presented a comprehensive account of how the research questions were addressed. The responses to all the study inquiries are substantiated by the evidence gathered from the literature reviews.

21.3.2.1 RQ1: What Is the Role of Middleware in a Shared Network Environment?

Middleware plays a crucial role in a shared network environment by enabling different applications and systems to communicate with each other seamlessly. In a shared network environment, multiple systems are connected to the same network and are expected to work together to achieve a common goal [26]. However, these systems may use different programming languages, protocols, and communication mechanisms, which can make it challenging to achieve interoperability between them. This is where middleware comes in. Middleware acts as an intermediary layer between different applications and systems, providing a common communication platform that enables them to interact with each other. The middleware layer provides a set of services that facilitate the exchange of data and messages between systems. These services include data transformation [27], routing, and orchestration, which allow different systems to exchange data with each other using different formats and protocols. One of the primary benefits of middleware in a shared network environment is its ability to decouple different systems. Decoupling means that systems can operate independently of each other and do not need to be aware of each other's existence. This means that even if one system goes down or undergoes changes, it does not affect the other systems on the network. Decoupling also enables asynchronous communication, where systems can send messages to each other without requiring an immediate response [28]. This can be useful in situations where systems have different processing speeds or are distributed across different locations. In addition to decoupling, middleware also provides security and reliability features that are essential in a shared network environment. Middleware can enforce security policies, authenticate users, and authorize access to resources. It can also provide features such as message queuing and transaction management, which ensure that messages are delivered reliably and consistently, even in the event of system failures or network disruptions. In summary, the role of middleware in a shared network environment is to enable different applications and systems to communicate with each other seamlessly. Middleware provides a common communication platform that facilitates the exchange of data and messages between systems. It enables decoupling, asynchronous communication, and provides security and reliability features that are essential in a shared network environment. By providing these services, middleware makes it possible for organizations to build complex systems that can easily integrate with each other.

21.3.2.2 RQ2: Which Facilities Are Provided by the Middleware in the Context of a Shared Network Environment?

Middleware acts as an intermediary between different software components in a shared network environment, enabling them to communicate with each other seamlessly. Middleware facilitates communication [26] in a shared network by providing a standardized interface for different applications, services, and devices to communicate with each other. Here are some of the key ways in which middleware can facilitate communication [29]:*Providing a common communication protocol, supporting multiple communication protocols, standardizing data formats, implementing message queuing, and providing API gateways.* Middleware provides a standard interface for communication, allowing software components developed using different programming languages, platforms, or operating systems to communicate with each other. Middleware provides interoperability [30] in a shared network by providing a standardized interface that enables different applications, services, and devices to communicate with each other, regardless of their underlying technology or platform. Here are some of the key ways in which middleware can provide interoperability [31].

Supporting multiple platforms, standardizing interfaces, implementing industry standards, providing translation services, and enabling legacy systems. Middleware provides a layer of security to shared network environments by enabling the implementation of access control policies, authentication, and authorization mechanisms. Figure 21.5 shows a block diagram of facilities provided by the middleware. Middleware provides a scalable platform for sharing resources, allowing the system to accommodate a growing number of users and devices. Middleware enables scalability [19] in a shared network by providing mechanisms for distributing load, replicating services, and managing resources. Here are some of the key ways in which middleware can enable scalability [32]: Load Balancing, replication, clustering, resource management, and elasticity.

Middleware provides security assurance [29] in a shared network by implementing authentication, authorization, and encryption to protect data and resources from unauthorized access. Here are some of the key ways in which middleware can provide security assurance [33]: Authentication, authorization, encryption, auditing, and integration with security frameworks. Middleware provides a platform for managing shared resources efficiently, reducing the risk of contention and improving overall system performance. Middleware increases efficiency [22] in a shared network environment by reducing the complexity of network design, streamlining communication, and improving the

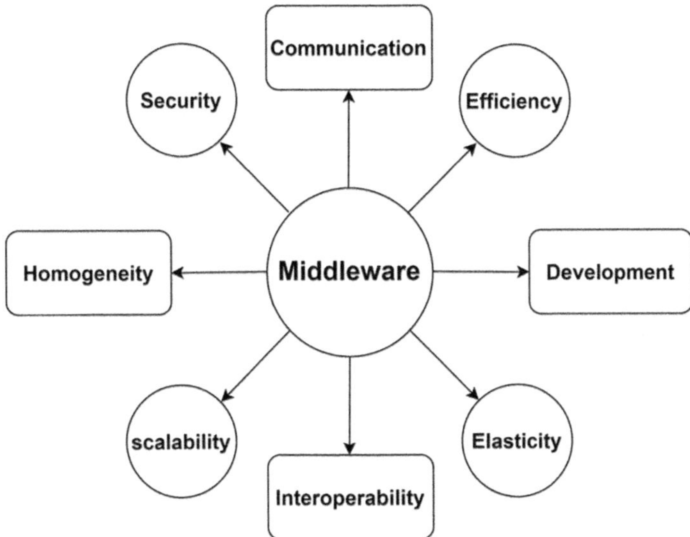

FIGURE 21.5 Middleware facilities.

scalability and reliability of networked systems. Here are some of the key ways in which middleware can increase efficiency [34]: Standardization, streamlined communication, scalability, reliability, and integration.

Middleware abstracts the underlying hardware and software components, providing a standardized interface for communication [35]. This simplifies the development of distributed applications and reduces the complexity of network management. Middleware plays a role in simplifying development [36] in a shared network by providing a standardized interface and reducing the complexity of networked systems. Here are some of the key ways in which middleware can simplify development: Standardization, reusability, modularity, abstraction, and integration.

Moreover middleware also provides a transparent mechanism [37] for accessing shared resources, ensuring that each user or device can access them in a controlled and secure manner. Middleware provides a platform for implementing fault-tolerant systems [38] in shared network environments, ensuring that the system can continue to operate even in the event of hardware or software failures. Middleware provides a platform for transforming data from one format to another, enabling interoperability [39] between different software components. Middleware enables legacy systems [40] to interact with modern applications and systems, extending their useful life and reducing the need for costly replacements. Middleware can be customized and configured to meet the specific needs of different networked systems, allowing for greater flexibility [21] and adaptability. Middleware can provide monitoring and management [41] tools to track network performance, identify and resolve issues, and optimize system resources. Support for different communication protocols. Middleware plays a critical role in enabling seamless communication and integration across a shared network, providing a common layer of abstraction and management for distributed systems.

21.3.2.3 RQ3: How Much Middleware Is Needy for a Shared Network?

Middleware plays a critical role in a shared network environment as it provides a standardized set of protocols [1] and interfaces that enable different software components to communicate and interact with each other seamlessly. The degree of need for middleware in a shared network environment depends on several factors, including the complexity [5] of the network, the number of applications and systems involved, and the nature of the data and services being exchanged.

In general, the more complex the network, the greater the need for middleware. A shared network environment may involve multiple applications and systems that are built using different programming languages [42], architectures [2], and platforms. Middleware provides a layer between these applications and systems, enabling them to exchange data and requests in a standardized way, regardless of their differences. As the number of applications and systems involved in the network grows, the need for middleware increases.

The nature of the data and services being exchanged also affects the need for middleware. If the data and services are relatively simple, with limited variations in format and structure, then middleware may not be as necessary. However, if the data and services are complex, with varying formats and structures, then middleware becomes critical in ensuring that the data can be transformed and integrated seamlessly.

Moreover, the need for middleware also arises when it comes to security and authentication. In a shared network environment, there is always a risk of data breaches and unauthorized access. Middleware can provide security features like authentication [25] and authorization, ensuring that only authorized users has access to data and services. It can also provide encryption and decryption facilities, ensuring that sensitive data is protected during transmission.

In summary, middleware is critical for a shared network environment, especially when the network is complex and involves multiple applications and systems with varying data and service structures [42]. Middleware provides standardized protocols and interfaces that enable seamless communication and interaction between different software components while ensuring security and authentication of data and services.

Middleware is essential in modern software development as it provides a layer of abstraction between different software components and systems, enabling them to communicate and work together seamlessly. Middleware helps to overcome the challenges of heterogeneity, scalability, and complexity that arise in distributed systems and shared networks. One of the primary benefits of middleware is its ability to provide interoperability between different software components and systems. Different software applications and systems may use different communication protocols, data formats, and interfaces, making it challenging for them to communicate and work together. Middleware provides a standardized interface and communication protocol that enables different software components and systems to communicate with each other efficiently.

21.4 DISCUSSION

The research paper on the role of middleware in a shared network environment explores the critical importance of middleware in enabling different software components to communicate and interact with each other seamlessly. The paper highlights the increasing complexity of shared network environments, with multiple applications and systems built using different programming languages [43], architectures [24], and platforms, and the need for middleware to provide a layer between these applications and systems.

The paper discusses the various facilities provided by middleware in a shared network environment, including communication and messaging, data transformation and integration [44], security [20] and authentication [25], performance and scalability, and service management and monitoring. The paper notes that middleware provides a standardized set of protocols and interfaces that enable applications to exchange data and requests in real time, regardless of the programming languages or architectures used.

The paper also emphasizes the critical role of middleware in ensuring the security and authentication of data and services in a shared network environment. With the increasing risk of data breaches and unauthorized access, middleware can provide security features like authentication and authorization, encryption and decryption facilities, and data routing facilities to ensure that only authorized users have access to data and services.

The paper highlights the need for middleware to optimize the performance of applications in a shared network environment by providing load balancing, caching, and other features. Middleware can also enable applications to scale up or down based on demand, ensuring that the system can handle large volumes of data and traffic [39]. It can also provide fault tolerance and disaster recovery facilities, ensuring that the system remains operational even in the event of a failure.

The paper concludes by highlighting the critical importance of middleware in a shared network environment, given the increasing complexity of networks and the need for seamless communication and interaction between different software components. Middleware provides a layer of standardization and security, enabling applications to exchange data and requests seamlessly, and ensuring that the network remains secure and reliable.

Overall, the research paper provides a comprehensive overview of the critical role of middleware in a shared network environment, emphasizing its importance in ensuring seamless communication and interaction between different software components while providing security, performance optimization, and fault tolerance facilities. The paper highlights the increasing complexity of networks and the need for middleware in ensuring their smooth operation, making it a valuable resource for researchers and practitioners alike.

21.5 CONCLUSIONS

In conclusion, middleware plays a vital role in a shared network environment, providing a standardized set of protocols and interfaces that enable different software components to communicate and interact with each other seamlessly. The facilities provided by middleware include communication and messaging, data transformation and integration, security and authentication, performance and scalability, and service management and monitoring. The need for middleware in a shared network environment depends on the complexity of the network, the number of applications and systems involved, and the nature of the data and services being exchanged. However, as the network becomes more complex and involves multiple applications and systems with varying data and service structures, the need for middleware becomes critical in ensuring seamless communication and interaction between different software components while ensuring security and authentication of data and services. Therefore, it is important to carefully evaluate the role of middleware in a shared network environment to ensure that the network operates efficiently, reliably, and securely.

REFERENCES

[1] V. Issarny, G. Bouloukakis, N. Georgantas, and B. Billet, "Revisiting service-oriented architecture for the IoT: a middleware perspective," presented at the *Service-Oriented Computing: 14th International Conference, ICSOC 2016*, Banff, AB, Canada, October 10–13, 2016, Proceedings 14, Springer, 2016, pp. 3–17.

[2] X. Li, M. Eckert, J.-F. Martinez, and G. Rubio, "Context aware middleware architectures: Survey and challenges," *Sensors*, vol. 15, no. 8, pp. 20570–20607, 2015.

[3] J. Bohuslava, J. Martin, and H. Igor, "TCP/IP protocol utilisation in process of dynamic control of robotic cell according industry 4.0 concept," presented at the *2017 IEEE 15th International Symposium on Applied Machine Intelligence and Informatics (SAMI)*, IEEE, 2017, pp. 000217–000222.

[4] R. Kirchgessner, A. D. George, and G. Stitt, "Low-overhead fpga middleware for application portability and productivity," *ACM Trans. Reconfigurable Technol. Syst. TRETS*, vol. 8, no. 4, pp. 1–22, 2015.

[5] M. A. Razzaque, M. Milojevic-Jevric, A. Palade, and S. Clarke, "Middleware for internet of things: a survey," *IEEE Internet Things J.*, vol. 3, no. 1, pp. 70–95, 2015.

[6] A. Gazis and E. Katsiri, "Middleware 101," *Commun. ACM*, vol. 65, no. 9, pp. 38–42, 2022.

[7] A. Gazis and E. Katsiri, "Middleware 101: What to know now and for the future," *Queue*, vol. 20, no. 1, pp. 10–23, Feb. 2022, doi: 10.1145/3526211.

[8] J. Yongguo, L. Qiang, Q. Changshuai, S. Jian, and L. Qianqian, "Message-oriented middleware: A review," presented at the *2019 5th International Conference on Big Data Computing and Communications (BIGCOM)*, IEEE, 2019, pp. 88–97.
[9] H. Ding, C. Zhang, X. Chen, J. Shi, and W. Wang, "Cloud-mom: a content-based real-time message-oriented middleware for cloud," presented at the *2018 IEEE 20th International Conference on High Performance Computing and Communications; IEEE 16th International Conference on Smart City; IEEE 4th International Conference on Data Science and Systems (HPCC/SmartCity/DSS)*, IEEE, 2018, pp. 750–757.
[10] A. Sterz, L. Baumgärtner, R. Mogk, M. Mezini, and B. Freisleben, "DTN-RPC: Remote procedure calls for disruption-tolerant networking," presented at the *2017 IFIP Networking Conference (IFIP Networking) and Workshops*, IEEE, 2017, pp. 1–9.
[11] Y. Tanaka, H. Nakada, S. Sekiguchi, T. Suzumura, and S. Matsuoka, "Ninf-G: A reference implementation of RPC-based programming middleware for Grid computing," *J. Grid Comput.*, vol. 1, pp. 41–51, 2003.
[12] M. Rubtcova and O. Pavenkov, "The common object request broker architecture of the information system," presented at the *International Conference on Advanced Computer Science and Information Technology (ICACSIT)*, Pune, India on 29th April, 2018.
[13] P. T. Eugster, B. Garbinato, and A. Holzer, "Middleware Support for Context-Aware Applications," in *Middleware for Network Eccentric and Mobile applications*, Springer Berlin Heidelberg, pp. 305–322, 2009.
[14] J. Park *et al.*, "GNU Data Language 1.0: a free/libre and open-source drop-in replacement for IDL/PV-WAVE," *J. Open Source Softw.*, vol. 7, no. 80, p. 4633, 2022.
[15] F. Halili and E. Ramadani, "Web services: a comparison of soap and rest services," *Mod. Appl. Sci.*, vol. 12, no. 3, p. 175, 2018.
[16] S. A. Hamid, R. A. Abdalrahman, I. A. Lafta, and I. Al Barazanchi, "Web services architecture model to support d systems," *J. Southwest Jiaotong Univ.*, vol. 54, no. 6, 2019.
[17] B. Petersen, H. Bindner, B. Poulsen, and S. You, "Smart grid communication middleware comparison," *SmartGreens Porto*, 2017.
[18] M. Aragão, P. Moreno, and A. Bernardino, "Middleware interoperability for robotics: A ros–yarp framework," *Front. Robot. AI*, vol. 3, p. 64, 2016.
[19] M. García-Valls, C. Calva-Urrego, J. A. de la Puente, and A. Alonso, "Adjusting middleware knobs to assess scalability limits of distributed cyber-physical systems," *Comput. Stand. Interfaces*, vol. 51, pp. 95–103, 2017.
[20] B. Bhushan, "Middleware and Security Requirements for Internet of Things," in *Micro-Electronics and Telecommunication Engineering*, D. K. Sharma, S.-L. Peng, R. Sharma, and D. A. Zaitsev, Eds., in Lecture Notes in Networks and Systems, vol. 373. Singapore: Springer Nature Singapore, 2022, pp. 309–321. doi: 10.1007/978-981-16-8721-1_30.
[21] J. Gascon-Samson, M. Rafiuzzaman, and K. Pattabiraman, "Thingsjs: Towards a flexible and self-adaptable middleware for dynamic and heterogeneous iot environments," presented at the *Proceedings of the 4th Workshop on Middleware and Applications for the Internet of Things*, 2017, pp. 11–16.
[22] S. Bharany *et al.*, "Efficient middleware for the portability of paas services consuming applications among heterogeneous clouds," *Sensors*, vol. 22, no. 13, p. 5013, 2022.
[23] K. Lingaraj, R. V. Biradar, and V. Patil, "Eagilla: An enhanced mobile agent middleware for wireless sensor networks," *Alex. Eng. J.*, vol. 57, no. 3, pp. 1197–1204, 2018.
[24] T. Devadithya, K. Chiu, K. Huffman, and D. F. McMullen, "The Common Instrument Middleware Architecture: Overview of Goals and Implementation," in *First International Conference on e-Science and Grid Computing (e-Science'05)*, Pittsburg, PA, USA: IEEE, 2005, pp. 578–585. doi: 10.1109/E-SCIENCE.2005.77.
[25] M. A. Christie *et al.*, "Managing authentication and authorization in distributed science gateway middleware," *Future Gener. Comput. Syst.*, vol. 111, pp. 780–785, 2020.
[26] A. Balador, N. Ericsson, and Z. Bakhshi, "Communication middleware technologies for industrial distributed control systems: A literature review," presented at the *2017 22nd IEEE International Conference on Emerging Technologies and Factory Automation (ETFA)*, IEEE, 2017, pp. 1–6.
[27] S. Bhowmik, M. A. Tariq, L. Hegazy, and K. Rothermel, "Hybrid content-based routing using network and application layer filtering," presented at the *2016 IEEE 36th International Conference on Distributed Computing Systems (ICDCS)*, IEEE, 2016, pp. 221–231.

[28] M. Treiber and H. Bernhardt, "NEVONEX—the importance of middleware and interfaces for the digital transformation of agriculture," *Eng. Proc.*, vol. 9, no. 1, p. 3, 2021.
[29] S. Demurjian Sr, K. Bessette, T. Doan, and C. Phillips, "Concepts and capabilities of middleware security," *Middlew. Commun.*, vol. 3, pp. 211–235, 2004.
[30] T. Coito *et al.*, "A middleware platform for intelligent automation: An industrial prototype implementation," *Comput. Ind.*, vol. 123, p. 103329, 2020.
[31] P. Verma, "A Brief study of middleware technologies: Programming applications and management systems," *Nov. Res. Asp. Math. Comput. Sci.* vol. 1, pp. 174–181, 2022.
[32] R. Priego, N. Iriondo, U. Gangoiti, and M. Marcos, "Agent-based middleware architecture for reconfigurable manufacturing systems," *Int. J. Adv. Manuf. Technol.*, vol. 92, pp. 1579–1590, 2017.
[33] Y. Zheng, J. Luo, and T. Zhong, "Service recommendation middleware based on location privacy protection in VANET," *IEEE Access*, vol. 8, pp. 12768–12783, 2020.
[34] D. Von Leon, L. Miori, J. Sanin, N. El Ioini, S. Helmer, and C. Pahl, "A performance exploration of architectural options for a middleware for decentralised lightweight edge cloud architectures," presented at the *IoTBDS 2018: Proceedings of the 3rd International Conference on Internet of Things, Big Data and Security*; Funchal, Madeira, Portugal, 19–21 March 2018, SciTePress, 2018.
[35] K. Michalakis, Y. Christodoulou, G. Caridakis, Y. Voutos, and P. Mylonas, "A context-aware middleware for context modeling and reasoning: a case-study in smart cultural spaces," *Appl. Sci.*, vol. 11, no. 13, p. 5770, 2021.
[36] J. Zhang, M. Ma, P. Wang, and X. Sun, "Middleware for the internet of things: A survey on requirements, enabling technologies, and solutions," *J. Syst. Archit.*, vol. 117, p. 102098, 2021.
[37] D. Wang, H. Yao, Y. Li, H. Jin, D. Zou, and R. H. Deng, "A secure, usable, and transparent middleware for permission managers on Android," *IEEE Trans. Dependable Secure Comput.*, vol. 14, no. 04, pp. 350–362, 2017.
[38] R. Belchior, A. Vasconcelos, M. Correia, and T. Hardjono, "Hermes: Fault-tolerant middleware for blockchain interoperability," *Future Gener. Comput. Syst.*, vol. 129, pp. 236–251, 2022.
[39] A.-T. Fadi and B. D. Deebak, "Seamless authentication: for IoT-big data technologies in smart industrial application systems," *IEEE Trans. Ind. Inform.*, vol. 17, no. 4, pp. 2919–2927, 2020.
[40] S. S. Albouq, A. A. Abi Sen, N. Almashf, M. Yamin, A. Alshanqiti, and N. M. Bahbouh, "A survey of interoperability challenges and solutions for dealing with them in IoT environment," *IEEE Access*, vol. 10, pp. 36416–36428, 2022.
[41] B. Saovapakhiran, W. Naruephiphat, C. Charnsripinyo, S. Baydere, and S. Ozdemir, "QoE-Driven IoT architecture: A comprehensive review on system and resource management," *IEEE Access*, 2022.
[42] L. Poirier, K. Fortun, B. Costelloe-Kuehn, and M. Fortun, "Metadata, Digital Infrastructure, and the Data Ideologies of Cultural Anthropology," in *Anthropological Data in the Digital Age: New Possibilities–New Challenges*, vol. 5, pp. 209–237, 2020.
[43] F. Hauser *et al.*, "A survey on data plane programming with p4: Fundamentals, advances, and applied research," *J. Netw. Comput. Appl.*, vol. 212, p. 103561, 2023.
[44] F. Siqueira and J. G. Davis, "Service computing for industry 4.0: State of the art, challenges, and research opportunities," *ACM Comput. Surv. CSUR*, vol. 54, no. 9, pp. 1–38, 2021.

22 Challenges and Frameworks in Cloud Security Governance

A Review

Farah Taj, Danish Shehzad, Faiqa Irum, Hinna Arqam, and Zaib Unnisa

22.1 INTRODUCTION

The security of the cloud is a shared duty of consumers of cloud services and cloud service providers. Cloud service suppliers must implement strong security measures, and clients must adopt adequate security controls to protect their data in the cloud [1]. Cloud providers must implement robust security measures. Failure to comply with these regulations can result in severe penalties and legal consequences. Cloud service governance is the set of policies, processes, and controls that organizations use to manage their cloud services. It involves establishing a framework for the procurement, deployment, and management of cloud services to ensure that they are secure, compliant, and aligned with business goals.

Effective cloud service governance requires a comprehensive approach that includes assessing cloud service providers, managing third-party risk, establishing clear roles and responsibilities, and implementing robust security controls. By effectively governing their cloud services, organizations can maximize the benefits of cloud computing while minimizing the risks. This may increase the need to enable businesses to link the security measures between users of mobile and cloud platforms. Businesses continually strive to demonstrate that they are following their guidelines for security compliance in areas where data is available. Operating systems and the apps that run on them that are shared in that environment's infrastructure should be more responsible for monitoring subscribers' non-pure cloud resources than the vendors that provide them. The enterprise's loss of control over its information assets, which indicates the need for a defined security governance policy, is the key security issue preventing the widespread adoption of cloud computing and creating resistance among practitioners [2]. Using the cloud concept, computing is extended outside of company walls. The specificities of this environment lead to the need for an assurance framework that will help organizations deal with security threats. This includes implementing encryption, access controls, network security, and physical security measures, as well as incident response and disaster recovery plans [3].

In addition, cloud providers are also subject to regulatory requirements, such as HIPAA, GDPR, and PCI DSS [5]. Cloud governance is a duty that falls under the purview of the company employing the cloud. The company needs to make sure that it is not being negligent and using malware to launch assaults on the cloud. It is required to follow all cloud-enforced rules. The company must also be subject to regulatory requirements, such as HIPAA, GDPR, and PCI DSS [4]. Cloud governance is a duty that falls under the purview of the company employing the cloud. The company needs to make sure that it is not being negligent and using malware to launch assaults on the cloud. It is required to follow all cloud-enforced rules.

DOI: 10.1201/9781003497851-22

The company must also train its staff on how to use the cloud properly. The organization must also have an authoritative figure that oversees all communications with the CSP. This person may be the Chief Cloud Officer (CCO), the Chief Information Security Officer (CISO), or a specific individual with that title (CCO) [6]. We'll go through a number of cloud governance topics below. Cyber risk is one of the cyber security topics being researched. More and more companies are now offering cyber risk insurance.

The significant contributions of this review paper are summarized below:

It reviews frameworks such as Security as a Service (SecaaS), Cloud Control Matrix (CCM), NIST Cloud Computing Reference Architecture(CCRA), and ISO 27001 discussed in this paper are different approaches for cloud security governance. These approaches have different strengths and weaknesses.

SecAAS is a model where security services are delivered through a cloud provider, and it can provide a cost-effective and flexible solution for organizations with limited resources or proficiency. However, organizations that require high levels of security or have unique compliance requirements may not find SecAAS to be sufficient. CCM is a set of controls and guidance that can be used to assess cloud service provider (CSP) security, and it can be used in an organization that provides a comprehensive view of cloud security risks and controls. However, CCM does not provide specific support for implementing security controls, and organizations may need to combine it with other frameworks or standards according to their local needs.

NIST Cloud Computing Reference Architecture (CCRA) is a risk-based framework that provides guidance on how to manage cybersecurity risks, and it can help organizations to develop an extensive security program that incorporates their overall risk management strategy. It is designed for large-scale cloud deployments; However, it may not be necessary or cost-effective for small-scale cloud implementations. It may also be not good for organizations with strict regulations and highly customized environments.

ISO 27001 is a comprehensive standard that provides a framework for information security management, and it can be used to develop a systematic and structured approach to cloud security governance. However, it may be more resource-intensive to implement compared to other frameworks, and organizations may need to invest significant time and resources in achieving certification.

In conclusion, it is suggested that organizations should carefully keep an eye on their specific needs, risk outlines, and compliance requirements before deciding which framework or approach to adopt for cloud security governance. No single framework or approach is a one-size-fits-all solution, and organizations may need to use a combination of frameworks and standards to achieve comprehensive cloud security governance.

The rest of the paper is structured as follows. In the next section, challenges are presented followed by models and frameworks of cloud computing governance along with their limitations. In the end, the conclusion is presented.

22.2 CHALLENGES

Some of the key challenges that organizations face in cloud security governance include the following challenges.

22.2.1 Lack of Visibility and Control

One of the primary challenges in cloud security governance is the lack of visibility and control over cloud resources. This is because cloud services are often delivered through a shared infrastructure, and organizations may not have direct access to the underlying hardware and software. One of the major challenges in cloud security. When an organization stores its data in the cloud, it may not have complete visibility or control over its data and infrastructure. This lack of visibility and control can lead to security breaches, data loss, and other security risks [7].

Some key components of an operational model for cloud security governance include risk assessment and management, incident management, access control, and monitoring. This lack of visibility and control can lead to security breaches, data loss, and other security risks [8].

22.2.2 Lack of Senior Management Participation

One of the frequent issues that cloud clients face is the absence of a security policy that has the support and influence of high management. Unfortunately, many businesses often write security rules without the input or influence of executives and with a heavy tactical element. As a result of this circumstance, executive tone and expectations for cloud security are poorly defined and communicated. Including enterprise executives in the conversation and set of tone and expectations for security that will feed a formal enterprise security policy is crucial to resolving this problem. Additionally, it is critical that CEOs accept full responsibility [9].

22.2.3 Missing Roles, Responsibilities, and an Operational Model

Roles and responsibilities define the tasks and responsibilities of individuals and teams involved in cloud security governance. Some common roles include the Cloud Security Officer, Cloud Security Architect, and Cloud Security Engineer. The Cloud Security Officer is responsible for overseeing the overall cloud security strategy, policies, and procedures, while the Cloud Security Architect is responsible for designing the security architecture of the cloud environment. The Cloud Security Engineer is responsible for implementing and maintaining the technical controls that protect the cloud environment. The operational model describes how cloud security governance is implemented in practice. This includes the processes, procedures, and tools used to manage cloud security risks. The operational model should be designed to align with the organization's business objectives, regulatory requirements, and risk appetite.

Companies must adopt the necessary security procedures and controls, frequently evaluate their security state, and refine their security governance methods [9].

22.2.4 Data Protection

Data protection is a critical aspect of cloud security governance. Organizations must ensure that their data is protected from unauthorized access, theft, and loss. This requires implementing robust security controls such as encryption, access controls, and data backups [10].

22.2.5 E-Compliance

Cloud security governance must also comply with various regulatory and industry standards such as GDPR, HIPAA, PCI DSS, and SOC 2. Compliance with these standards requires organizations to implement appropriate security controls and demonstrate compliance through regular audits and assessments [11].

22.2.6 Vendor Management

Cloud security governance also involves managing third-party vendors who provide cloud services. Organizations must ensure that their vendors adhere to the same security standards and policies that they have in place. This requires regular audits, assessments, and vendor risk management [12].

22.2.7 Security Incident Response

Cloud security governance must also address security incident response. Organizations must have a well-defined plan for responding to security incidents such as data breaches or cyberattacks. This requires having a dedicated team, incident response procedures, and regular testing and training. Ultimately, cloud security governance necessitates a comprehensive strategy that considers every facet of cloud security [13].

22.3 MODELS AND FRAMEWORKS OF THE CLOUD SECURITY GOVERNANCE

There are several different models and frameworks that organizations can use to guide their cloud security governance efforts. Frameworks and models can be extremely helpful in the security of the cloud. Here are some of the ways in which frameworks and models can help. Frameworks and models provide standardization in cloud security practices, ensuring that organizations follow a consistent and best-practice approach to securing their cloud environments. This can help organizations identify and address security gaps and vulnerabilities in a structured and comprehensive manner. Frameworks and models provide guidance on how to identify, assess, and manage risks associated with cloud computing. This enables organizations to prioritize security controls and resources based on the level of risk posed by specific threats. Here are some of the most commonly used models.

22.3.1 Cloud Control Matrix (CCM)

A cybersecurity framework called the Cloud Control Matrix (CCM) from the Cloud Security Alliance (CSA) offers a uniform method for evaluating cloud service provider's security postures (CSPs). In order to assess the security of cloud services across a number of domains, including governance, risk management, and compliance, the CCM framework contains a set of security controls and objectives. The 133 security controls of the CCM architecture are divided into 17 domains [14]. These domains address a variety of security issues, such as:

1. Compliance and Audit
2. Human Resources
3. Data Security and Information Lifecycle Management
4. Information Management and Data Governance
5. Interoperability and Portability
6. Legal and Electronic Discovery
7. E-discovery, cloud forensics, and security incident management are included in.
8. Supply chain management, accountability, and transparency
9. Key administration and encryption
10. Security for Applications
11. Identity and Access Management, number (IAM)
12. Security for Infrastructure and Virtualization 12.Mobile Security
13. Business Continuity and Disaster Recovery
14. Management
15. Network Security
16. Security Operations

The CCM framework is widely used by organizations and third-party auditors to assess the security posture of CSPs and to ensure that cloud services are deployed securely. The model provides a comprehensive and standardized approach to cloud security, enabling organizations to compare different CSPs and make informed decisions about their cloud security requirements. The CCM framework also supports the implementation of the Shared Responsibility Model, which is a common approach to cloud security that outlines that each of these domains includes specific controls that CSPs should implement to protect their customers' data and ensure the security of their cloud services [15]. For example, the Compliance and Audit domain includes controls related to compliance with regulatory requirements, while the Data Security and Information Lifecycle Management domain includes controls related to encryption, data classification, and data retention policies. The framework provides a comprehensive and standardized approach to assessing the security posture of cloud service providers. By using the CCM framework, organizations can evaluate the security of cloud services based on a common set of security controls and objectives, and ensure that their cloud services are deployed securely.

22.3.2 NIST Framework (CCRA)

The National Institute of Standards and Technology (NIST) has developed a comprehensive framework for cloud security, known as the (CCRA). This framework is designed to help organizations understand and address the unique security challenges associated with cloud computing. The NIST CCRA framework consists of five primary components.

Cloud Consumers. The organization that makes use of cloud services is known as cloud consumers. This might be a person, a company, or a government organization [16].

Cloud Provider. The organization that offers cloud services is known as a cloud provider. A hybrid cloud provider, a private cloud provider, or a public cloud provider might be this.

Cloud Broker. A cloud broker is a middleman who links cloud providers and customers. Service level agreement (SLA) management, security management, and cost management are just a few of the services the broker may offer. The cloud carrier is the entity that provides the network infrastructure for cloud services. This includes the physical infrastructure, such as routers and switches, as well as the logical infrastructure, such as virtual private networks (VPNs) and software-defined networking (SDN) [17].

Cloud Auditor. The cloud auditor is an independent third party that assesses the security and compliance of cloud services. The auditor can provide assurance to cloud consumers that the cloud provider is meeting their security and compliance requirements. The NIST CCRA framework also includes a set of security and privacy controls that organizations can use to assess the security posture of cloud providers. These controls are based on the NIST Special Publication 800-53, which is a set of security and privacy controls that are widely used in the US federal government. The various cloud deployment models that the NIST CCRA architecture is intended to be flexible and adaptive to. Service (IaaS). The framework also includes guidance on risk management, incident management, and disaster recovery in cloud environments. Overall, the NIST CCRA framework provides a comprehensive approach to cloud security that organizations can use to understand and address the unique security challenges associated with cloud computing. The framework is widely used by organizations, auditors, regulators, and industry associations to evaluate the security of cloud services and promote best practices for cloud security [18].

22.3.3 Limitations of CCM

Cloud Control Matrix (CCM) is a useful option for organizations due to a better understanding of the security posture of their cloud service providers and take proactive steps to mitigate potential security risks. But there are a few drawbacks that organizations should be aware of:

1. Limited Scope: It may not be sufficient to address all security risks associated with the use of cloud-based services.
2. Complexity: The CCM is sometimes difficult to adopt due to its complexity.
3. Lack of Customization: The CCM is a standardized framework that cannot be customized to meet the unique security requirements of individual organizations. This can make it challenging for organizations to tailor the framework to their specific needs.
4. Compliance Challenges: The CCM is not a compliance standard in and of itself. While it can be used to assess the compliance of cloud service providers with various regulatory and industry-specific standards, organizations may face challenges in demonstrating compliance with these standards.
5. Maintenance and Updates: The CCM is a living framework that is regularly updated to reflect changes in the cloud computing landscape. Keeping up with these updates and maintaining compliance with the latest version of the framework can be a significant challenge for organizations.

Table 22.1 explains and presents the further details.

Best Features of CCRA

One of the key features of the NIST CCRA framework is its focus on risk management. The framework provides guidance on identifying, assessing, and managing risks associated with cloud computing. This includes guidance on risk assessment methodologies, threat modeling, and vulnerability assessment. The NIST CCRA framework also includes guidance on incident management and disaster recovery in cloud environments. This includes guidance on incident response planning, business continuity planning, and disaster recovery planning. The framework emphasizes the importance of having a well-defined incident management and disaster recovery plan in place to ensure that organizations can quickly respond to and recover from security incidents in cloud environments. The NIST CCRA framework also includes guidance on compliance and privacy in cloud environments. The framework provides guidance on complying with various regulatory requirements, such as HIPAA and PCI-DSS, as well as privacy laws, such as GDPR. The framework emphasizes the

TABLE 22.1
Different Framework

	Cybersecurity Framework	
Framework	***Issued by***	***Industry***
NIST CSRA	NIST	Operational of the critical infrastructure + general
CSA CCM	CSA	Cloud service Provider
ANS/ISA	ANSI	Industrial based cloud governance f
HITRUST CSF	HITRUST	Healthcare service providers

importance of understanding and complying with the relevant regulatory and privacy requirements when using cloud services. Finally, the NIST CCRA framework is widely recognized as a leading standard for cloud security. The framework is used by organizations, auditors, regulators, and industry associations around the world to evaluate the security of cloud services and promote best practices for cloud security. The framework is also regularly updated to reflect changes in the cloud computing landscape and emerging security threats [19].

22.3.4 Limitations of NIST CCRA

The NIST Cloud Computing Reference Architecture is a standardized framework for the design, deployment, and management of cloud computing security environments. While it can be applied to a wide range of scenarios, here may be some situations where it is not appropriate:

1. Highly regulated industries: Some industries, such as healthcare and finance, are subject to strict regulations regarding data privacy and security. The NIST Cloud Computing Reference Architecture may not fully address all of the regulatory requirements in these industries.
2. Legacy IT systems: Organizations with legacy IT systems may face challenges when adopting cloud computing, and the NIST Cloud Computing Reference Architecture may not be well-suited for these scenarios.
3. Small-scale deployments: The NIST Cloud Computing Reference Architecture is designed for large-scale cloud deployments, so it may not be necessary or cost-effective for small-scale cloud implementations.
4. Hybrid cloud environments: The NIST Cloud Computing Reference Architecture is focused primarily on public and private cloud deployments, and may not fully address the unique requirements of hybrid cloud environments that combine both public and private clouds.
5. Highly customized environments: Some organizations may require highly customized cloud environments that do not fit well within the standardized framework provided by the NIST Cloud Computing Reference Architecture.

22.3.5 Security as a Service (SECAAS)

Security as a Service (SECAAS) is a cloud-based security model in which security services are delivered and managed by a third-party service provider. SECAAS providers offer a range of security services, including threat detection and response, vulnerability management, data encryption, and identity and access management.

SECAAS is becoming increasingly popular among organizations due to its many benefits, including cost savings, scalability, and flexibility. With SECAAS, organizations can outsource their security needs to a third-party provider, allowing them to focus on their core business operations. Additionally, SECAAS providers often have access to the latest security technologies and expertise, allowing them to provide a higher level of security than organizations could achieve on their own. However, there are also some potential risks associated with SECaaS, such as the possibility of data breaches or service disruptions. To mitigate these risks, organizations should carefully evaluate potential SECaaS providers, and ensure that they have robust security measures in place, including data encryption, access controls, and incident response plans. Overall, SECaaS is an important development in cloud security governance, and organizations should carefully consider the benefits and risks of this approach when developing their security strategies [11].

In addition to the benefits and risks mentioned above, there are some other important considerations to keep in mind when implementing SECaaS:

Integration with existing security infrastructure: Before adopting SECaaS, organizations should consider how it will integrate with their existing security infrastructure Integration is important to ensure that all security systems work together seamlessly and that there are no gaps in security coverage. Compliance with regulations: Depending on the industry and location, organizations may be subject to various regulations and standards related to data privacy and security. Before implementing SECAAS, organizations should ensure that the service provider can meet all necessary regulatory requirements.

Service-level agreements (SLAs). When working with a SECAAS provider, organizations should ensure that there is a clear service-level agreement (SLA) in place that outlines the provider's responsibilities, performance metrics, and compensation in the event of a security breach or service disruption [20].

Data ownership and access. Organizations should be aware of who owns the data and where it is stored. It is also important to ensure that only authorized personnel have access to sensitive data. Organizations should regularly monitor and evaluate the performance of their SECaaS provider, to ensure that they are meeting all necessary security requirements and that the service is providing the expected level of protection [21].

Best Features of SECAAS model. In summary, while SECaaS can provide many benefits for organizations, it is important to carefully evaluate the risks and considerations mentioned above and implement implementing appropriate technical controls, and ensuring compliance with relevant regulations and industry standards. Organizations should ensure that their cloud providers have robust security controls. They should also monitor their cloud environment regularly for unauthorized access, malware, and other security incidents [22].

22.3.6 Limitations of SECAAS

Security as a Service (SecaaS) can be a useful tool for many organizations due to outsourcing and its greater flexibility, scalability, and regulatory compliance. However, there may be situations where it may not be feasible for the organizations.

Some examples of environments where SecaaS may not be the best fit include:

1. Cost Considerations: While SecaaS can be a cost-effective solution for many organizations, it may not be the most cost-effective option for all organizations.
2. High-Security Environments: Organizations that deal with highly sensitive data, such as classified government information or intellectual property, may not be comfortable with outsourcing their security needs to third-party providers. In such cases, organizations may need to maintain their own security infrastructure to ensure the highest level of security.
3. Limited Provider Availability: Depending on the location and industry, some organizations may not have access to SecaaS providers that meet their specific security needs or compliance requirements.

22.3.8 ISO 27001

ISO 27001 is a worldwide recognized framework for information security management, and it is also used as a framework for cloud security governance [23]. The framework provides a well-structured architecture for managing and protecting information assets in the cloud environment. ISO 27001 can help organizations to maintain a secure cloud environment by providing a systematic and risk-based approach to information security management [24]. The standard offers a set of requirements that must be fulfilled for an organization to achieve certification. This includes the following features:

- It provides an Information Security Management System (ISMS): This is a framework of policies, procedures, and processes introduced to manage and safeguard the organization's information assets. It comprises identifying the scope of the cloud environment, setting ups security objectives, and defining roles and responsibilities for establishing cloud security.
- Risk assessment and management: organizations following the ISO 27001 require to identify and assess information security risks, and to apply controls to mitigate those risks. It includes identifying and assessing the risks associated with cloud service providers, data breaches, and other security threats.
- Security controls: The standard figures out a set of security controls that organizations can apply to protect their information assets. These controls are divided into 14 domains, including access control, cryptography, and incident management. In the cloud context, organizations can use these controls to ensure the safeguarding of cloud data, applications, and assets.
- Continuous improvement: organizations following this standard continuously monitor and improve their ISMS, including the cloud security controls that are in place. This includes conducting regular audits and reviews, as well as identifying chances for improvement and taking corrective measures where necessary.

22.3.9 Limitations of ISO 27001

While the ISO 27001 framework is a comprehensive approach to managing information security risks associated with cloud-based services, there may be situations where it may not be the best fit to use this framework for cloud security governance:

1. Short-term Projects: Organizations that are implementing short-term cloud projects may not need to implement the full ISO 27001 framework, as the costs and efforts involved may spoil the benefits.
2. Small Organizations: Small organizations with limited resources may find it difficult to implement and maintain the ISO 27001 framework, as it requires a significant investment of time and resources.
3. Cloud Services with Limited Risks: Organizations that use cloud-based services with limited risks, such as storage of non-critical data, may not need to implement the ISO 27001 framework, as they can adopt a more lightweight framework to cloud security governance.
4. Cloud Services without Sensitive Data: Organizations that use cloud-based services for non-sensitive data, such as public-facing websites, may not need to implement the full ISO 27001 framework, as the level of risk associated with these services may be relatively low.
5. Regulatory Requirements: ISO 27001 may not be an acceptable framework in such cases where regulatory requirements may dictate the use of specific frameworks for cloud security governance,

22.4 CONCLUSION

Cloud security governance might emphasize the importance of effective security governance for organizations that rely on cloud services. Effective cloud security governance requires a comprehensive approach that considers people, processes, and technology. It involves defining policies and procedures for managing cloud security risks, implementing appropriate security controls, and ensuring compliance with relevant regulations and industry standards.

The relevance of cyber security was emphasized while discussing the governance components of the cloud. The hazards associated with the cloud have also been covered, as well as issues of analyzing and certifying the cloud. The cloud strategy of a company using the cloud is one of the key issues that must be addressed. A company strategy must be integrated with such a strategy. We further contend that in order to develop the plan, commercial and legal experts must collaborate with cloud and security specialists. Also, it is crucial to clearly identify the person in charge of cloud governance. In addition to the advantages and hazards discussed above, the following factors should be considered while integrating SECAAS with current security infrastructure.

To effectively manage these risks, organizations must adopt a comprehensive approach to cloud security governance. This includes defining policies and procedures, implementing appropriate technical controls, and ensuring compliance with relevant regulations and industry standards. Organizations should ensure that their cloud providers have robust security controls. They should also monitor their cloud environment regularly for unauthorized access, malware, and other security incidents.

There is no single best security model for cloud computing, as the most effective security model will depend on various factors such as the specific use case, the type of cloud environment (public, private, or hybrid), and the level of security and compliance requirements. However, there are some general principles that can guide the development of a strong security model for cloud computing:

- Defense in depth: This principle involves layering multiple security measures to protect against different types of threats. For example, a defense-in-depth approach might involve using firewalls, intrusion detection systems, encryption, and access controls.
- Identity and access management: Implementing strong authentication and authorization processes is essential to prevent unauthorized access to cloud resources. This might include multi-factor authentication, role-based access control, and regular access reviews.
- Data protection: Protecting data is critical in cloud computing, as data can be accessed from anywhere and at any time. This might involve using encryption for data at rest and in transit, data loss prevention (DLP) technologies, and regular data backups.
- Compliance and governance: Cloud providers and customers must adhere to various regulations and standards, such as HIPAA, PCI DSS, and GDPR. A strong security model should include processes for compliance and governance, such as regular audits and assessments.
- Continuous monitoring and incident response: Regular monitoring of cloud resources is necessary to detect and respond to security incidents quickly. This might include using security information and event management (SIEM) systems and having an incident response plan in place.

Ultimately, a strong security model for cloud computing will involve a combination of these and other security measures, tailored to the specific needs and requirements of the organization. It is important to conduct a thorough risk assessment and regularly review and update the security model to stay ahead of emerging threats.

REFERENCES

[1] G. Gupta, P. R. Laxmi, and S. Sharma, "A survey on cloud security issues and techniques."

[2] M. Ahmed and M. Ashraf Hossain, "Cloud computing and security issues in the cloud," *International Journal of Network Security & Its Applications*, vol. 6, no. 1, pp. 25–36, Jan. 2014, doi: 10.5121/ijnsa.2014.6103.

[3] V. Maria Antoniate Martin, K. Kavitha, and A. Proffessor, "Cloud computing security issues and challenges 1," JETIR, 2019. [Online]. Available: www.jetir.org

[4] D. Yimam and E. B. Fernandez, "A survey of compliance issues in cloud computing," *Journal of Internet Services and Applications*, vol. 7, no. 1, Dec. 2016, doi: 10.1186/s13174-016-0046-8.

[5] C. Bryce, "Security governance as a service on the cloud," *Journal of Cloud Computing*, vol. 8, no. 1, Dec. 2019, doi: 10.1186/s13677-019-0148-5.

[6] S. B. Maynard, M. Onibere, and A. Ahmad, "Defining the Strategic Role of the Chief Information Security Officer," *Pacific Asia Journal of the Association for Information Systems*, vol. 10, pp. 3, 2018.

[7] S. Ahmad, S. Mehfuz, and J. Beg, "Cloud security framework and key management services collectively for implementing DLP and IRM," *Mater Today Proceedings*, vol. 62, pp. 4828–4836, Jan. 2022, doi: 10.1016/J.MATPR.2022.03.420.
[8] J. Archer and D. Cullinane, "Security guidance for critical areas of focus in cloud computing V2.1," 2009. [Online]. Available: http://www.cloudsecurityalliance.org/guidance/csaguide.v2.1.pdf
[9] O. M. Yigitbasioglu, "The role of institutional pressures and top management support in the intention to adopt cloud computing solutions," *Journal of Enterprise Information Management*, vol. 28, no. 4, pp. 579–594, Jul. 2015, doi: 10.1108/JEIM-09-2014-0087.
[10] K. Jakimoski, "Security techniques for data protection in cloud computing," *International Journal of Grid and Distributed Computing*, vol. 9, no. 1, pp. 49–56, 2016, doi: 10.14257/ijgdc.2016.9.1.05.
[11] D. Yimam and E. B. Fernandez, "A survey of compliance issues in cloud computing," *Journal of Internet Services and Applications*, vol. 7, no. 1, Dec. 2016, doi: 10.1186/s13174-016-0046-8.
[12] J. Opara-Martins, R. Sahandi, and F. Tian, "Critical analysis of vendor lock-in and its impact on cloud computing migration: a business perspective," *Journal of Cloud Computing*, vol. 5, no. 1, Dec. 2016, doi: 10.1186/s13677-016-0054-z.
[13] S. Achar, "Cloud computing forensics," *International Journal of Computer Engineering and Technology*, vol. 13, pp. 1-0. 2022, doi: 10.17605/OSF.IO/9N64K.
[14] G. Carrera, "Building a comprehensive cloud security audit program,"*EDPACS*, vol. 66, no. 1, pp. 15–18, 2022, doi: 10.1080/07366981.2021.2004689.
[15] A. S. Saidah and N. Abdelbaki, "A new cloud computing governance framework," 2014, pp. 671–678.
[16] V. Kaushik, P. Bhardwaj, and K. Lohani, "Game of definitions—do the NIST definitions of cloud service models need an update? A remark," *Lecture Notes in Electrical Engineering*, vol. 936, pp. 653–666, 2022, doi: 10.1007/978-981-19-5037-7_47/COVER.
[17] X. Li, L. Pan, and S. Liu, "A survey of resource provisioning problem in cloud brokers," *Journal of Network and Computer Applications*, vol. 203, p. 103384, Jul. 2022, doi: 10.1016/J.JNCA.2022.103384.
[18] Z. Yi, L. Wei, H. Yang, X. A. Wang, W. Yuan, and R. Li, "An improved secure public cloud auditing scheme in edge computing," *Security and Communication Networks*, vol. 2022, 2022, doi: 10.1155/2022/1557233.
[19] S. A. Safar Al Ghamdi, K. Than Win, and E. Vlahu-Gjorgievska, "Information security governance challenges and critical success factors: Information security governance challenges and critical success factors: Systematic review Systematic review," 2020. [Online]. Available: https://ro.uow.edu.au/eispapers
[20] "Cloud computing: Business benefits with security, governance and assurance perspectives an ISACA emerging technology white paper," 2009. [Online]. Available: www.isaca.org
[21] R. P. Padhy, M. R. Patra, and S. C. Satapathy, "Cloud computing: Security issues and research challenges," 2011. [Online]. Available: http://www.nist.gov/
[22] S. Girs, S. Sentilles, S. A. Asadollah, M. Ashjaei, and S. Mubeen, "A systematic literature study on definition and modeling of service-level agreements for cloud services in IoT," *IEEE Access*, vol. 8. Institute of Electrical and Electronics Engineers Inc., pp. 134498–134513, 2020. doi: 10.1109/ACCESS.2020.3011483.
[23] K. Beckers, I. Côté, S. Faßbender, M. Heisel, and S. Hofbauer, "Erratum to: A pattern-based method for establishing a cloud-specific information security management system (Requirements Eng, 10.1007/s00766-013-0174-7)," *Requirements Engineering*, vol. 18, no. 4. p. 397, Nov. 2013. doi: 10.1007/s00766-013-0176-5.
[24] M. I. Tariq and V. Santarcangelo, "Analysis of ISO 27001:2013 controls effectiveneb for cloud computing," in *ICISSP 2016 - Proceedings of the 2nd International Conference on Information Systems Security and Privacy*, SciTePress, 2016, pp. 201–208. doi: 10.5220/0005648702010208.

23 LTE/Wi-Fi Coexistence in Unlicensed Band

Irfan Ud Din, Fatima Zahid, Uns Bin Younas, Umar Farooq, and Ommair Hameed

23.1 INTRODUCTION

23.1.1 LTE-LAA Technology

The number one goals of LTE-LAA technology are to provide a sincere coexistence mechanism with the current Wi-Fi networks operating in the unlicensed spectrum so that the LTE community won't have to reduce the throughput performance of Wi-Fi nodes and to provide a sincere coexistence mechanism with unique LTE-LAA networks deployed by amazing operators so that a similar throughput standard performance can be carried out. The CSMA/CA method serves as the foundation for MAC Wi-Fi layer.

To prevent a collision, the Wi-Fi device must first sense the channel before sending any data; if the channel is free, the device may then send the data. Otherwise, the Wi-Fi equipment is allowed to transmit, and a control response (ACK) frame is used for detection. On the other hand, there is no such frame and no LBT mechanism in the licensed LTE to detect the collision. The MAC layers of Wi-Fi and LTE differ significantly, as was previously discussed, and this creates some difficulties for the coexistence of these two technologies in the unlicensed spectrum. The fundamental issue is that if LTE and Wi-Fi coexist on the same unlicensed band without any kind of fair mechanism, LTE transmission would have an impact on Wi-Fi performance because it prevents Wi-Fi transmission due to its continuous nature. Wi-Fi, on the other hand, uses a random backoff and channel sensing technique to cohabit with other networks.

It is necessary to periodically check to feel the channel (listen) before broadcasting (speak) in these markets where LTE transmission uses an LBT algorithm in the unlicensed spectrum. Therefore, in order to transmit, a device or BS must detect the energy level at a specific time that corresponds to the Clear Channel Assessment (CCA) period. According to 3GPP TR 36.889, the LBT process uses Energy Detection (ED) to evaluate the channel's availability, making it a key component of equitable coexistence between Wi-Fi and LAA in the unlicensed spectrum. According to 3GPP TR36.889, "the capability of a LAA network not to damage Wi-Fi networks active on a carrier more than an additional Wi-Fi network operating on the same carrier, in terms of throughput and latency" is the definition of fair coexistence between LTE and Wi-Fi in the 5 GHz band.

In order to obtain a comparable performance between various LAA deployed by various operators in terms of throughput and latency, the design of LAA should take into account a number of factors, such as a fair and effective coexistence mechanism with Wi-Fi.

DOI: 10.1201/9781003497851-23

23.2 BACKGROUND

This paper presents the concept to utilize the unlicensed spectrum by means of deploying other technology over these unlicensed bands to coexist with Wi-Fi, radar, and Bluetooth. On the opposite hand, this coexistence between LTE and Wi-Fi technologies faces many boundaries and challenges over those bands.

The coexistence mechanism is studied by way of deploying distinctive scenarios of LTE. The first scenario is by using the usage of LTE Unlicensed duty-cycling (LTE-U), while the second is by means of the use of LTE Licensed-Assisted Access (LTE-LAA). In particular, simulation outcomes the usage of NS-3 simulator for the throughput and latency for specific coexistence deployments are supplied. The simulation outcomes show that the coexistence mechanism among LTE-LAA and Wi-Fi inside the 5 GHz band outperforms that of LTE-U with Wi-Fi.

The exponentially developing call for Wi-Fi packages and cell services, it's far suspected that the licensed spectrum isn't sufficient. Examining the operational effects of lengthy term evolution on the performance of Wi-Fi in an unlicensed spectrum band and devising answers to minimize such an effect has the main consideration of the community research network.

23.3 OVERALL PERFORMANCE EVALUATION

Consequently, overall performance evaluation of the new technology of wireless cell networks is an extreme problem to be taken into consideration in destiny studies inside the uplink in addition to within the downlink to satisfy the continuous demand for higher statistics rate, better utilization of network assets. The number one-person information connections are served by using the LTE outer quarter, whereas the excellent-effort person statistics connections are served through Wi-Fi internal sector. Assuming maximum capacity of the cellular network, the admission control decision is based on a minimum bit rate that must be met for a request to be admitted to the network. Besides, primarily based on the kind of service, a concern degree is given to each carrier.

The LTE sector serves higher priority requests, but decrease priority requests are forwarded by the LTE region to be served via the Wi-Fi inner zone as a result, the stability of the visitors over the whole community will be carried out. Hence, the site visitors is distributed among both technology, and therefore, the burden can be balanced all over the entire community. The models are built upon popular ns-three simulator, were designed in near session with industry professionals, and were established through calibration methods and towards lyrical proposals and experimental structures.

We check the optimizer in numerous situations and look into the effect of numerous parameters. Our outcomes reveal that the optimizer is able to predict the first-rate possible parameters for 3GPP fairness and therefore the proposed approach proves useful for tuning 5G NR-U parameters in the course of their coexistence with WiFi. Recently, mobile information traffic has swiftly grown which leads to many challenges particularly inside the radio spectrum wishes.

Long Term Evolution (LTE) in unlicensed band. Thus, operating Long Term Evolution (LTE) in unlicensed bands is turning into an attractive vicinity of research. LTE and Wi-Fi have successful designs inside the mobile and Wi-Fi networks operating within the certified and unlicensed spectrum bands from a time. This aims to behavior a scientific literature evaluation to find all the problems about truthful coexistence of the LTE and WiFi in the unlicensed spectrum band. The paper can even spotlight the proposed mechanism and analyze the effect of some of these techniques to provide the fair coexistence of those wireless technology with synthesis evaluation of previous paintings. According to a current study commissioned through the Wi-Fi Alliance (WFA), among 2020 and 2025 users worldwide are probable to experience a spectrum shortfall. As an end result of the increasing demand for traffic and bandwidth, cellular operators are increasingly interested in deploying complementary access using unlicensed spectrum. There has been a unique interest these days for accessing the 5 GHz band, traditionally mainly used by Wi-Fi technologies, with Long Term Evolution (LTE). This has generated the definition of novel LTE-based totally get right of entry to technology able to running in unlicensed spectrum, even as coexisting with other technology.

The partner editor coordinating the assessment of this manuscript and approving it for publication was Yi Zhang. Of unlicensed spectrum above 6 GHz has attracted a whole lot of interest by using industry, law and standardization bodies. 3rdGenerationPartnershipProject (3GPP), in Release 16, has centered on a take a look at item on New Radio (NR) in unlicensed band, under and above 6 GHz [5], which ended in TR 38.889. A new Work Item on this difficulty is ongoing. With an emphasis on the 5GHz band, technologies that access the unlicensed spectrum can be divided into two main categories, depending on the radio access technology used:

1. Technology basedon integration of LTE and Wi-Fi radio hyperlinks and the use of Wi-Fi to access the unlicensed spectrum, and 2) technology the usage of LTE Radio Access Network (RAN) in unlicensed spectrum. The integration of LTE and Wi-Fi radio hyperlinks has been proposed on the grounds that 3GPP Release 13 [6]. Examples of these technologies are: LTE-WLAN Aggregation (LWA) and LTE-WLAN Radio Level Integration with IPsec Tunnel (LWIP). As for unlicensed LTE technology, using LTE RAN to access the unlicensed spectrum, their main challenge is the fair coexistence with other wireless technologies running inside the identical band.
2. While LTE is designed to have a special get entry to channel and carry out in uninterrupted and synchronous style, the present unlicensed technology perform in a decentralized, asynchronous manner employing protocols commonly primarily based on carrier sensing to be able to gain an honest usage of the spectrum. Some of the challenges of such coexistence state of affairs are defined in [7, 8].

23.4 DESIGN OF UNLICENSED LTE

Therefore, a crucial requirement for the design of unlicensed LTE is that it has to coexist with different technology, like Wi-Fi, on a "fair" and "friendly" foundation [2, 9], by way of extending its synchronous layout. In a few markets, like Europe and Japan, a Listen-Before-Talk (LBT) characteristic for Clear Channel Assessment (CCA) earlier than gaining access to the 5 GHz unlicensed channel is needed, at the same time as in others, such as the USA, China, India and Korea, there is no such requirement. For markets that don't require LBT, the commercial consortium LTE Unlicensed (LTE-U) Forum specified a proprietary solution for unlicensed LTE primarily based on Release 12, which is referred to as LTE-U [10]. On the other hand, to meet LBT requirement, 3GPP has produced in Release 13 [2] Licensed-Assisted Access (LAA) specification, for Supplemental DownLink (SDL) in unlicensed band.

In Release 14, the uplink operation changed into also defined, inside the context of Enhanced LAA (eLAA) [11] and new capabilities are beneath definition within the Further Enhanced LAA (feLAA) Study Item [12]. Several merchandise have already been offered at Mobile World Congress 2017 and 2018 through agencies inclusive of Qualcomm, to reach the 1 and a couple of Gigabit LTE, respectively, with and without aggregation of unlicensed bands. In unique, in lots of business deployments around the sector, presently LAA is used as supplemental downlink to deliver Gigabit LTE, paving the manner to 5G.

On the other hand, lately, the Multefire alliance [4] proposed an answer for unlicensed LTE that operates in a very stand-by myself manner in unlicensed spectrum by capitalizing on 3GPP Release thirteen and 14 LAA. In this work, we recognize LAA and LTE-U for the reason that they represent the maximum promising and substantial LTE-based totally unlicensed technology. This is due to the fact those technology use the same RAN in each certified and unlicensed spectrum, which permits a unified mobility, authentication, security, and control. Additionally, on account that they leverage Carrier Aggregation (CA) with the licensed carrier, they assure extensive-area insurance and the Quality of Service (QoS) traditional of the licensed service.

We carry out an in-depth evaluation, take a look at of these technologies, and compare their performance in the direction of their customers and in terms of coexistence. There is a common perception that LAA, due to its LBT function, is generally considered superior to LTE-U in terms of coexistence performance with Wi-Fi. We discuss in this paper that the fact isn't always so honest and we attempt to deeply understand the boundaries and strengths of each of the two technologies. A key undertaking to evaluate these technologies is that despite the large body of simulation effects by way of industry [2, 3] and inside the literature [13–15], the simulators are not publicly available, the two technologies were evaluated in a standalone style, and that they have not been in comparison but over the identical scenarios and the equal simulation or test systems.

As a consequence, the obtained results are nor reproducible, neither comparable, and gadget performance metrics are presented without a good deal info discovered approximately the underlying fashions and assumptions. Few analytical models had been proposed inside the literature to study channel access of both technologies [16], [17], and [18]. However, these fashions rent one of a kind assumptions, which restriction their functionality to assess the effect of some of the key coexistence parameters. In order to perform a coexistence examine and comparison of LAA and LTE-U technologies, we have constructed an in-depth simulation platform, strictly complying with LTE-U Forum and LAA 3GPP specifications, extending the famous open source network simulator ns-three [19]. This permits access to the total configuration of the system (i.e. from the application to the network interface) and the reproducibility of outcomes. The simulator that we have built basedonns-3,lets in to reproduce each 3GPP and WFA assessment procedures, and differently from any other simulator used in literature or in 3GPP studies, permits a full protocol stack simulation and a cease-to-cease overall performance evaluation.

23.5 PROPOSED MODEL

3 The proposed models had been validated via calibration campaigns, following 3GPP strategies, and against analytical answers and experimental platforms to be had in literature. Based on this simulation platform, we've carried out an extremely detailed simulation campaign, analyzing many aspects affecting coexistence, and evaluating LAA and LTE-U coexistence performance. This has allowed us to reach meaningful conclusions, that are mentioned at some stage in the paper. The paintings on this simulator have been supported with the aid of the WFA and with the aid of a small cellular seller, SpiderCloud Wireless, intensively working in unlicensed spectrum, and so it has been designed in near consultation with enterprise. The proposed models have been proven by calibration campaigns, 3GPP-compliant methods, and comparisons to experimental systems and analytical solutions that have been published in the literature.

4 Based on this simulation platform, a very accurate simulation campaign was run to explore several coexistence-affecting aspects and compare the performance of LAA and LTE-U. As a result, we were able to draw several important findings that will be discussed in the study. This simulator was developed in close collaboration with business thanks to funding from the WFA and assistance from a small mobile vendor, Spider Cloud Wireless, who actively uses unlicensed spectrum.

23.6 IMPLEMENTATION OF LAA AND LTE-U

The implementations of LAA and LTE-U fashions in conjunction with the documentation, and the simulation scenarios offered on this paper are publicly available at [20] to facilitate results reproducibility and further collaborative trends. The outline of the paper is organized as follows. The work presented in [7] proposed a QoS aware mechanism which has a fairness constraint. It has been shown in [8] that, in at least the commonplace scenarios, 3GPP fairness is satisfied. Given that 3GPP equity won't be carried out in every community state of affairs [4], it might be beneficial to determine the optimal 5GNR-U parameters such that fairness criterion is met.

As carefully as feasible. To the greatest of our knowledge, constrained paintings have been posted for coexistence of 5G NR-U with WiFi, and there may be no work in literature to music an arbitrary variety of 5G NR-U parameters to gain 3GPP fairness. Our key contributions on this paper are as follows:

We look at the problem of achieving 3GPP equity as an optimization problem with a non-linear goal function and constraints. We display how the Sequential Quadratic Programming set of rules [9] can be used to remedy the problem via tuning 5G NR-U parameters and obtain the "fairest" state of affairs wherein WiFi and 5G NR-U throughput are the closest viable.

We check out the effect of diverse 5G NR-U parameters on the 3GPP criterion and demonstrate the validity of the proposed method. We count on that nodes comply with an unbiased time homogeneous backoff technique with an identical constant-country opportunity of a hit transmission of head-of-line (HOL) packets as defined in [10]. The assumption holds true as long as the difference in the number of sensing slots between the nodes is relatively small. Though the definition of a fit in 5G NR-U differs from LTE-LAA, we counterpart whilst computing the conserving times$L$ of gNBs [11]. A WiFi node, can recollect the 5G NR-U slot period (()) rather than the LTE-LAA transmission rate. On the other hand, a 5G NR-UW node will transmit$RW$for a time upon accessing the channel transmits a () bits payload at () Mbps length of most channel occupation time (MCOT).

We consider some preferred assumptions discovered in literature [3, 4, 10]. The nodes are saturated and the channel is noiseless, i.e., there's no random blunders. The hidden node problem is assumed to no longer occur. Also, every node is believed to have an infinite. With the evolution of wi-fi programs and offerings, the spectrum scarcity has come to be a tough trouble. Unfortunately, the value and the supply of the licensed spectrum is likewise a tough trouble [1]. Therefore, it's important to find a technique to have more spectrum bands. One of those solutions is to utilize the unlicensed spectrum greater efficiently by way of occupying those bands with different wireless technologies.

The unlicensed bands are occupied by using a few Wi-Fi technologies.

LTE era has been developed to function in unlicensed bands to present higher throughput, higher overall performance in dense deployment, and extra capacity [3]. On the alternative hand, the coexistence of LTE with Wi-Fi in these loose bands creates many demanding situations since there's a major distinction between the LTE and Wi-Fi MAC layers.

In Wi-Fi, the MAC layer is primarily based on the Carrier Sense Multiple Access with Collision Avoidance (CSMA/CA) mechanism. Thus, the node senses the channel, and if it is free, the transmission takes place. Otherwise, the node selects a random backoff timer, and the transmission begins once the timer counts down to zero. While in LTE, there's no sensing scheme. As an end result, the coexistence of LTE with Wi-Fi in the unlicensed bands can critically degrade the overall performance of Wi-Fi because the Wi-Fi node sends its own records after checking the supply of the channel.

23.7 OBJECTIVES

This study's main objective is to conduct a thorough comparison of LTE-U and LTE-LAA when they are deployed alongside Wi-Fi in the five GHz band. It has been established that sharing a channel without a managed device is unfair when using Wi-Fi and LTE nodes absence of a controlled device. The Advanced Long Term Evolution (Advanced LTE) and IEEE 802 standards for 4G wireless broadband devices have been recommended, according to the Three Era Partnership Project (3GPP).

Deliver a thorough survey, execution evaluation, and distinction of LAA and LTE-U in a wide range of structure while ensuring 3GPP and Wi-Fi Alliance (WFA) recommendations in this study.

Despite the fact that the market confirms LAA as the primary unlicensed LTE technology, properly prepared LTE-U performance may also provide comparatively better Wi-Fi coexistence in the exact location. Comparably, the anticipated actions of the LAA LBT technique may reveal the untruth, depend on the flow of traffic, or require complex execution strategies, balanced in Wi-Fi. This study also includes a framework requirement for TCP, which is a problem that is rarely mentioned in writings about the coexistence of LTE and Wi-Fi.

This research effort proposes a contemporary model of existence between LTE and Wi-Fi technology so that the give-up-individual will benefit from the integration of each technology and profit from the highest quality best of offerings for surrender-customers. In this analysis, motivation and LTE service aggregation with the unlicensed band coexistence trends and standard overall performance evaluation of each technology are documented.

23.8 ISSUES IN LTE/WI-FI COEXISTENCE IN UNLICENSED BAND

LTE and Wi-Fi are the most advanced Wi-Fi technologies available today in terms of coexistence, and numerous coexistence concerns are highlighted in this context. According to the researchers' simulation study, the coexistence situation is detrimental to Wi-Fi in comparison to LTE. It has been advised to evaluate the performance of a multilayer mobile device scenario. The simulation results provided by the author show that the use of Wi-Fi doesn't significantly impact LTE performance. On the plus side, Wi-Fi performance generally suffers. When the two technologies are combined, the network's performance benefits from the overall good community performance. As a result, any admissions shifted from the community to the access point will improve overall performance as well as network performance. The cohabitation of LTE and WiFi technologies is thought to be decorated by the multiband armed (MBA) algorithm.

This method is based entirely on the idea of making better use of Wi-Fi channels that could be assumed to be perfect for LTE generation. The study refers to these great channels as "white area," and as a result, the overall performance metrics offered to demonstrate the efficacy of this set of rules are entirely dependent on this idea. Even while academics working at the 4G and Wi-Fi generation aggregate in the literature, fresh research and artwork are suggested for coexistence between WiFi-6 and 5G. The effectiveness of this integration on cellular performance is presented, and it is demonstrated that this idea can be applied to 5G technology in the future in addition to 4G technology.

A hybrid adaptive channel access (HyACA) mechanism is suggested in specific research studies as a way to improve LTE/Wi-Fi performance while coexisting.

A dynamic channel switch (DCS) capability and an adaptive nearly easy-sub frame (AABS) capacity are advised to manage the channel get-in method in order to ensure channel access equality. The recommended mechanism enables higher throughput and creates higher equity. Compared to the conventional solution of LTE data offloading, LTE-U can provide greater hyperlink average performance, medium access control, mobility control, and big insurance. Although LTE-U has numerous benefits, it also has several technical issues, such as throughput deterioration, interference, and inefficient spectrum usage. These technologies, with a focus on the 5GHz band, can be split into two main categories based on the radio access generation utilized to access the unlicensed spectrum:

1. Technologies that are entirely based on integrating LTE and Wi-Fi radio links and using Wi-Fi to gain access to the unlicensed spectrum, and
2. Technologies that use LTE Radio Access Network (RAN) in the unlicensed spectrum.

23.9 CONCLUSION

We conclude that two LTE-U technologies are compared and contrasted along with the difficulties they have encountered. LBT Category 4 has to be improved in future work. The simulation structures, the unique application options for both LTE and Wi-Fi, and many other factors, such as the considered trading patterns, come into play in the coexistence standard presentation, in addition to the specific knowledge of the channel get approach to techniques that have long been significant. When a Wi-Fi community operates concurrently on the same channel, the performance of the Wi-Fi will be considerably impacted, but the overall performance of the LTE is minimally impacted. The reason is that each technology uses unique procedures to regulate the medium access at the MAC layer as a result, the difference in MAC protocols is what causes performance reduction in each technology. We therefore draw the conclusion from these findings that the coexistence performance depends on more than only the access mechanism. The apparent finding that LAA is a better neighbor to Wi-Fi than LTE-U is not always accurate. For example, the received interference level, the eNB's ability to detect interference, or the traffic pattern may cause one technology to perform better than the other, making it impossible to say that one technology is superior to another in terms of coexistence.

REFERENCES

1. Bojović, B., Giupponi, L., Ali, Z., & Miozzo, M. (2019). Evaluating unlicensed lte technologies: Laa vs lte-u. *IEEE Access, 7*, 89714–89751
2. Alhulayil, M., & Lopez-Benitez, M. (2018, July). Coexistence mechanisms for LTE and Wi-Fi networks over unlicensed frequency bands. In *2018 11th International Symposium on Communication Systems, Networks & Digital Signal Processing (CSNDSP)* (pp. 1–6). IEEE.
3. Mustafa, S., Alam, K. A., Khan, B., Ullah, M. H., & Touseef, P. (2019, July). Fair coexistence of LTE and WiFi-802.11 in unlicensed spectrum: A systematic literature review. In *Proceedings of the 3rd International Conference on Future Networks and Distributed Systems* (pp. 1–10).
4. Zreikat, A. I., & Elbasi, E. (2021, January). Downlink interoperability model of LTE/Wi-Fi integration/coexistence with performance evaluation. In *2021 IEEE 11th Annual Computing and Communication Workshop and Conference (CCWC)* (pp. 0373–0379). IEEE.
5. Methley, S., & Webb, W. (2017, February). Wi-Fi spectrum needs study, final report to Wi-Fi alliance 3rd ed. [Online]. Available: http://www.wi-fi.org.
6. Study on Licensed-Assisted Access to Unlicensed Spectrum, document TR 36.889 V13.0.0, Release 13, 3GPP, 2015, June.
7. Evolved Universal Terrestrial Radio Access (E-UTRA) and Evolved Universal Terrestrial Radio Access Network (E-UTRAN) and Multe Fire (MF); Overall description; Stage 2, document TS 36.300 V1.0.1, Release 1, MulteFire Alliance,2017, May.
8. Cavalcante, M., Almeida, E., Vieira, R. D., Chaves, F., Paiva, R. C., Abinader, F., Choudhury, S., Tuomaala, E., & Doppler, K. (2013). Performance evaluation of lte and wi-fi coexistence in unlicensed bands. In *Vehicular Technology Conference (VTC Spring), 2013 IEEE 77th.* (pp. 1–6). IEEE.
9. Huang, Y., Chen, Y., Hou, Y. T., Lou, W., & Reed, J. H. (2017). Recent advances of LTE/WiFi coexistence in unlicensed spectrum. *IEEE Network, 32*, 107–113.
10. Zreikat, A. I. (2015). Performance evaluation of downlink LTE-Advanced CELL by MOSEL-2 language. In *Proc. of 29th European Conf. on Modelling and Simulation, ECMS 2015, May 26th-29th*, 2015, Technical University of Sofia, Albena (Varna), Bulgaria (pp. 662–668). (ISBN: 9780-9932440-0-1). DOI: 10.7148/2015-0662-0668.
11. New SI: Study on NR-based Access to Unlicensed Spectrum, document RP-170828, Release 15, 3GPP, RAN Meeting 75, Dubrovnik, Croatia, 2017, March.
12. Pateromichelakis, E., Bulakci, O., Peng, C., Zhang, J., & Xia, Y. (2017). LAA as a key enabler in slice-aware 5G RAN: Challenges and opportunities. *IEEE Communications Standards Magazine, 2*(1), 29–35.

13. Morris, T. P., White, I. R., & Crowther, M. J. (2019). Using simulation studies to evaluate statistical methods. *Statistics in Medicine, 38*(11), 2074–2102.
14. Nelson, D., Springel, V., Pillepich, A., Rodriguez-Gomez, V., Torrey, P., Genel, S., . . . Weinberger, R. (2019). The IllustrisTNG simulations: Public data release. *Computational Astrophysics and Cosmology, 6*, 1–29.
15. Okuda, Y., Bryson, E. O., DeMaria Jr, S., Jacobson, L., Quinones, J., Shen, B., & Levine, A. I. (2009). The utility of simulation in medical education: What is the evidence? *Mount Sinai Journal of Medicine: A Journal of Translational and Personalized Medicine, 76*(4), 330–343.
16. Ferraro, F., Pfeffer, J., & Sutton, R. I. (2005). Economics language and assumptions: How theories can become self-fulfilling. *Academy of Management Review, 30*(1), 8–24.
17. Fletcher, K. (2010). Slow fashion: An invitation for systems change. Fashion Practice, 2(2), 259–265.
18. King, P. M., & Kitchener, K. S. (2004). Reflective judgment: Theory and research on the development of epistemic assumptions through adulthood. *Educational Psychologist, 39*(1), 5–18.
19. Blazquez, S., Algaba, J., Míguez, J., Vega, C., Blas, F., & Conde, M. (2024). Three-phase equilibria of hydrates from computer simulation. I. Finite-size effects in the methane hydrate. *The Journal of Chemical Physics, 160*(16), 45–56.
20. Chen, Y., Guo, S., He, Y., Luo, Y., Chen, W., Hu, S., . . . Su, S. (2023). Simulation and design of an underwater LiDAR system using non-coaxial optics and multiple detection channels. *Remote Sensing, 15*(14), 36–47.

24 Evolution of Next-Generation Firewall System for Secure Networks

Asim Noor, Noshina Tariq, Farrukh Aslam Khan, and Muhammad Ashraf

24.1 INTRODUCTION

Over the past few years, there has been a significant increase in the utilization of the Internet and its applications, which can be attributed to the rapid evolution of social networking and online business [1]. Internet users have also increased exponentially, and the dynamics of the Internet have changed melodramatically. Traditional firewalls are insufficient to defend networks from contemporary cyber threats. Because they operate primarily at the transport and network layers, traditional firewalls can only check packet headers and IP addresses. However, modern cyber-attacks are becoming increasingly sophisticated, and they frequently employ sophisticated evasion techniques and encrypted traffic to circumvent traditional firewalls [2]. The advanced capabilities of Next Generation Firewalls (NGFWs) enable them to identify and inhibit various cyber threats, including viruses, malware, and ransomware, in real time. NGFWs use deep packet inspection to scrutinize the contents of network traffic at the application layer, which enables them to identify and block malicious traffic that traditional firewalls may not detect. Additionally, NGFWs can detect and prevent attacks that exploit application or operating system vulnerabilities.

NGFWs can timely identify even the most sophisticated threats and provide additional benefits, including scalability and adaptability, which are essential for the security of today's networks. NGFWs can integrate with other security technologies, including intrusion detection systems (IDS) and security information and event management (SIEM) systems, to provide an all-encompassing security solution. In addition, they can be deployed in both hardware and software forms, providing organizations with a deployment model that can be tailored to their specific requirements. NGFWs include sophisticated security capabilities compared to traditional firewalls. By integrating deep packet inspection, sophisticated threat detection techniques, SSL inspection, and granular access control, NGFWs offer a more comprehensive approach to network security [3]. They can increase user and network visibility and identify and alleviate a variety of threats, including malware, phishing, and data exfiltration. Similarly, firewalls can help stop internal assaults on the blockchain in addition to external threats by limiting access to specific areas of the network and implementing stringent authentication and authorization standards. This can support preserving the privacy and security of sensitive data stored on the blockchain [4].

The dangers and challenges to network security increase with the advancement in technology. NGFWs, which provide sophisticated security capabilities beyond standard firewalls, have become an essential network security solution. This study intends to review the development of firewalls from traditional to NGFWs, explain their key features, implementation and use on a broad scale, possible impacts on network performance, and future NGFW research. In light of developing cyber threats, the article helps to increase the knowledge of NGFWs in terms of their capabilities and

DOI: 10.1201/9781003497851-24

significance for network security. It offers helpful information on the functions, uses, and future directions of NGFW research. In addition to outlining the advanced security capabilities that NGFWs provide, such as deep packet inspection, increased threat detection, SSL inspection, and granular access control, this paper provides an overview of the development of firewalls from conventional to NGFWs [5].

Overall, this survey provides a timely and comprehensive examination of the evolution of NGFWs, highlighting their significance and potential for future research. By shedding light on critical issues and considerations, our work aims to provide insights to help organizations make informed decisions about their network security strategies. The significant contributions of this work encompass:

1. The critical characteristics of NGFWs, including their capacity to recognize and thwart various threats, scalability and flexibility, and interaction with other security technologies, are provided in this study [6].
2. The study focuses on the significance of extensive NGFW deployment and employment, and the need for ongoing monitoring and upgrading to maintain their efficacy.
3. The survey investigates the impacts of SSL inspection on network performance, a crucial aspect to consider while adopting NGFWs.
4. The study outlines potential research possibilities for NGFWs, including using cloud-based NGFWs, incorporating Artificial Intelligence (AI) and Machine Learning (ML), and integrating other security technologies to increase their efficacy [7].

The structure of this paper is presented as follows: The security objectives and several forms of advanced attacks are presented in Section 24.2. The NGFW, various layer applications, and benefits over conventional firewalls are covered and are highlighted in Section 24.3. In Sections 24.4 and 24.5, the evolution and up-to-date innovation in NGFWs are provided, respectively. Recommendations for future NGFWs have been presented in Section 24.6. Finally, Section 24.7 presents a conclusion and a look ahead.

24.2 SECURITY REQUIREMENTS AND TYPES OF ATTACKS

This section discusses the security requirements and types of attacks in a network environment.

24.2.1 Security Requirements

The criteria for network security cover several significant topics, such as confidentiality, privacy, integrity, availability, and accountability (see Figure 24.1).

24.2.2 Types of Attacks over Network

Firewalls primarily deal with network and TCP/ IP layers and block suspicious IP addresses and Internet protocol ports to defend against traditional cyber-attacks. These attacks are primarily visible and opportunistic (see Figure 24.2). Network cyber-attacks are evolving to be targeted, focused, and stealthier against applications and associated sensitive data. Furthermore, present threats have been moved up to the OSI layers, which must be guarded through new-generation firewalls [8]. Presently, Next Generation firewalls have the proficiency to scrutinize all data over the Internet but are simultaneously unable to measure, assimilate and detect advanced attack methodologies. Numerous systems led to **"Next-Generation Firewalls."** Some of the primary reasons for a network attack are listed below.

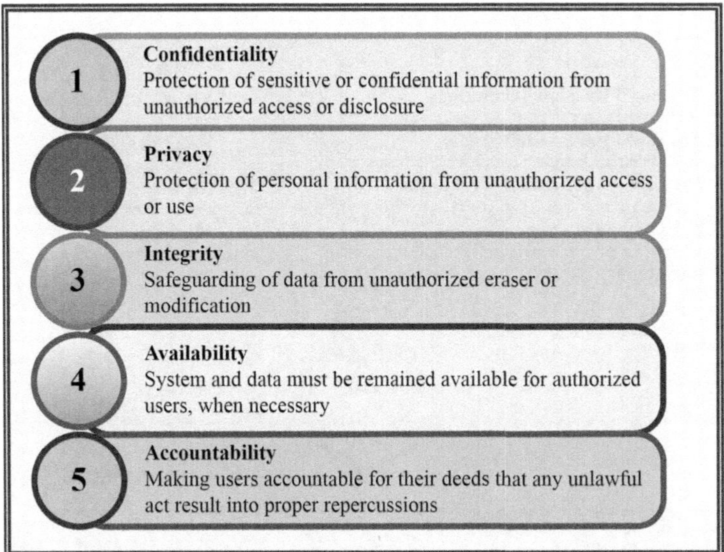

FIGURE 24.1 Security requirements.

1. **Advanced Evasion Techniques (AETs)**: Sometimes, network attack is based upon a combination of various evasion methods, and a new type of attack technique is created to dodge multiple network layers of security protocols. AETs can disguise malicious data by segregating it into smaller parts/frames and transmitting smaller frames across rarely used network protocols. AET attacks mostly happen silently, cannot be detected by traditional firewalls, and remain undetected [9].
2. **Advanced Persistent Threats (APTs):** APTs are deliberate, well-rehearsed, and protracted attacks. They can enter the network and avoid detection for a long time by utilizing social engineering techniques to get beyond typical firewalls [10].
3. **Targeted Cyber Attack Techniques (TCATs)**: In these attacks, the entire network system is targeted instead of attacking individuals or systems. Although these cyber-attacks are not widely known, they are anticipated to target and jeopardize the network infrastructure actively [11].
4. **Insider Threats**: Insider hazards are dangers from within an organization, unintentionally or intentionally. Since they already have permission to access the network; conventional firewalls cannot detect or prevent insider attacks [3].
5. **Encrypted Traffic**: Encrypted traffic is fast becoming the standard for Internet conversations. Encryption offers protection and privacy, but it also presents a problem for conventional firewalls because they cannot read the content of the encrypted data [8].
6. **Zero-day Exploits**: Because the vendor is unaware of the vulnerability in the software or hardware, there is no patch for a zero-day exploit. Attackers may use these weaknesses to break into a network without being noticed by conventional firewalls [3].
7. **IoT Devices**: These devices frequently lack adequate security safeguards and are easy to use [1].

2.3 Types of Firewalls

To defend computer networks from intrusions and unauthorized access, firewalls of various forms are frequently used. A few of the most typical kinds of firewalls are tabulated below in Table 24.1.

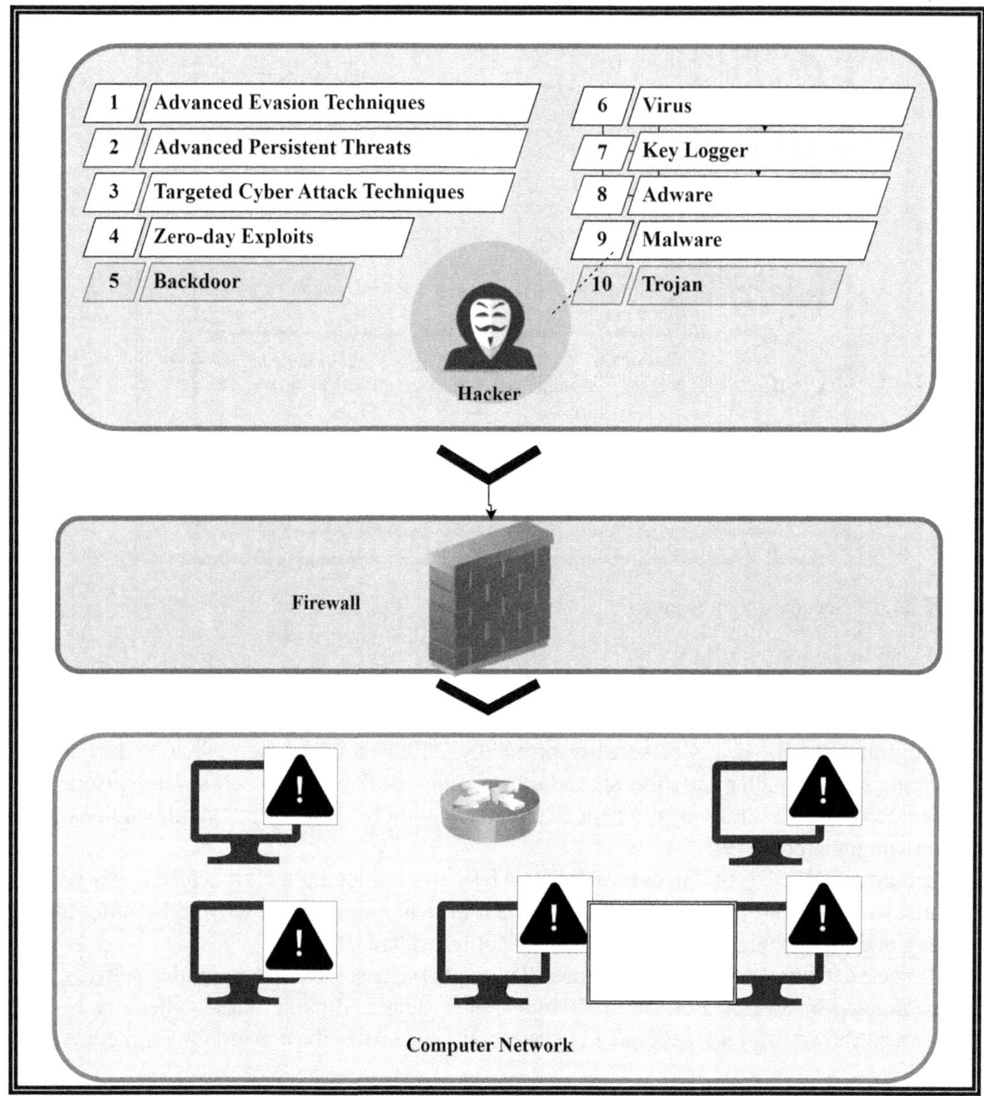

FIGURE 24.2 Type of attacks over the network.

24.3 NEXT-GENERATION FIREWALLS (NGFWS)

In Advanced Threats Patterns, attackers/hackers critically examine their target, inspect network security, and find a means to sidestep security protocols. In the early 1990s, the first-generation firewalls were developed. They were not intended to examine network data/traffic in which the Internet is required to connect to user applications/other sensitive information. However, there are multiple threats over networks that are growing exponentially. Due to rapidly evolving network threat scenarios, the overall efficacy of first-generation firewalls is limited. Later, major network stakeholders paved the waypoints for the Next Generation Firewalls (NGFWs) progression [3]. In this regard, brief details have been summarized in Table 24.2.

Legacy firewall systems typically provide packet filtering and stateful inspection but lack advanced security features like deep packet inspection, application awareness, and intrusion prevention. These systems also typically have limited or no user identification capabilities and offer the

TABLE 24.1
Types of Firewalls

Type	Description
Firewall with Packet Filtering [12]	A collection of predetermined rules to filter packets coming into and leaving the network. According to the set rules, it examines each packet's source, destination, and port information before allowing or blocking it.
Stateful Inspection Firewall [13]	A more sophisticated form of packet filtering that examines data packets and the connection status between the systems at the source and the destination. Every network connection is monitored, determining whether incoming packets are part of an active connection. It aids in thwarting harmful activity and blocking unauthorized traffic.
An application Level Gateway [14]	A proxy firewall is a network device that inspects and filters traffic at the application level while operating at the network stack's application layer. Analyzing and filtering packets based on application-level protocols like HTTP, FTP, and SMTP offers more precise control over network traffic.
Circuit-level Gateway Firewall [3]	A circuit-level gateway checks the transmission control protocol (TCP) handshake between two systems at the network stack's transport layer. It checks that the handshake process is valid and that the traffic comes from an authorized source; it does not examine the data packets' contents.
Virtual Private Network (VPN) Firewall [15]	Data transmission between a remote user and a private network is protected by a virtual private network (VPN) firewall. To prevent data from being intercepted and to provide secure remote access, it uses a VPN to establish an encrypted tunnel between the user and the network.
Next-Generation Firewall (NGFW) [5]	An advanced firewall that combines conventional firewall functionality with extra security capabilities like intrusion prevention, application-level control, and advanced malware detection is known as a next-generation firewall. To recognize and stop malicious traffic, NGFWs combine signature-based detection, behavioral analysis, and machine learning techniques.

TABLE 24.2
Characteristics of Next-Generation Firewall

Features	Legacy Firewall System	Next Generation Firewall System
Packet Filtering	Yes	No
Stateful Inspection	Yes	Yes
Deep Packet Inspection	No	Yes
Application Awareness	No	Yes
Intrusion Prevention	No	Yes
User Identification	No	Yes
VPN	No	Yes
Quality of Service	Limited	Advanced
Bandwidth Management	Basic	Advanced
Threat Intelligence	No	Yes
Cloud Integration	Limited	Advanced
Real-time data handling	No	Yes
Scalability	Limited	Advanced
Overall Solution	Inefficient	Efficient

essential quality of service and bandwidth management functionality. Legacy firewalls are typically managed using a command line interface (CLI) or a simple graphical user interface (GUI). NGFW systems, on the other hand, offer advanced security features like deep packet inspection, application awareness, and intrusion prevention. They also provide user identification capabilities, advanced quality of service, and bandwidth management functionality and are typically managed using an intuitive GUI. Additionally, NGFWs typically integrate with cloud-based security solutions and leverage threat intelligence feeds to detect and prevent attacks better [16].

24.4 EVOLUTION OF FIREWALL

To comprehend the advancement of NGFWs, it is essential to know various types of firewalls. The evolution of firewalls has been categorized based on their primary features, the scope of operation in the OSI layer, design style, and capabilities (see Figure 24.3).

The NGFW differs significantly from its predecessors as it employs advanced techniques such as deep packet inspection, which combines Intrusion Prevention Systems (IPS) and other sophisticated network security measures. Traffic flow management remains a fundamental aspect of all NGFWs. Managing First Generation Firewalls has become a considerable security risk, highlighting the need for more robust firewalls with advanced traffic inspection capabilities, easy management, and enhanced accessibility to secure networks against present and future threats [17]. To address the limitations of First-Generation Firewalls, Next Generation Firewalls must possess the following essential characteristics, as given in Table 24.3.

24.5 TOPICAL INNOVATION IN NEXT-GENERATION FIREWALL

Implementing Palo Alto's NGFWs significantly changes application usage, user behaviour, and complex network infrastructure. This evolution has exposed vulnerabilities in traditional port-based

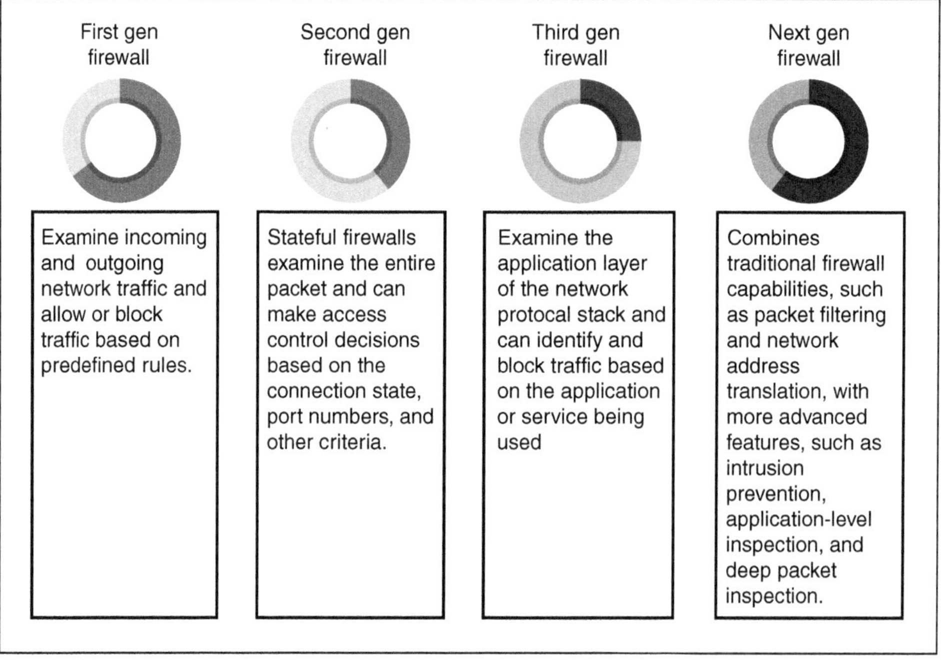

FIGURE 24.3 Evolution of firewalls.

TABLE 24.3
Characteristics of Next-Generation Firewalls

Characteristics	Description of NGFWs
Application Awareness	NGFWs are capable of identifying and controlling applications that traverse the network, allowing granular control.
Deep Packet Inspection (DPI)	NGFWs use DPI to analyze packet payloads, enabling them to identify malware and other advanced threats.
Intrusion Prevention System (IPS)	NGFWs include an IPS that detects and prevents known and unknown network-based attacks.
User Identification	NGFWs can identify users and apply access policies based on user identity.
Quality of Service (QoS)	NGFWs can prioritize network traffic based on business needs, ensuring critical applications always have the necessary bandwidth.
Virtual Private Network (VPN)	NGFWs support secure remote access to the corporate network via VPNs.
Centralized Management and Reporting	NGFWs provide centralized management and reporting interface, simplifying administration and enabling better visibility.
Cloud Integration	NGFWs can integrate with cloud-based security solutions, providing a layered defense against threats.
Threat Intelligence	NGFWs leverage threat intelligence feeds to detect better and prevent attacks.
Automation	NGFWs can automate routine tasks like policy deployment and update management, freeing up IT staff to focus on more strategic initiatives.

TABLE 24.4
Salient Characteristics of Palo Alto

Requirements	Traditional Firewall System
Next-Generation	Offers advanced security features such as deep packet inspection, intrusion prevention, user identification, and application awareness.
Multi-Layered Security	Provides multi-layered security, including network, application, and user-based policies, to protect against modern cyber threats.
Single Pass Architecture	It uses a single-pass architecture that reduces latency and allows for faster processing and inspection of the network traffic.
Centralized Management	Offers centralized management through the Panorama management platform, allowing administrators to manage multiple firewalls from a single interface.
Threat Intelligence	Integrates with threat intelligence feeds, such as WildFire, provides proactive threat prevention and response capabilities.
Cloud Integration	Integrates with cloud-based security solutions, such as Prisma Cloud, to provide enhanced protection for cloud-based environments.
Scalability	Provides flexible and scalable deployment options, including virtual and physical appliances to meet the needs of organizations of all sizes.

network security. As a result, the expansion of data centres, network virtualization, and mobility has led to a need for protection against advanced persistent threats like botnets [18] and targeted cyber-attacks. The Palo Alto NGFW is the foundation of next generation security platforms that tackle these emerging threats [3]. A summary of noteworthy features of NGFWs is tabulated below in Table 24.4.

The NGFW inspects all data traffic, including applications and potential threats, and associates them with the user. The Palo Alto Next Generation security platform identifies unfamiliar threats, minimizes the time to respond to incidents, and simplifies security deployment [19].

24.6 RECOMMENDATIONS FOR FUTURE NGFW SYSTEMS

Following are some suggestions for future research and development of NGFW systems:

1. **Integration with other security technologies**: NGFWs ought to be integrated with other security tools, including SIEM systems, threat intelligence feeds, and intrusion prevention systems. Better threat identification and response will be possible, assisting organizations in staying ahead of developing dangers [20].
2. **Continuous monitoring and updating:** To ensure NGFWs are operating effectively and offering the best level of security, they should be continuously monitored and updated. To keep NGFWs current with the most recent threat landscape, routine software updates, threat intelligence feeds, and security assessments should be carried out.
3. **Machine learning and artificial intelligence:** AI-powered NGFWs can better detect and respond to threats in real time by analyzing massive amounts of data and spotting patterns. Machine learning and Artificial Intelligence (ML/AI) based NGFWs can use ML/AI to enhance threat detection and response capabilities.
4. **Cloud-based NGFWs:** Cloud-based NGFWs are necessary as more businesses transfer their workloads to the cloud. These NGFWs are made exclusively for cloud environments. Flexible, scalable, and able to connect with other cloud-based security technologies are all essential characteristics of cloud-based NGFWs.
5. **User education and awareness:** NGFWs are only a small component of a thorough security plan. Businesses must inform users of optimal security practices and the most recent dangers. It can assist in ensuring that NGFWs are functioning correctly and stopping users from unintentionally bringing vulnerabilities into the network.

24.7 CONCLUSION

Next-Generation firewalls (NGFWs) offer sophisticated security features surpassing conventional firewalls. NGFWs provide a more thorough network security method by integrating deep packet inspection, sophisticated threat detection techniques, SSL inspection, and granular access control. They can provide more visibility into network activity and user behaviour, and detect and stop various threats, such as malware, phishing, and data exfiltration. NGFWs can be installed on-premises or in the cloud and are versatile and adaptable. Although NGFWs can cost more than the standard firewalls, they provide a higher level of protection and call for less additional hardware or software to perform advanced functions. NGFWs are crucial for businesses defending their networks and data from complex cyber assaults. In conclusion, NGFWs should develop and adapt to keep up with the evolving threat environment. Future research in NGFWs should concentrate on several key areas, including user education and awareness, continuous monitoring and updating, machine learning and AI applications, cloud-based NGFWs, and integration with other security technologies.

REFERENCES

1. N. Tariq, A. Qamar, M. Asim and F. A. Khan.: Blockchain and Smart healthcaresecurity - a survey. *Procedia Computer Science*, vol. 175, pp. 615–620, (2020).
2. A. Hahn.: Operational technology and information technology in industrial control Systems. Cyber security of SCADA and other industrial control systems, vol. 66, no. 1, pp. 51–68, (2016).

3. K. Neupane, R. Haddad and L. Chen.: Next generation firewall for network security - A survey. In: *Southeast Conference*, pp. 1–6, IEEE, (2018).
4. N. Tariq, M. Asim, F. A. Khan, T. Baker, U. Khalid, A. Derhab.: A blockchain-based multi-mobile code-driven trust mechanism for detecting internal attacks in internet of things. *Sensors*, vol. 21, no. 1, Art. no. 1, pp. 23, (2021).
5. J. Liang and Y. Kim.: Evolution of firewalls - Toward securer network using next generation firewall. In: *IEEE 12th Annual Computing and Communication Workshop and Conference (CCWC)*, pp. 0752–0759, (2022).
6. B. Daya.: Network security - History, importance, and future. Department of Electrical and Computer Engineering, University of Florida, (2013).
7. F. A. Khan and A. Gomaei.: A comparative study of machine learning classifiers for network intrusion detection. *5th International Conference on Artificial Intelligence and Security (ICAIS)*, (2019).
8. M. V. Pawar and J. Anuradha.: Network security and types of attacks in network. *Procedia Computer Science*, vol. 48, pp. 503–506, (2015).
9. E. Korhonen.: Advanced Evasion Techniques - Measuring the threat detection capabilities of up-to-date network security devices, vol. 2, pp. 53–66, (2012).
10. N. A. S. Mirza, H. Abbas, F. A. Khan and J. Al Muhtadi.: Anticipating Advanced Persistent Threat (APT) countermeasures using collaborative security mechanisms. (ISBAST), vol. 1, pp. 129–132, (2014).
11. H. Al-Mohannadi, Q. Mirza, A. Namanya, I. Awan, A. Cullen and J. Disso.: Cyber-attack modeling analysis techniques - An overview. In: *IEEE 4th international conference on future internet of things and cloud workshops (FiCloudW)*, pp. 69–76, (2016).
12. C. Sheth and R. Thakker.: Performance evaluation and comparative analysis of network firewalls. In: *IEEE International Conference on Devices and Communications (ICDeCom)*, pp. 1–5, (2011).
13. C. Roeckl and C. M. Director.: Stateful inspection firewalls - Juniper Networks. White Paper, (2004).
14. C. Basile and A. Lioy.: Analysis of application-layer filtering policies with application to HTTP. In: *IEEE / ACM Transactions on Networking*, (2013).
15. Y. P. Kosta, U. D. Dalal and R. K. Jha.: Security comparison of wired and wirelessnetwork with firewall and virtual private network (VPN). In: *IEEE International Conference on Recent Trends in Information, Telecommunication and Computing*, pp. 281–283, (2010).
16. F. Malecki.: Next-generation firewalls-Security with performance. *Network Security*, vol. 2012, pp. 19–20, (2012).
17. D. Freet and R. Agrawal.: Network security and next-generation firewalls. In: *Proceedings of International Conference on Technology Management (ICTM)*, (2016).
18. A. Derhab, R. Alawwad, K. Dehwah, N. Tariq, F. A. Khan, and J. Al-Muhtadi.: Tweet-based bot detection using big data analytics. *IEEE Access*, vol. 9, pp. 65988–66005, (April, 2021).
19. S. Gold.: The future of the Firewall. *Network Security*, vol. 2011, pp. 13–15, (2011).
20. N. Tariq, M. Asim, and F. A. Khan.: Securing SCADA-based critical infrastructures: Challenges and open issues. *The 16th International Conference on Mobile Systems and Pervasive Computing (MobiSPC 2019)*, Halifax, Canada, (August 19–21, 2019).

25 Security and Performance Comparison of Window and Linux

A Systematic Literature Review

Johar Mumtaz, Attiq ur Rehman, Hamayun Khan, Irfan Ud Din, and Imran Tariq

25.1 INTRODUCTION

Operating systems provide standard services for implementing processes such as storage, scheduling, and deadlock, while also offering a programming environment that facilitates efficient program execution. Today, there are numerous types of operating systems available, including Apple's Mac OS, Microsoft's Windows, the community-driven Linux, and Google's Android. Windows is the most popular operating system in the market, offering numerous advantages. It is user-friendly due to its attractive GUI [1] Users are currently migrating from Window to Linux due to security lacking. In our country, the growth of Linux has been emerging rapidly in recent years. Windows and Linux have been competing for the operating system market since 1993, each with their own strengths and weaknesses. Our research aims to identify the major distinctions between the two systems and their potential impact on end-user adoption. We plan to analyze several key factors, including cost, security, configurability, and user-friendliness, to provide a comprehensive comparison of Linux and Windows [13]. As per the market data for April 2020, Linux ranks third according to [2]. The lack of attention and adoption of Linux by enterprises and individuals has resulted in a significant shortage of Linux experts, posing a real challenge. IT experts who intend to implement the evaluated workloads should not limit their considerations to the purchasing expenses of the technologies under investigation [3]. The scarcity of skilled Linux professionals has led to monthly salaries of Linux engineers being three times higher than those of Windows developers, due to the higher demand. This highlights the pressing need for training and development of Linux expertise. Moreover, Linux is considered to be more secure compared to Windows' high security features and capability to fight against the malware cyber-attacks as Linux's modular kernel architecture allows programs to function independently, which reduces the risk of infected or flawed programs propagating their damage to other programs in the kernel [4]. The adaptability of equipment isn't given by window and it isn't exceedingly reliable. Currently, there is a significant gap in Linux training and proficiency in our country, which is hindering its development. Ongoing research on Linux-based malware analysis faces a challenge in detecting variants of Linux malware utilized in IoT/embedded settings. Moreover, due to the support of diverse architectures by Linux malware, it becomes intricate to determine the specific architecture employed by the Linux malware while conducting malware analysis [5]. In terms of research and development (R&D) efforts, only 53% is focused on Linux, with the majority (31%) being dedicated to Windows, and a smaller percentage (16%) allocated to UNIX [6]. Referring to Figure 25.1, it illustrates the flow of the paper.

DOI: 10.1201/9781003497851-25

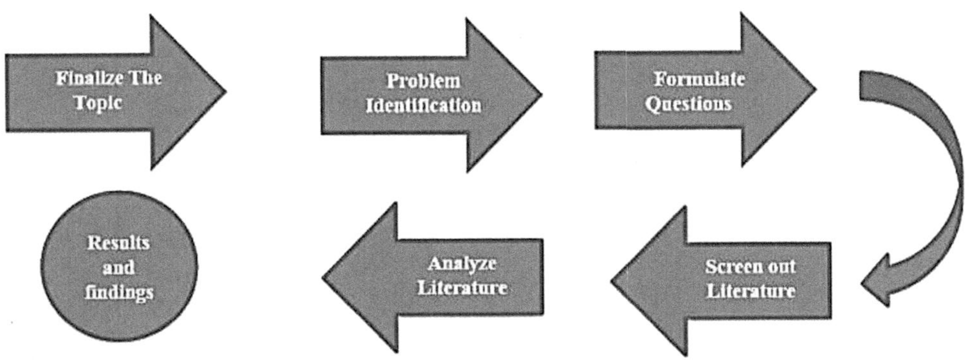

FIGURE 25.1 Literature review work flow.

Windows OS was introduced in 1985 and has since become a dominant and all-encompassing type of software, holding a market share of approximately 90% over competing operating systems [7]. Notable characteristics of Windows include:

- Symmetrical multiprocessing
- Multiple operating environments
- Virtual memory, primitive scheduling, portability, extensibility
- Integrated catching,
- Client server
- Integrated catching

Linux is a cost-free Unix-like operating system that can be installed on computing devices. It is characterized by its various distributions, including Debian, Ubuntu, and CentOS, and can be downloaded, modified, executed, and redistributed without any charges. Linux features a potent packet filtering firewall and operates solely on binary files, ensuring encrypted data, and maintaining privacy and security. It finds most of its use in mobile devices such as Android. Some of Linux's salient features include:

- Security and Live CD/USB
- Graphical user interface
- Multiple user option
- Multitasking and portability

In the Windows OS, users typically need to search for programs online or in stores and install them using separate installers, which can lead to uninstallation issues and virus risks. By contrast, many Linux distributions have adopted the concept of "software repositories" that simplify program installation. Users can use the add/remove programs tool to search for and install programs, and the package manager will handle the necessary dependencies and uninstallation. This streamlines the installation process, keeps the system free of unwanted software, and reduces the risk of data theft [3]. The characteristic of Windows- and Linux-based operating systems are illustrated in Table 25.1.

25.2 LITERATURE REVIEW

The Windows operating system is built upon a two-layer architecture comprising of User mode and Kernel mode. Both these layers consist of several modules. The User mode contains two essential

TABLE 25.1
Characteristics of Windows and Linux

OS	Source code Availability	User Modification	Usability (User-Friendly)	Security	Cost
Windows	Proprietary (not freely available)	User cannot modify in OS independently	Window provides ease of use and nice GUI	Security improved on new versions but still not much secure	The cost of software and OS is high as it is proprietary.
Linux	Open source (freely available)	User can easily modify the OS as source code freely available	Widely used by programmers and gained much attention in a few years	Very secure and safe to use	community based and freely or less costly available

subsystems, namely the Integral subsystem and the Environment subsystem. The Integral subsystem comprises fixed system support processes like login process and session manager, service processes like print spooler and task scheduler service, security subsystem for access management and security tokens, and user applications. On the other hand, the Environment subsystem acts as a bridge between user applications and the kernel functions of the OS.

On the other hand, the Kernel mode possesses full control over a computer system's resources and hardware. It runs code within a protected memory area and is composed of the Executive, microkernel, kernel mode drivers, and hardware abstraction layer (HAL). Numbers of inquiries have been carried out related to Linux and window working frameworks [8]. Windows is a widely used operating system that boasts excellent compatibility with a variety of programs. It has a reputation for reliability and speed, making it a popular choice for productivity-related tasks. While Linux also has its strengths, such as enhanced security, stability, and efficiency, it is particularly adept at bringing new technologies to less powerful machines. The review examined how the migration from Windows to Linux can be facilitated using Wine, a distinct implementation of Windows API and architecture [9]. It also explored how common issues of cloud computing could be resolved using leveraging POSIV (Portable Operating System Interface for vStar Cloud), a specific framework [10] displayed a investigate over usage and assessments of application about information in working framework Linux and window. They proposed taking after condition for Fourier change whereas displaying estimations of application of information in two OS Linux and Windows and Linux.

$$X(f) = F\{x(t)\} = \int \infty\infty - dt \tag{25.1}$$

X(t) represents a time-space flag, X(f) represents Fast Fourier Transform, ft denotes the rate for evaluating. Xinyu et al. (2006) suggested a method for measuring and analyzing risks in a specific window. Christian M. Garcea Arelano et al. (2006) demonstrated how autonomic computing has been utilized to reduce complexity in IBM DB2 Universal Database planning for Linux and Windows (DB2 UDB). Hurst Ref. [11] presented a method for using computers running Windows in a campus computer lab, based on the instructional mode of colleges. Ref. [12] projected a classification of Linux proficiency into four grades, along with tailored educational curricula for professionals at different skill levels. Ref. [13] presented an advanced modification of Casadeus-Masanel and Ghemawaa's research, providing a comparative analysis of Linux and Windows in the context of separate operating systems. A comprehensive analysis of existing work has been summarized in Table 25.2.

TABLE 25.2
Overview of Literature Review

#	Paper	Feature
1	M. G. DElia and V.Pacielo, 2011 [9]	Present a study comparing the performance of data applications in Linux and window operating systems, including implementation and evaluation measures.
2	Chrisian M. Garcia Arellano et al., 2006 [18]	Present how IBMDB2 Database for window & Linux (DB2 UDB) has been utilized in conjunction with autonomic computing to simplify administration and reduce complexity.
3	Ms NH.Gere et al, 2017 [19]	Introduced a virtual operating system to facilitate the transition from window to Linux.
4	Song Xinyu et al., 2006 [20]	Xinyu Song, Michael Stinsen, Roger Lee, and Paul Albee suggested a method for assessing and measuring the risks inherent in a particular window feature in 2006.
5	Z hang Lufei and Chen Zoning, 2017 [8]	Explained how POSIV can be utilized to resolve common cloud-related issues.
6	Meyun Kong et al., 2010 [7]	Developed a system of categorizing the skills of Linux professionals into four grades, and offered comprehensive educational programs for individuals at varying levels of expertise.
7	Soloviv et al., 2008	Introduced a revised mathematical model that was developed based on the research of Casadeus-Masanel and Ghemewa.
8	Hurst Severni et al., 2008 [24]	Outlined a strategy for integrating Window computers into college computer labs in a manner that aligns with the educational paradigm utilized by the institution.
9	Yue Zho and Jinyo Yan et al., 2012 [15]	Proposed an exploratory technique for evaluating the performance of Linux compared to window 7 execution of TCP.
10	Rue Le et al. (2012) [16]	Rue Le et al. (2012) proposed an approach for window-like virtual memory administration by unlocking the uncommon consistent space

25.3 METHODOLOGY

The researchers put forth a virtual operating framework that utilized a pre-existing virtual component for executing .exe files, with the aim of enabling access to both Linux and Window at the same time [19]. Their system comprised the following sections:

Global Terminal (Users can run command from window and Linux within the same interface using the Global Terminal).

Access Partition of Hard Drive (The hard disk segmentations access using access the Partition of Hard Drive module, without the need for any type of virtual Operating system).

exe Executor (This utilized wine to execute a window 32-bit file on a Linux platform).

Ref. [14] proposed an approach for Windows-like virtual memory administration by unlocking the uncommon consistent space. They also utilized the Windows heap management method, which connects a block of memory by listing and working on it. They proposed the Recreation Windows Synchronization Instrument and built a multi-object hold-up and release component. Ref. [15] proposed an exploratory technique for evaluating the performance of Linux compared to Windows 7 execution of TCP (Figure 25.2).

25.3.1 Metric

The two primary measurements that were taken into consideration were throughput and equality.

Throughput could be a clear purposeful of the lion's share clog controlling calculation. This is often introductory metric for assessing execution of TCP. The throughput which a great put is segregated by us and common tenure is utilized by us. [23]. Both the collective behavior of all flows

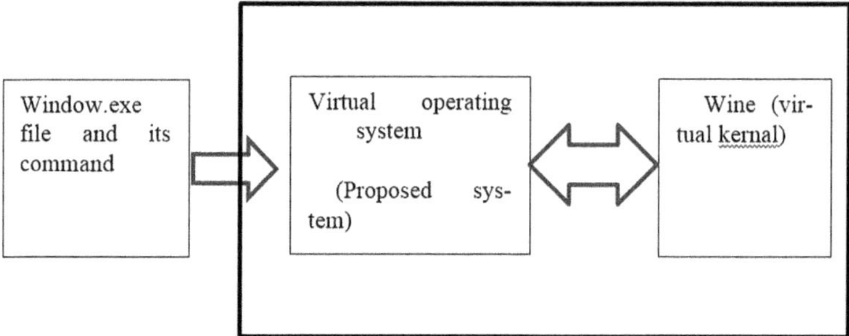

FIGURE 25.2 System design.

and the individual behavior of each flow are taken into account. The average throughput, aggregated from all flows, was measured in each test and is defined by Equation 25.2.

$$U: = \Sigma^{i=1}{}_{n}\, \overline{x_1} \quad (25.2)$$

Here talk about nth no of flow, average count of through-put on the test aiming for add up to no of n streams.

To ensure fairness in TCP, it is important that any new protocol is allocated an equitable amount of bandwidth, similar to that of a comparable TCP stream. Fairness can be defined in various ways and can be classified into two categories: inter-protocol fairness and intra-protocol fairness. In our evaluation, we focus on two types of TCP fairness: inter-implementation fairness, which compares fairness among different sender implementations, and intra implementation fairness, which compares fairness among senders using the same TCP implementation and operating system. Ref. [16] used Jain's fairness index as a metric to measure the minimum-maximum fairness.

25.3.2 Test Cases

Test cases arrangement were arranged for assessing execution of usage of TCP in real web. In every scenario, interfaces are connected to a single receiver at the same time, so that the bottleneck interface is distributed among n entities of streams and alternative conditions of systems. In every scenario the extent of the round trip time is 26ms to 471ms. Taking after are experimental cases utilized by [17].

In first scenario 1 flows on Linux vs. 1 flows on Windows 7. A sole TCP stream is initiated either across Windows 7 or Linux. Once 60 seconds elapse, an additional TCP stream is established over a different sender with a distinct operating system. This triggers a contest between the diverse implementations of TCP utilized by Windows and Linux [21].

In second scenario n flows on Window 7 A solitary TCP stream is established either on Window7 or on Linux. Once a minute has passed, another TCP stream is established over a distinct sender with an unlike operating system. This initiates a rivalry between the diverse implementations of TCP utilized by Windows and Linux.

In third scenario n-flows on Linux. This scenario is equivalent to Case 2, where multiple TCP streams are established on Linux, illustrating the competition among Linux's TCP flows.

25.3.3 Test Setting

In their 2012 study, Yue Zho and Jinyau Yan utilized two identical nodes as senders located in Beijing's Communication University. One node operated on Windows 7, while the other operated

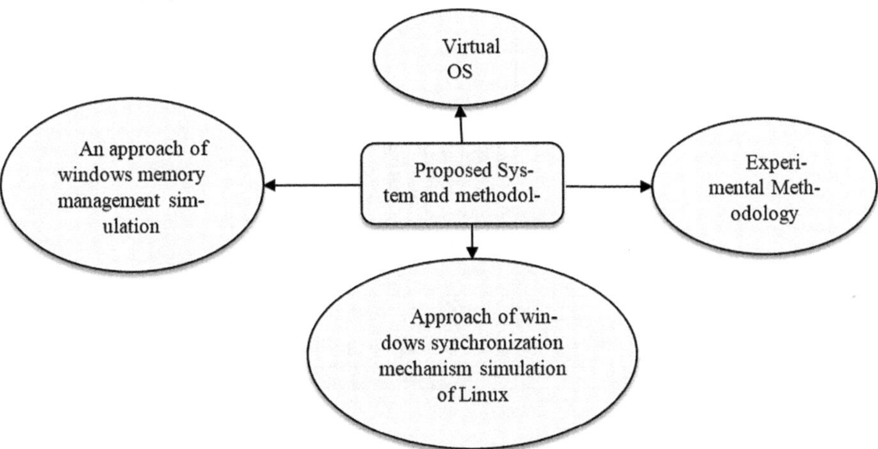

FIGURE 25.3 Overview of methodology.

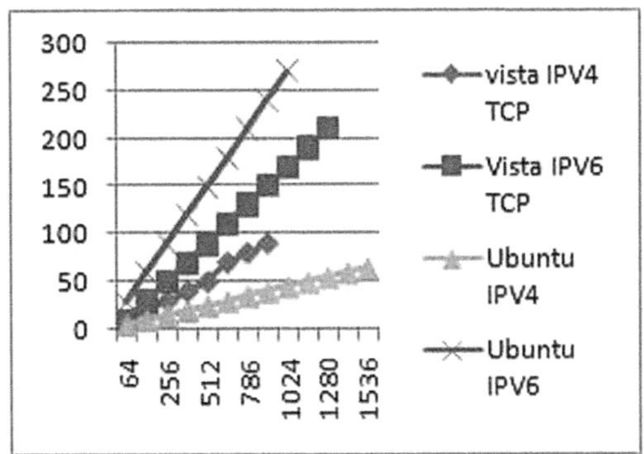

FIGURE 25.4 Through-put of TCP.

on Fedora Linux. To measure the TCP parameters, they employed the TCP_Data option with the getsockopt() function on Linux, while the API was utilized (Figure 25.3).

25.4 DISCUSSION AND RESULTS

Below is a chart that displays their results for calculating packet throughput using our values, with the TCP throughput being measured (Figure 25.4).

The chart below displays the findings of Shanel Naryan, Peng Shang, and Na Fan research, showing the calculated packet throughput based on our values for packet size. The throughput of UDP is measured in the chart (Figure 25.5).

Li Xin conducted tests on a laptop equipped with an Intel 1.5 GHz CPU, 512 MB memory, and running on the RedHAT 9.0 operating system. Each verification were performed on the Linux kernel source code, version 2.6.11.5, with the confirmation of the intervention of several control functions.

Table 25.3 displays the results of experiments carried out by Mohd Anuar Mat Isa and fellow researchers [22]. The experiment involved using a tool on Windows Vista SP1 x64 to measure the performance of its baseline process, which included measuring the execution time for specific file extensions (*DLL and *EXE) known for being vulnerable to common malicious attacks, such as

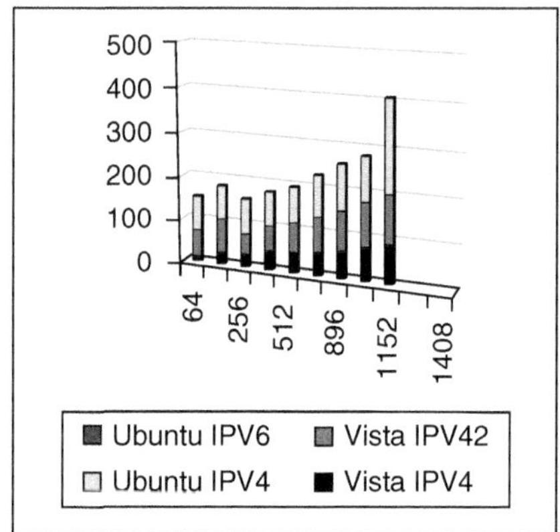

FIGURE 25.5 UCP throughput.

TABLE 25.3
Mouhd Anar Mat Issa Performance Measurements

	Base Line	Verification On	Self-Healing
Windows	377.944 (IM)	148.638	22.152
System32	309.522 (BC)		
Windows	103.537 (IM)	52.339	21.045
SysWOW64	128.264 (BC)		

IM = Integrity Measurement.
C = Backup & Compress Files.

viruses and malware. The System32 folder was compared to the SysWOW64 folder, and the results indicated that the number of vulnerable files in the System32 folder was almost triple that of the SysWOW64 folder. To further test the system's vulnerability, we launched malicious attacks on specific files in both folders, and the results showed significant differences in execution time and folder depth. Our proposed solution is an Integrity Verification Architecture (IVA)-based system that provides a mechanism for verifying and self-healing of disk or partition files of the operating system and its application files, to detect and remove threats such as online key loggers and spyware. The IVA-based system provides more confidence and trust in the application's security. The experiment results are also influenced by confounding variables, such as device speed and I/O drivers, which were taken into account during the testing process.

25.5 CONCLUSION

Both Linux and Windows have advantages and disadvantages. Linux is renowned for its stability, security, and performance, whereas Windows is known for its ease of use and extensive software library. The decision between the two operating systems ultimately depends on the user's requirements and preferences. For those who prioritize system performance. A Comparative Study of Linux and window is a research paper aimed at an in-depth analysis of the features, capabilities, and performance of two popular operating systems, Linux and Windows. Due to differences in the design concepts of the two operating systems, the implementation and results are different. A previous issue was discussed regarding the similarity in semantics between different OS, which can be a limitation for window and requires migration to Linux with minimal changes. This problem was solved by simulating the implementation of window application tools. Another previous research solved this issue by simulating synchronization strategies for Linux. Virtual private networks (VPNs) have various drawbacks, which were eliminated in one research by introducing SMA.

REFERENCES

1. Giri, N.H., Nandgaonkar, V.N., and Gosavi, Rahul. (2017). Virtual operating system for Windows to Linux migration. *2017 International Conference on Energy, Communication, Data Analytics and Soft Computing (ICECDS)*, pp. 2125–2127. doi: 10.1109/ICECDS.2017.8389825.
2. Desktop operating system market shareworldwide, [online]. Available: https://gs.statcounter.com/os-market-share/desktop/worldwide
3. Al-Rayes, Hadeel. (2012). Studying main differences between android & linux operating systems. *International Journal of Electrical & Computer Sciences*, 12, 46–4.
4. IvyPanda. (2022, June 21). Making informed user decisions: Windows v. Linux. https://ivypanda.com/essays/a-comparative-research-on-windows-vs-linux/
5. Hwang, Chanwoong, Hwang, Junho, Kwak, Jin, and Lee, Taejin. (2020). Platform-independent malware analysis applicable to windows and linux environments. *Electronics*, 9, MDPI AG, 793. https://doi.org/10.3390/electronics9050793.
6. Adekotujo, Akinlolu, Odumabo, Adedoyin, Adedokun, Ademola, and Aiyeniko, Olukayode. (2020). A comparative study of operating systems: Case of Windows, UNIX, Linux, Mac, Android and iOS. *International Journal of Computer Applications*, 176(39), 1623
7. Meiyun, Kong, Jun, Li, Fengming, Wang. (2010). Study on educational mode of Linux majors in colleges. *International Conference on Artificial Intelligence and Education (ICAIE)*, (pp. 623-626), 2010. doi: 10.1109/ICAIE.2010.5641146.
8. Lu-fei, Zhang, and Zuo-ning, Chen. (2017). vStarCloud: An operating system architecture for Cloud computing. *2017 IEEE 2nd International Conference on Cloud Computing and Big Data Analysis (ICCCBDA)*, pp. 271–275.
9. D'Elia, Maria Grazia, and Paciello, Vincenzo. (2011). Performance evaluation of LabView on Linux Ubuntu and Window XP operating systems. *2011 19thTelecommunications Forum (TELFOR) Proceedings of Papers*, Belgrade, Serbia, pp. 1494–1498. doi: 10.1109/TELFOR.2011.6143840.
10. Li, Yee-Ting, Leith, Douglas, Shorten, Robert. (2007). Experimental evaluation of TCP protocols for high-speed networks. *IEEE/ACM Transactions on Networking*, 15, 1109–1122. doi: 10.1109/TNET.2007.896240.
11. Adekotujo, Akinlolu Solomon, Ademola, Adedokun, Odumabo, Adedoyin, Aiyeniko, Olukayode. (2020, July). A comparative study of operating systems, the case of window, UNIX, Linux, Mac, Android and iOS. *International Journal of Computer Applications*, 176(39), 16–23.
12. Thubaasini, P. et al. (2009). Efficient comparison between windows and Linux platform applicable in a virtual architectural walkthrough application. *International Symposium on Symbolic Computation in Software Science,* Springer.
13. Awan, Muhammad Talha, and Khan, Kashaf. (2022). Linux vs. Windows: A comparison of two widely used platforms. *Journal of Computer Science and Technology Studies*, 4(1), 41–53. doi: 10.32996/jcsts.2022.4.1.4.
14. Li, Rui, Yang, Nanjun, and Ma, Shilong. (2012). An approach of windows synchronization mechanism simulation on Linux. *2012 13th International Conference on Parallel and Distributed Computing, Applications and Technologies*, pp. 442–445. doi: 10.1109/PDCAT.2012.39.
15. Zhou, Yue, and Yan, Jinyao. (2012). Experimental evaluation of TCP implementations on Linux/Windows platforms. *2012 21st International Conference on Computer Communications and Networks (ICCCN)*, pp. 1–5. doi: 10.1109/ICCCN.2012.6289279.
16. Li, Rui, Yang, Nanjun, and Ma, Shilong. (2012). An approach of windows memory management simulation on Linux. *2012 Third World Congress on Software Engineering*, pp. 143–146. doi: 10.1109/WCSE.2012.34.
17. Khan, Umaima. (2020, February). Comparative study of Linux and Windows. *International Journal of Academic Research in Business, Arts & Science*, 2(2), 53–70. Zenodo. doi: 10.5281/zenodo.3692081.
18. Garcia-Arellano, Christian, Lightstone, Sam, Lohman, Guy, Markl, Volker, Storm, Adam. (2006). Autonomic features of the IBM DB2 universal database for Linux, UNIX, and Windows. *IEEE*

Transactions on Systems, Man, and Cybernetics, Part C: Applications and Reviews, 36, 365–376. doi: 10.1109/TSMCC.2006.871572.

19. Giri, N. H., Nandgaonkar, V. N., and Gosavi, R. (2017). Virtual operating system for Windows to Linux migration. *2017 International Conference on Energy, Communication, Data Analytics and Soft Computing (ICECDS)*, Chennai, India, 2017, pp. 2125–2127, doi: 10.1109/ICECDS.2017.8389825.
20. Xinyue Song, Stinson, Michael, Lee, Roger, and Albee, Paul. (2006). An approach to analyzing the Windows and Linux security models. *5th IEEE/ACIS International Conference on Computer and Information Science and 1st IEEE/ACIS International Workshop on Component-Based Software Engineering, Software Architecture and Reuse (ICIS-COMSAR'06)*, Honolulu, HI, USA, pp. 56–62, doi: 10.1109/ICIS-COMSAR.2006.18.
21. Lu-fei, Zhang, and Zuo-ning, Chen. (2017). vStarCloud: An operating system architecture for Cloud computing. *2017 IEEE 2nd International Conference on Cloud Computing and Big Data Analysis (ICCCBDA)*, Chengdu, China, 2017, pp. 271–275. doi: 10.1109/ICCCBDA.2017.7951923.
22. Isa, Mohd Anuar Mat, Hashim, Habibah, Manan, J. -l. A., Mahmod, Ramlan, and Othman, Hanunah. (2012). Integrity Verification Architecture (IVA) Based Security Framework for Windows Operating System. *2012 IEEE 11th International Conference on Trust, Security and Privacy in Computing and Communications*, Liverpool, UK, pp. 1304–1309, doi: 10.1109/TrustCom.2012.189.
23. Floyd, Saly. (2008, March). Metrics for the evaluation of congestion control mechanisms. RFC 5166, doi: 10.17487/RFC5166.
24. Severini, Horst, Neeman, Hnery, Franklin, Chris, Alexander, Joshua, and Sumanth, J. (2008). Implementing Linux- enabled Condor in windows computer labs. *Paper presented at the 2008 IEEE Nuclear Science Symposium Conference Record*, pp. 873–874.

For Product Safety Concerns and Information please contact our EU representative GPSR@taylorandfrancis.com Taylor & Francis Verlag GmbH, Kaufingerstraße 24, 80331 München, Germany

Batch number: 10392095

Printed by Printforce, the Netherlands